P9-APG-700

This volume addresses the problems arising from pollutants that all too commonly contaminate the indoor environment, including biological sources such as bacteria, fungi and moulds, common combustion products, radon and other sources of radiation, solvents used in industry and the home, asbestos and dust pollution. The aim is to provide a balanced account of the health risks associated with these major pollutants and to quantify the scale of the problem on a pollutant-by-pollutant basis. Each chapter covers exposure levels, sources of pollution and routes of uptake, health effects, control measures, and regulatory guidelines.

INDOOR AIR POLLUTION: PROBLEMS AND PRIORITIES

INDOOR AIR POLLUTION:

problems and priorities

G. B. LESLIE
Independent Consultant in Toxicology

F. W. LUNAU
Independent Consultant in Occupational Hygiene

CAMBRIDGE
UNIVERSITY PRESS

Published by the Press Syndicate of the University of Cambridge
The Pitt Building, Trumpington Street, Cambridge CB2 1RP
40 West 20th Street, New York, NY 10011-4211, USA
10 Stamford Road, Oakleigh, Melbourne 3166, Australia

First published 1992
Reprinted 1993

Printed in Great Britain by Antony Rowe Ltd, Chippenham, Wiltshire

A catalogue record of this book is available from the British Library

Library of Congress cataloguing in publication data

Indoor air pollution: problems and priorities / [edited by] G.B.
Leslie, F.W. Lunau.
p. cm.
Includes index.
ISBN 0-521-38510-5 (hardback)
1. Indoor air pollution. I. Leslie, G. B. (George B.)
II. Lunau, F. W.
RA577.5.I542 1991
615.9′02—dc20 91-21196
CIP

ISBN 0 521 38510 5 hardback

GE

CONTENTS

CONTRIBUTORS

P. APPLEBY, BSc(Hons), CEng, MCIBSE, MASHRAE, is a building services engineer with experience in consultancy, contracting, lecturing and research, and is currently Managing Director of London-based Building Health Consultants Ltd, specialising in the diagnosis, prevention and alleviation of building-related illnesses and indoor environment problems.

DR G. R. BETTON is a veterinary pathologist specialising in the field of toxicologic pathology with particular interest in carcinogenesis. He has been Head of Pathology at Smith Kline and Beckman UK and at Unilever and is presently Head of Pathology at ICI, Pharmaceuticals Division at Alderley Park, Cheshire.

DR R. C. BROWN is a senior scientist with the UK Medical Research Council. He is interested in all types of environmental hazard but with an especial interest in pulmonary carcinogenesis and its relation to smoking and other exposures. His primary research interest is the biological effects of mineral fibres which he has been studying for 15 years both at the MRC Toxicology Unit and at the now defunct Pneumoconiosis Unit.

DR P. SHERWOOD BURGE is a consultant physician with a particular interest in occupational lung and environment diseases in Birmingham and Solihull. His main research has been occupational asthma and alveolitis and in problems of indoor air quality. He trained in clinical immunology with Professor Pepys at the Brompton Hospital and lung medicine at the Brompton Hospital, and in occupational medicine at the London School of Hygiene with Professor Corbett McDonald.

DR G. CLOUGH, is a consultant environmental physiologist with special interests in environmental control and monitoring, the control of disease transmission and the maintenance of gnotobiotic environments.

G. W. CROCKFORD, BSc, MIBiol, MIOH, is a Fellow of the Ergonomics Society and is a Registered Occupational Hygienist. He is an Honorary Consultant in Occupational Hygiene at the Robens Institute of Industrial and Environmental Health at the University of Surrey, UK.

DR L. H. HAWKINS, BSc, PhD, is an environmental physiologist specialising in Health and the Indoor Environment. He is Senior Lecturer in Applied Physiology and Head of the Occupational Health Unit in the Robens Institute of Environmental and Occupational Health and Safety at the University of Surrey, UK.

DR J. A. HOSKINS is a scientist with the Medical Research Council in their Toxicology Unit. He is a chemist by training who has studied both in England and Australia and has been involved in medical research for over 20 years. His especial interests are in metabolic management and control, particularly the disruption to this caused by mineral fibres.

DR J. B. KURTZ is consultant virologist to the Oxford Health Authority and Honorary Clinical Lecturer, Oxford University. He has been involved in research into environmental aspects of *Legionella pneumophila* since 1978.

G. B. LESLIE, FRCPath is a consultant toxicologist who was, for 13 years, Head of Toxicology at Smith Kline & French, UK.

PROFESSOR T. LINDVALL is on the staff of the Karolinska Institute in Stockholm.

F. W. LUNAU, MSc, DFH, C Eng. Dip. Occ. Hyg., MIEE, MIOH is a Chartered Engineer and a Registered Occupational Hygienist. As a consultant Occupational Hygienist, his work has covered a wide variety of assignments in the engineering, electronics, chemical and food industries. Latterly he has specialised in office environmental problems. He is a former Executive Secretary of the British Occupational Hygiene Society and is the President of the International Association for Indoor Air Quality. He was awarded the Bedford Memorial Prize of BOHS in 1967.

DR P. NICHOLLS, BSc, PhD, is Professor of Pharmacology and Head of the Division of Pharmacology in the Welsh School of Pharmacy at the University of Wales. He has a long-standing research interest in the study of occupational diseases caused by inhalation of vegetable dusts.

DR A. POOLE, BSc, PhD, a graduate of the Universities of Wales and Surrey, and a Member of The Royal College of Pathologists, has had wide experience in both academic and industrial areas of toxicology. His interests lie in understanding mechanisms of toxicological and carcinogenic effects. He is currently employed as project manager, in Health and Environmental Sciences, the Dow Chemical Company.

DR D. E. RAY was born in Oxfordshire in 1952 and studied at King's College London, obtaining a BSc in Biochemistry and Physiology followed by a PhD on the neurotoxicity of hyperbaric oxygen. His current research at the Medical Research Council Toxicology Unit is on the physiological response to natural and synthetic neurotoxins.

DR J. M. SNEDDON, BSc, PhD, is Head of the School of Pharmacology at Sunderland Polytechnic, UK. He was awarded the Sandoz Prize of the British Pharmacology Society in 1973 and has been an editor of the *British Journal of Pharmacology* since 1968. He was the recipient of the Visiting Scientist Award at the Alberta Heritage Foundation for Medical Research at the University of Alberta, Canada from 1981 to 1984. His research interests are in cardiovascular and respiratory diseases.

D. F. WEETMAN is Professor of Pharmacology at Sunderland Polytechnic, where he has occupied various academic positions for more than 20 years. Amongst his many interests and activities in the medical sciences, he specialises in communication between cells, experimental design and all aspects of drug development.

1

INTRODUCTION

F. W. LUNAU and G. B. LESLIE

This book is aimed at the informed non-specialist who has to deal with questions of indoor air pollution. Although most of the contributors are British, the book is designed to be internationally relevant. It is hoped that it will be of value to occupational health physicians and nurses, industrial hygienists, personnel and safety officers, ventilation engineers, architects and building maintainers and operators.

The contributors to this book are all internationally recognised experts in their various fields. They have endeavoured to present, for each category of indoor air pollution an assessment of exposure levels and uptake into the body as well as a consideration of the health effects and the possibilities for the control of the pollutants. The references are restricted to the most important papers and reviews on each subject.

Indoor air quality problems depend in large part on climate and the economic requirements and abilities of developed and developing countries. Those living in the more affluent countries tend to forget that the bulk of the World's urban population have to live with much more heavily polluted outside air. Historically this distinction did not always exist. In the West, urbanisation led to horrific air pollution which was perhaps best exemplified by the infamous 'pea-souper' fogs for which London was notorious. These fogs occurred during the winter when sulphurous smoke, mainly from domestic coal fires, mixed with fog in conditions of thermal inversion. This choking fog often stayed for days at a time and at its worst caused thousands of acute deaths. Road traffic became impossible and businesses had to close until the fog lifted. Visibility was sometimes reduced to 2 to 3 metres. The situation in London changed radically with the introduction of the Clean Air Act which prohibited the use of soft coal in the 'smokeless' zones. Variations of the 'pea-soup' fog phenomenon, however, prevail in many parts of the world today.

With such severe external pollution, virtually no attention was paid to the quality of indoor air. People escaped into their homes to avoid the noxious exterior and thought or cared little about the presence of pollutants in their indoor environment. However, as the problems of outdoor air pollution have lessened in developed countries, more attention is now being given to the problems of indoor environments. Buildings in colder developed countries now usually have central heating, while in warmer developed countries there is air conditioning in homes, offices and public places. These systems may limit the intake of fresh air, particularly where air interchange rates with the outside have become lower due to the need to conserve energy both for economic and environmental reasons. Thus, the problems of indoor air pollution have become more pronounced.

In the urban areas of developing countries with warm climates few people can afford air conditioning and ventilation is largely natural via open windows, perhaps assisted by fans. In these situations the quality of internal air is largely dependent on the quality of the external environment. Developing countries usually have less stringent regulations or at least less rigidly enforced regulations controlling the pollution of outdoor air. In the developing countries, cities which experience cold winters very often have domestic heating by coal and the combustion products of coal are their major source of outdoor air pollution. In addition to coal combustion, other sources of outdoor pollution are industry and road traffic. The outdoor pollutants strongly influence the indoor environment, though there are also indoor sources. For example, the fuel used for cooking purposes and the smoke from cooking often contribute considerably to indoor air pollution.

Another factor influencing indoor air quality is the development of new materials and products which produce novel air pollutants or at least pollutants which were quantitatively unimportant a few decades ago. For example, modern building materials, insulation, bonding resins, fabrics, cleaning materials, personal care products, and pesticides now often expose us to high levels of a wide spectrum of chemicals, some of which have been demonstrated to be highly toxic in humans or even carcinogenic in experimental animals. In addition the air conditioning systems that we use in offices and public buildings can give rise to indoor air pollution. They may contain biological products such as moulds, fungi and bacteria and they may distribute dust, fibres and various pollutants.

We are now living in an era where indoor air pollution is seen to be increasingly important. Perhaps in many instances too much importance is attached to the potential risks to health from indoor air pollutants. All too often our perception of risk is determined not by what is real and measurable but by the extent of media presentation of the issue. Nevertheless, the issue is important, for indoor air quality can influence health. The chapters in this book show this clearly.

In the developing countries the greatest problems of indoor air quality come from the external air and their highest priorities should be the control of the sources of external pollution. In affluent countries, much more research is needed on the quantification of the risk from the known pollutants and on the means to ameliorate these risks. These requirements are, fortunately, becoming recognised by public and industrial health physicians, occupational hygienists, toxicologists, ventilation engineers and building designers.

It is immediately evident that there are several distinct types of problems. Most obvious and most dramatic is sudden death following exposure to a pollutant. Such events are, fortunately, rare, at least in the developed countries of the West and, when they occur, they attract considerable publicity as is witnessed by recent episodes of legionellosis. In Korea and parts of China, however, several thousand people die every winter following exposure to lethal levels of carbon monoxide. The traditional domestic heating system in this part of the world uses a form of underfloor duct through which the hot smoke and fumes from coal combustion pass. Cracks in the floor lead to leakage of the products of incomplete combustion such as carbon monoxide and NO_x. The problem is exacerbated by supplementary heating from often inefficient kerosene stoves and by insufficient ventilation. Even in Western Europe there are several hundred acute carbon monoxide deaths every winter.

The problem of legionellosis has been identified primarily in affluent countries and can be tackled by better design and maintenance of air cooling systems. The problems of North East Asia with carbon monoxide can only be resolved by the adoption of improved systems of domestic heating which are often beyond the financial resources of many people in this part of the world.

The other types of indoor air pollutants exposure to which can be potentially lethal are much less easy to determine since, in general, exposure often occurs many years or even decades before the onset of symptoms of the disease. One problem is that available scientific techniques do not allow one to say with certainty that various pollutants are a health hazard. We are all familiar with the concept of occupationally induced diseases. Epidemiology has identified as risk factors for workers exposure at high concentration and prolonged periods to heavy metals, asbestos, coal and other dust, various solvents, ionising radiation, etc, but it is only now becoming apparent that exposure to various pollutants in the home, the office, public buildings, transport and places of entertainment may also have long-term health consequences. Epidemiology has been of much less value in these situations since the risk ratios are very low, probably less than 2 in most cases, and the diseases which are reportedly associated with exposure are well known to be multifactorial in causation. This, of course, means that there will be a wide array of confounding factors to be taken into consideration. Because of the long time of onset of symptoms for diseases such as cancer of the lung it may be very difficult to obtain satisfactory quantitative evidence of exposure to the various risk factors. This is the situation prevailing with the purported association of exposure to environmental tobacco smoke (ETS) and lung cancer.

Of course, we need not rely on epidemiology in every case. Radon and its daughters represent for most people by far their most important source of exposure to ionising radiation. The level of suggested risk of lung cancer from exposure to radon and its daughters is not derived from epidemiological studies of the at-risk population but is based upon extrapolations from observed lung cancers occurring in people exposed to very high levels of radiation. It is difficult to make such extrapolations with any degree of accuracy since the body has mechanisms which can repair low levels of damage. Thus, estimates of risk to people living or working in areas with high levels of ground emanation of radon vary considerably. It is now widely accepted that 10 ± 5% of the lung cancer frequency in the temperate zones is induced by radon. To put this in perspective, adjustment for the somewhat lower average levels of radon in the UK gives an annual death rate of about the same as that due to road traffic accidents. Recently (*Nature*, 1990) the UK National Radiation Protection Board increased its estimate of lung cancers caused by radon by a factor of two. Since lung cancer in non-smokers is a rare disease, this level of risk from radon is sufficient to account for most of the cases of lung cancer occurring in non-smokers.

Some animal data are also available. Formaldehyde has been shown to cause nasal cancer in experiments on rats. Although formaldehyde exposure is common in some occupations there is no epidemiological evidence of any association with human cancer. Formaldehyde can, of course, cause irritation and sore eyes and nose, and chronic exposure has been known to cause sensitisation and allergy.

Most of the cancers which have been reported to be associated with exposure to indoor air pollutants are cancers of the lungs. The majority of our evidence about the association between exposure and cancer is epidemiological and often very controversial. In very few cases do we have any evidence on causation. Unfortunately the epidemiological literature contains numerous examples of publications showing associations between various types of exposure with serious outcomes which have subsequently been shown to be unsustainable. Recently Bailar (1985) discussed two papers which appeared in one issue of a journal. One of the papers showed that treatment with synthetic steroid hormones reduced the risks of cardiovascular disease in women and the other paper showed that such treatment increased such risks. The use of spermicidal creams has been reported to be associated with an increased risk of foetal deformities. The manufacturers of the cream withdrew it from the market and were sued by parents of deformed babies for many millions of dollars. Subsequently the USA Food and Drugs Administration exonerated the product and it returned to the market. There is a great dilemma facing the editors and referees of epidemiological papers. Over-willingness to publish unsupported papers may cause great panic when the media oversimplify the issues and cause people to fear the consumption of coffee, beer, mustard, spices and herbs or various food additives. On the other hand, failure to publish could allow continued exposure to dangerous agents.

Over the last decade, debate has raged over the risk of lung cancers in non-smokers exposed to other people's tobacco smoke. There is a wide range of epidemiological data available. This subject is dealt with in Chapter 12. At present it seems reasonable to accept the conclusions of the WHO International Agency for Research on Cancer (IARC, 1986) that 'the data are consistent with a low level of risk or with no risk'. It certainly seems most implausible that the relatively high level of risk (30% to 40% increase) reported in some epidemiological studies could be due to the low levels of exposure of non-smokers. Despite this basic implausibility on dosimetric grounds the US Surgeon General (1986), the National Research Council (1986) and the fourth report of the UK Independent Committee on Smoking and Health (1988) have all concluded that the risk of lung cancer from ETS is real. One might speculate that these committees might have been willing to moderate their scepticism in order to put pressure on smokers to indulge their habit less frequently.

A recent report that the keeping of pet birds (Holst *et al.*, 1988) is associated with a six- or seven-fold increased risk of lung cancer seems to have attracted relatively little interest although this level of risk is 20 times higher than that reported for exposure to environmental tobacco smoke which has attracted immense attention. The work of Holst merits thorough further independent investigation and, if confirmed, requires wide publicity.

Other studies have identified several indoor generated air pollutants which are considered serious risk factors for lung cancer. Asbestos is well known to be a very serious risk factor for mesothelioma, the cancer of the linings of the lungs. It can also increase the risk of lung cancer but, probably nowadays, there is rarely a sufficient level of exposure in the home or the office to cause any great risk – at least in the developed countries. Recently developed man-made mineral fibre substitutes for asbestos are not known with certainty to be without risk but probably seldom reach high enough concentrations to be dangerous.

There may be some risk factors for lung cancer in polluted outdoor air. In Britain, before the Clean Air Acts of the 1950s, London and other big cities experienced notorious 'pea-souper' fogs caused by combustion of coal for domestic heating and for industry. The British had higher rates of lung cancers than did most other countries for which reliable data were available. British who emigrated to New Zealand or South Africa had much higher rates of lung cancers than those born in these countries. We cannot, therefore exclude the possibility that outdoor air pollution may be a risk factor for lung cancer either through exposure to respirable carcinogenic material or perhaps subsequent to bronchitis or damage caused to the lungs by pollutants such as SO_2. There are parts of North East Asia where such 'pea-soup' fogs still occur and where chronic bronchitis is very common.

A final point worth mentioning here is the problem of allergies to indoor air pollutants. One of the most common and disabling forms of ill health due to exposure to indoor air pollutants are the allergic disorders such as asthma. These disorders have been referred to in Chapters 4, 10 and 11 as being induced by moulds, fungi, animal danders, house dust mites and their excretion, vegetable dust, etc. Chemical agents can also be causes of asthma.

Asthma is a disease which seems to be increasing in frequency worldwide. There have been in the last decade or so reports from countries as far apart as New Zealand and Finland that the incidence of childhood asthma has almost doubled. This effect has been reported in cold climates and hot climates, in affluent countries and developing countries and the causes are not at all clear.

In the UK the most important indoor allergen is the house dust mite (or its excreta) and it is commonly supposed to be the major domestic cause of asthma. It may be that changing conditions in homes and offices worldwide have contributed to the success of the house dust mite and thus have increased the incidence of asthma, but much research needs to be done in this field.

One of the important conclusions to be drawn from reading this book is that research and education are vitally and urgently needed into many aspects of indoor air quality.

One of the past problems has been that practitioners in one part of the overall science of indoor air quality have not been fully aware of the expertise and problems of practitioners in allied fields. It is hoped that this book will, at least, give the reader an awareness of the work and knowledge of experts in various disciplines. Without the development of this inter-relationship, progress in improvement of indoor air quality will not be as rapid as it might otherwise be.

Finally, it may be of interest to the reader that a new multidisciplinary international learned society has recently been formed[1] in order to bring together experts from all the various sciences and technologies involved in these problems.

Note

1 Indoor Air International, Postfach 2, CH-4467, Rothenfluh, Switzerland.

References

BAILAR, J.C. (1985). When research results are in conflict. *New England Journal of Medicine*, **313**, 1080–1.

HOLST, P.A.J., KROMHOUT, D. & BRAND, R. (1988). For debate: Pet birds as an independent risk of lung cancer. *British Medical Journal*, **297**, 1319–21.

IARC (1986). Monographs on the evaluation of the carcinogenic risk of chemicals to humans: tobacco smoking. **38, 314**, WHO, IARC, Lyon, France.

INDEPENDENT SCIENTIFIC COMMITTEE ON SMOKING AND HEALTH (1988). Fourth Report. London: Her Majesty's Stationery Office.

NATIONAL RESEARCH COUNCIL (1986). *Environmental Tobacco*

Smoke: Measuring Exposures and Assessing Health Effects. Washington, DC, National Academy Press.

UNITED STATES SURGEON GENERAL'S REPORT (1986). *The Health Consequences of Involuntary Smoking*. Rockville, Maryland: US Department of Health and Human Services.

2

THE PERCEPTION OF INDOOR AIR QUALITY

F. W. LUNAU

Poor indoor air quality may, rarely, have clinically identifiable adverse consequences. Less rarely it may have sub-clinical health effects. All these may, or may not be, identified by the particular occupant as connected with the indoor environment to which he or she is exposed. However, there are many other factors which cause the sum of the sensations leading to the total perception of the indoor environment.

The factors which affect perception are shown in Figure 2.1.

Within the limits of this chapter, it is not possible to deal fully with all these factors. Attention is therefore concentrated on the two most important ones – thermal comfort and air quality – since these are generally the two most important factors in determining perception of the indoor environment.

The thermal environment

This is a very powerful factor in the occupant's perception of overall satisfaction with the environment. In the author's experience, most cases of dissatisfaction with the immediate environment have an element of thermal discomfort associated with them.

The following inter-related factors are the important ones in the total picture.

- Air temperature
- Mean radiant temperature (mrt)
- Rate of air movement
- Relative humidity
- Insulation value of clothing worn
- Metabolic rate of task being performed

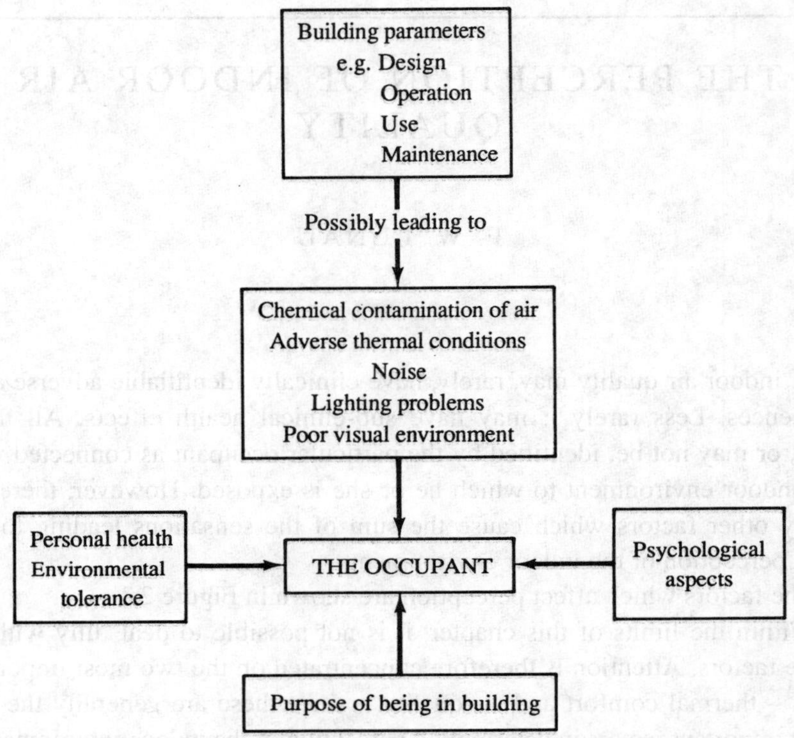

Fig. 2.1. Factors in the perception of air quality.

Fanger (1972) has produced an equation correlating these factors. This equation is complex since the heat transfer mechanisms are complicated. However, to ease the use of this thermal comfort equation, Fanger has reduced it to a series of graphs and tables, from which it is possible to predict the number of subjects who will be satisfied or, more importantly, dissatisfied with a particular set of thermal conditions. His results show that there are no, or only minor, differences in requirements for thermal neutrality between the sexes, or among persons of different ethnic origin or age. (Thermal neutrality is defined as the subject desiring to be neither hotter nor colder.) Two interesting points emerge. First, in a large population exposed to a particular set of conditions, it will be impossible to satisfy more than 96% of that population; generally a satisfaction rate of 90% is a suitable target figure. Second, whilst acclimatised persons working in tropical countries have become used to accepting heat-induced discomfort, their choice of conditions for thermal neutrality is the same as that for those persons from temperate climates.

There are over-riding limits for the first four individual factors given above.

Air temperature

There is conflicting evidence as to whether a varying air temperature is better than an absolutely static one. However, excessive temperature fluctuations will certainly cause complaints. Sprague and NcNall have postulated this limit for excessive complaints as

$$(\Delta t)^2 \cdot (\text{cph}) < 4.6$$

where Δt = peak to peak amplitude of the air temperature change (°C) and cph is the cycling frequency per hour.

Mean radiant temperature (mrt)

This should normally be maintained within 3 °C of the air temperature. In circumstances where the mrt is lower than the air temperature, the atmosphere tends to be regarded as 'stuffy' whilst, if the mrt is above the air temperature, then the atmosphere is regarded as 'fresh'. There are also various aspects of the asymmetry of the radiant heating to be considered. In most situations this is not particularly important except that the existence of large cold surfaces may cause draughts by the high local air movements involved.

Rate of air movement

Movement of air past the body will obviously result in extra cooling of it. Except in summer, rates of air movement greater than 0.3 metres per second (m/s) should not be allowed. This airspeed requires an increase of air temperature of 1 °C to compensate for it. When a person is warm, cooling by air movement is pleasant, when the person is cold then the extra cooling is unpleasant and the environment is felt to be draughty. In this latter circumstance velocity fluctuations may also be of importance. The part of the body exposed to air movement is also important, the back of the neck being the most sensitive area.

Relative humidity

Purely from a thermal comfort point of view, relative humidities in the range 40–70% are best. However, other factors influence the desirable range. 70% is getting near the level at which mould growth increases and house dust mites start to increase rapidly. Increases in either of these will have an adverse influence on the incidence of asthma. In winter, humidification will often be required to maintain a value in excess of 40% and this in itself introduces difficulties of possible fungal growth in the ventilating system. It is probably most sensible to try to keep within the range 40–60% during most of the year, whilst accepting that relatively short-term excursions below and above the target range will not produce too great an incidence of complaints.

Evaluation of thermal comfort conditions

Whilst the full application of the principles of Fanger is undoubtedly the best way to proceed, there are other indices which have some utility. The most useful practical index is probably that of dry resultant temperature (t_{res}). This is approximately equal to

$0.5\ (t_a + t_r)$

Where t_a = air temperature

t_r = mean radiant temperature

This relationship only holds good for air speeds in the region of 0.1 m/s and relative humidities in the range 40–70%. For sedentary workers wearing temperate region clothing, a t_{res} in the range 19–23 °C will be found to satisfy most people, although generally the number of dissatisfied will increase towards the lower end of the range.

Other sensory aspects

Skin temperature is only one of the sensory aspects which affect people's perception of their environment. The other main ones are chemical, such as olfactory and taste effects as well as irritancy.

Olfactory effects

In the matter of perception of indoor air quality by smell, two aspects are of importance. Firstly, there is the presence or absence of smell and, secondly, there is the acceptability or otherwise of any such smell. With regard to presence of a smell, various lists have been published, which use somewhat different criteria. WHO (1987) list two different criteria:

Detection threshold level (the lowest concentration that can be detected in 50% of tests).

Recognition threshold level (the lowest concentration that can be identified in 50% of tests).

Nuisance is dealt with differently and WHO (1987) define its threshold value as:-

Nuisance threshold level

That concentration at which not more than a small proportion of the population ($<5\%$) experience annoyance for a small part of the time ($<2\%$).

Generally WHO (1987) in their Guidelines use the detection threshold level which is more onerous than the recognition threshold level.

Use of these criteria in assessing indoor air quality is very difficult. The problem has been tackled in various ways, all of which have advantages and drawbacks. In order of complexity these are:

Use of untrained observers

This approach is exemplified by the American ASHRAE Standard 62–89. (ASHRAE, 1989) One of the criteria set out in this standard for acceptable air quality is that it must be acceptable to at least 80% of an untrained panel immediately on entry. The purpose of this qualification is that olfactory fatigue occurs after quite short exposure times and the percentage of 'satisfied' people would rise with time. The reproducibility of a test such as this is low and it can only be regarded as a rough screening test to aid in a decision as to whether more thorough investigation of a suspect building is warranted.

Use of trained observers

This is a technique that has been pioneered by Fanger (for instance, Fanger 1988) using a similar approach to the one he has used in his work on the thermal environment. This approach has been quantified by the introduction of two new units.

The olf

This is the emission rate of air pollutants arising from 1 standard person.

The decipol

This is the unit of pollution. In a closed area, assuming steady state conditions and complete mixing, the concentration of a pollutant depends on the source strength and the dilution caused by the ventilation. The perceived air

Table 2.1. *Typical examples of pollution concentration in decipol*

Decipol level (approx.)	Example
0.05	Outdoor air, mountain
0.01	Outdoor air, town
1	Healthy building
10	Sick building

Based on a study of 15 offices, the source contribution has been found (Table 2.2).

Table 2.2. *Average pollution sources in 15 Danish offices (average floor area 230 m², average population 17)*

Source	olf level
Occupants	17
Building and furnishings	28
Smoking	35
Ventilation system	56

pollution as determined by a trained panel, is defined as that concentration of bioeffluents that would cause the same dissatisfaction as the actual air pollution. 1 pol is the pollution caused by 1 olf ventilated by 1 l/s. The practical unit is the decipol which is 1 olf ventilated by 10 l/s.

Fanger (1988) has reported an investigation comparing the dissatisfaction of trained observers as a function of the perceived air pollution. 15% are dissatisfied on first entry into a space with pollution of 1 decipol. This rises to 50% at a level of 6 decipol. He has also quantified decipol levels for some typical cases (see Table 2.1).

Whilst a powerful research tool, this approach obviously has practical difficulties for its application on a wide scale because of the need for a trained panel. Perhaps its greatest utility is for the identification and quantification of sources of emission. This enables total emission within a space to be minimised. However, it is difficult to see how this approach can be written into legislation or guidance because of the difficulty in specifying and checking the performance of the various panels that would have to be used. To do this requires an objective method of assessment rather than a subjective one. The following two methods go some way towards reaching this objective.

Table 2.3. *Tentative dose-response relation for discomfort resulting from exposure to solvent-like volatile organic compounds*

Total concentration mg/m3	Irritation and discomfort	Exposure range
<0.20	No irritation or discomfort	The comfort range
0.20–3.0	Irritation and discomfort possible if other exposures interact	The multifactorial exposure range
3.0–25	Exposure effect and probable headache possible if other exposures interact	The discomfort range
>25	Additional neurotoxic effects other than headache may occur	The toxic range

Use of total volatile organics. Molhave (Seifert, 1990) postulates that volatile organic compounds (VOCs) cause problems of perceived air quality by reason of irritant reactions and the other sensory reactions discussed in this section. He suggests that the perceived air quality is primarily determined by the total VOC content. His restrictions on this approach are:

(a) It is restricted to the perceived unspecific stimulation. It cannot be used to predict effects such as CNS effects, carcinogenic effects, etc.

(b) The compounds must be comparable in chemical properties, molecular weight and vapour. This includes those volatile organic compounds (VOCs) commonly found in indoor air.

(c) The best method of measurements is by gas chromatography using a flame ionisation detector calibrated against toluene.

This tentative dose-response relationship is given in Table 2.3.

Considerable further experimental work must be done to define this concept but if it can be confirmed it has the merit of being a reasonably inexpensive diagnostic tool.

Use of specific VOCs. Seifert (1990) proposes an overall limit of 0.3 mg/m3 for total VOCs with classes of VOC having specific limits (see Table 2.4).

Table 2.4

Alkanes	0.1 mg/m^3
Aromatic hydrocarbons	0.05 mg/m^3
Terpenes	0.03 mg/m^3
Halocarbons	0.02 mg/m^3
Esters	0.02 mg/m^3
Aldehydes and ketones[a]	0.02 mg/m^3
Other	0.05 mg/m^3

[a]Formaldehyde is excluded.
The concentration of any one component should not exceed 50% of its class concentration nor 10% of the total VOC concentration.

Like the proposals of Molhave these concepts require to be further refined. One difficulty is that testing of this nature will be extremely expensive.

The future

The foregoing sections highlight the complex nature of the problem. This is not unexpected since there is a vast range of response of the human body and mind to the kinds of stimuli that we are considering.

The biggest research problem must be to devise methods of measurement of the main parameters. This has been successfully done in the case of the thermal environment, where the main factors have had relatively simple methods of measurement developed for them and the effect on the body quantified. There is, however, one area in this section that warrants further research; this is the psychological one of who is in control of the environment. It is often stated that satisfaction is greatest when there is the opportunity to exercise control of his/her thermal environment. It is difficult to know whether this is because of the failure of centrally controlled systems to control the environment correctly in all parts of a building, particularly when external conditions are changing rapidly, or whether it is a psychological response to not being in control of one's destiny. The work of Fanger shows that a small percentage of people will always be dissatisfied by a given set of thermal conditions but the apparent merits of local control are usually thought to affect a much greater percentage than that. In these days of trying to reduce CO_2 emissions from burning fossil fuels to reduce the greenhouse effect, it is necessary to conserve energy as much as possible. It is therefore necessary to consider whether the apparent advantages of local control are worth the extra energy consumption over an effective centralised control system which most probably will use less energy.

The perception of the indoor environment due to air pollution is even more complex and less well understood than that of the inter-related thermal environment. The first steps have been taken to introduce objective measurement of the chemicals thought to influence perception and to link the results of such measurement to the effect on people. Whether more extensive work will support the various hypotheses is unknown. A relatively cheap chemical analytical method that reflects perception would be a valuable diagnostic tool. However, it seems to the writer that the main value of the various methods discussed lie in their use in determining emission rates from building materials and furnishings and linking these to the effect on perception. Any easy quantification method for the perception of groups or building occupants will continue to be a very difficult and likely unattainable goal. Thus the investigation of buildings where there are problems will continue to be difficult unless a specific cause can be quickly identified. This, of course, has always been the case where buildings with true sick building syndrome have been investigated. However, a more detailed knowledge of the factors influencing perception will help the building designer to avoid creating problem buildings.

References

General reading

ASHRAE Standard 62–89 (1989). Ventilation for acceptable indoor air quality. *ASHRAE* Atlantic Ga., USA.

FANGER, P.O. (1972). *Thermal Comfort*. New York: McGraw-Hill.

FANGER, P.O. (1988). Perceived quality of indoor air. *Indoor and Ambient Air Quality*. London: Selper Press.

McINTYRE, D.A. (1980). *Indoor Climate*. Barking, UK: Applied Science Publishers Ltd.

SEIFERT, B. (1990). Regulating indoor air. **5**, 15-33. *Proceedings of the fifth International Conference on Indoor Air Quality*. Conference Secretariat, Ottawa, Canada.

SPRAGUE, C.H. & McNALL, P.E. (1970). The effects of fluctuating temperature and relative humidity on the thermal sensation of sedentary subjects, *ASHRAE Transactions*. **1**, 76.

WORLD HEALTH ORGANISATION (1987). *Air Quality Guidelines for Europe* Copenhagen: WHO.

3

LEGIONELLA

J. B. KURTZ

In 1976, 182 cases of pneumonia occurred amongst American Legionnaires attending their convention in a Philadelphia hotel. Twenty-nine died. The cause of this 'Legionnaires' disease' was subsequently identified as a bacterium and was called *Legionella pneumophila*. Other outbreaks of pneumonia, some predating the convention and a non-pneumonic flu-like illness, Pontiac Fever, named after an outbreak in Pontiac, Michigan in 1968, were soon recognised to have been caused by the same bacterium.

L. pneumophila was the first member of the family Legionellaceae to be isolated and characterised. It has been subdivided serologically into at least 14 serogroups. Further members of the family have also been discovered and now there are over 20, some of which have also caused human disease (*L. micdadei* and *L. feeleii* are mentioned below). The diseases caused by the Legionellaceae are covered by the general term 'legionellosis'.

Pneumonia is the major disease. It is usually caused by *L. pneumophila* serogroup 1, which therefore is considered to contain the most virulent strains. Although Legionnaires' disease only accounts for about 2% of cases of community acquired pneumonias in Britain it has caused considerable public concern for two reasons. First, the illness may be severe and is associated with a high mortality in those who are elderly or have some pre-existing illness. Secondly, the source of infection is environmental, it is not spread from person to person. After an incubation period of 2–10 days, patients develop pneumonia but in some people gastro-intestinal symptoms (e.g. nausea, vomiting and abdominal pain) or signs of central nervous system disturbance (e.g. loss of memory, confusion, stupor) may be more prominent. The chest X-ray is abnormal in nearly all those affected.

Pontiac Fever is the other illness caused by Legionellaceae and is possibly a sensitivity reaction to the organism. It has a high attack rate, in contrast to Legionnaires' disease (which has 1% or less) and affects previously healthy people. It is not fatal and causes no abnormalities on the chest X-ray. The incubation period is usually about 2 days after which there is a sudden onset of a flu-like illness with fever, head and muscle aches, and a cough. It lasts for 2–5 days. Secondary spread to other people does not occur.

Both illnesses are caused by inhalation of minute water droplets (aerosols) containing the organism. Whether such an aerosol is likely to cause an infection depends on many factors: the number of bacteria present; the virulence of the strain – at least one aspect of which is its survival time in the aerosol; the size of the water droplets (less than 5 μm diameter) small enough to be inhaled into the lungs and the exposure time and susceptibility of the host. It is therefore not possible to give an absolute value to the upper limit of bacteria below which exposure would not cause disease.

The natural habitat of the Legionellaceae is aquatic where they live in symbiosis with other micro-organisms (algae, protozoa, bacteria, etc) and their distribution is worldwide. Man commonly comes into close contact with the organism in the environment of buildings whose water systems become colonised. Water entering the building from the mains or other supplies is the most likely seed source and, provided that conditions somewhere in the water system are favourable for the growth of these bacteria, they will multiply. Following this local amplification step, dissemination in a suitable aerosol is all that is required to put those exposed at risk. Well-recognised factors that contribute to amplification are stagnant water as when a water system is shut down or only in intermittent use or the pipework contains 'dead legs', warm temperatures, 20 °C–40 °C, at which bacterial growth occurs; the presence of organic material that provides appropriate nutrients for growth. All these factors encourage the development and growth of the biofilm – a slimy layer of microbes and their by-products – of which Legionellae may be a part.

A survey of hotels and hospitals in England and Wales carried out in the mid-1980s showed *L. pneumophila* to be very widespread. It was found in at least one site in 53 of 104 (53%) of the former and in 28 of 40 (70%) of the latter buildings, most commonly in domestic hot water systems. Also noted was a positive correlation between the presence of the bacterium and the size of the building. Cooling towers, in those buildings which had recirculating cooling systems, also frequently harboured the organism. 44% of those tested were infected. Today, with better maintenance this figure can be expected to be less.

With such widespread distribution of Legionellae, why is there not more legionellosis? The answer must be that it is relatively uncommon for all the factors necessary to cause disease to come together. An amplification site where adequate numbers of a virulent strain can multiply; a mechanism of dissemination of an aerosol; atmospheric conditions that allow the survival of the organism from the time and place the aerosol is generated to that when it is inhaled. As mentioned above, the state of health of those exposed is also important. This was clearly shown in the 1985 outbreak at the Stafford District General Hospital (UK). The attack rate of Legionnaires' disease was about 5% amongst the approximately 1800 outpatients attending during the outbreak, whereas there were no reports of cases of pneumonia in the hospital staff although 42% of those tested had antibodies to *L. pneumophila* – evidence of recent exposure.

Four outbreaks are outlined below which illustrate various environmental features typically associated with clusters of disease.

1. Pontiac fever affected 317 assembly plant workers at an automobile plant in Ontario, Canada in August 1981. The plant was naturally ventilated through open doors and windows. The attack rate gradient was consistent with airborne spread of the infectious agent along the North West to South East wind direction axis. The proportion of workers affected also decreased with the distance of their work places from some grinding machines that used a water/oil mixture for lubricating and cooling their processes (when in operation the jet of fluid hitting the moving tools creates an aerosol). These grinding machines and their recirculating coolant system had been shut down for the week preceding the outbreak. This provided stagnant conditions in the coolant, ideal for bacterial growth. Two days after this section of the plant was brought back into operation the first cases of Pontiac fever occurred. A new legionella species, *L. feeleii,* was isolated from the coolant fluid and those workers who were ill had raised antibodies to this organism. The addition of a biocide to the coolant ended the outbreak.
2. At the Wadsworth Medical Center, California, USA from May 1977, two months after the new hospital opened, until February 1980 there was an average of 4.5 cases of Legionnaires' disease per month in hospitalised patients. This continued despite the introduction of continuous chlorination (>3 ppm free residual chlorine) of the cooling tower in November 1978 after it had been

found to be contaminated with *L. pneumophila*. In March 1980 the monthly total jumped to 26. This cluster of infections had been preceded by a test of the hospital's emergency domestic water services during which the pump broke down. This resulted in a temporary but precipitous drop in water pressure throughout the hospital. After this pressure 'shock' the tap water was turbid and discoloured 'black' or 'brown' for several weeks. Environmental investigations at that time found that the same *L. pneumophila* serogroups (1 and 4) which had been isolated from patients were present in 80% of the tap water samples and 50% of the shower waters tested. An experimentally produced pressure 'shock' produced similarly discoloured water which had a legionella count ten-fold higher than the 'pre-shock' clear tap water. This suggests that the 'shock' must have disturbed the biofilm inside the water system. Hyperchlorination of the water supply (2 ppm) commenced in July, 1980 following which the outbreak ended. Although it is not clear how all the patients became infected, the whole water system of the hospital was heavily contaminated.

3. In April 1985 there was an outbreak of Legionnaires' disease at the Stafford District Hospital (UK) affecting 101 people of whom 28 died. A case control study showed that there was a significant association between visiting the hospital outpatients' department (OPD) and becoming infected. The OPD was air-conditioned by a cooling water system and tower which served not only the ground floor OPD but also operating theatres on the first floor and maternity wards on the second floor. There was a common fresh air inlet above rooftop level for this 'cylinder' which served separate air ducts for each floor. The conclusion of the Committee of Inquiry into this outbreak was that this cooling water system had become heavily contaminated with a virulent strain of *L. pneumophila* serogroup 1. Aerosols from the cooling tower entered the cylinder air ducts both via the fresh air inlet and also through a gap, approximately 3" wide, in the floor of the cooling tower which created a common inlet for air to both the cooling tower and cylinder. Once in the air ducts the infected aerosol was distributed via the air-conditioning system to all three floors. Infection was most severe in those particularly vulnerable people attending the out-patients department. Contaminated aerosols

were also carried down the outside of the building to create a potential source of infection to people in the immediate vicinity of the hospital.

There were areas in the plant where construction differed from the design intent and this may have resulted in its faulty operation which led to the multiplication and dissemination of legionella. Why the outbreak started on the 9th or stopped on the 19th of April is uncertain. During that time, however, the ambient temperature and relative humidity were high for the time of year. The temperature rose from 5.8 °C to 14 °C between 27th and 30th March and fell again from 16.5 °C to 8 °C between 18th and 20th April.

4. In January 1988 an outbreak of Pontiac fever affected more than 170 people including children celebrating the New Year in a hotel at Lochgoilhead on the west coast of Scotland. The attack rate was about 90%. The hotel had a modern indoor leisure complex adjoining it with a swimming and whirlpool spa area, a curling rink and lounge and toilet facilities. Moist air from the pools was dehumidified, heated and recirculated in that area, but, despite this, the bathing environment was very warm and humid because of the negligible air change. In an investigation after the outbreak, high numbers (1.4×10^4/l) of *Legionella micdadei* were isolated from the whirlpool spa water. High or rising antibody levels to this organism were also found in the majority of those infected which indicates that it had caused the illness. It was also found that the automatic bromination system which disinfected this whirlpool spa by the addition of bromine was faulty. All the guests who were ill had been in the modern complex part of the hotel – the number increased because the toilet facilities of the old part were temporarily out of order. They had therefore been exposed to an aerosol which was generated in the pool area but which spread through the frequently opened door to the lounge and in which conditions for survival of the organism were ideal.

In Examples 1, 2 and 4 given above, outbreaks resulted from the indoor air being contaminated by locally generated aerosols. Showers, taps, whirlpool spars or jets of water impinging on a surface have all caused illness in people closely exposed to these sources. Other sources of localised aerosols include fountains, car washes, greenhouse sprinklers, high-speed dental drills, room and incubator humidifiers and mechanical cleaning. Aerosols can also cause infection at a distance from where they are generated, carried on air currents or through air ducts. Evaporative condensers and cooling towers are the usual sources of aerosols generated at a distance as in the outbreaks at Pontiac, Michigan in 1968, at the Stafford District General Hospital in 1985 (Example 3) and one at the BBC, London in 1988. Distant spread could also be responsible for many of the sporadic infections that account for about half the cases reported but for which no obvious source can be found (Bhopal *et al.* 1991).

Outbreaks or a sudden increased incidence of illness as occurred in the Wadsworth Memorial Hospital (Example 2) have often been associated with *an event*, in this instance the failure of a pump. Such *events* may be seen as disturbing a steady state where a system may be sufficiently contaminated but conditions for dissemination of an infected aerosol are not suitable or vice versa. Other *events* that have resulted in outbreaks by completing the requirements discussed above include: lowering of the hot water temperature (e.g. deliberate or accidental lowering of the calorifier thermostat; removal of asbestos lagging around pipes; heavy demand so that contaminated pre-heated water (40 °C) in holding tanks was heated insufficiently to kill the bacterium by the calorifier before entering the piped water system), introduction of stagnant water into the circulation (e.g. from 'dead legs', auxiliary cooling towers, or storage tanks), alterations in meteorological conditions (e.g. temperature inversion – when the drift from a cooling tower will spread horizontally and down, rather than up), and changes in the condition of the water in recirculating systems (e.g. rapid accumulation of dust and debris in the water by scrubbing dirty air in the vicinity of building sites).

Given the worldwide presence of legionella species in natural habitats, prevention of infection in man must centre around control of the multiplication of the organism in situations where people might be exposed. This involves two main aspects, design and maintenance. In the design of buildings, sites where colonisation and multiplication can occur must be minimised or eliminated. Unnecessarily long pipe runs should be avoided. Both pipes and storage tanks for hot and cold water should be insulated so that the water in them is maintained above or below the temperature range ideal for legionella multiplication. Calorifiers should be designed to minimise the possibility of temperature stratification which can lead to tepid areas where bacterial growth can occur. Areas of stagnation can occur in both the pipework and water holding tanks. Dead legs and blind loops should be minimised and unnecessarily large storage tanks should not be installed; storage capacity should not exceed 24 hours of demand. Materials should be used for washers, joints, pipework and the packs of cooling towers that do not support the growth of bacteria. Information about these materials is given in the *Water Fittings and Materials Directory.* Where air conditioning systems are used, the cooling towers must be sited downwind of, and as far away as possible from, the air inlet ducts and be fitted with efficient draft eliminators which only allow 0.001% of the circulating water to escape. The use of air-cooled rather than water-cooled systems should also be considered. Further recommendations of design and commissioning of new buildings have been made in the Second Report of the Committee of Inquiry into the Stafford outbreak.

Proper operation and maintenance is essential but has too often been neglected. Recent improvements have been made, not least because those responsible for the management of a building and its water systems could be legally liable if it was the source of an infection. An operational engineer should be in overall command and responsible for all systems. They must be well trained and be familiar with the equipment under their care and its functioning and be provided with the relevant operating manuals. Codes of Practice have been produced by several organisations. That produced in 1988 by the UK Department of Health (DOH) for health care premises is one of the most comprehensive. Such codes provide a programme for the operation and maintenance of a system and should include a schedule of procedures which must be carried out at regular intervals and a log book in which dates of inspection, results of tests and remarks are written. For example, hot water services should ensure that water is stored at 60°C and at the draw-off point is not less than 50 °C within one minute of full flow of the tap. Calorifiers ought to be drained four times each year to minimise the accumulation of sludge and be thoroughly cleaned once every two years when they are dismantled for statutory inspection. Again, standby or backup pumps should be used at regular intervals to reduce the danger of stagnation. There is also a Health and Safety Executive Guidance Note (HSE, 1991) which contains essential advice.

For recirculating water systems, twice yearly cleaning is advisable. Some form of regular biocide treatment is also necessary, in addition to scale and corrosion inhibitors, to reduce microbial growth which would otherwise proliferate and lead to fouling of the system and loss of efficiency. A wide variety of biocides are available for this purpose and are discussed in a Department of Health Report (1989). It gives the properties of an ideal biocide, which should include: good activity at low concentrations and under a variety of chemical and physical conditions, against all micro-organisms; its concentration should be easy to measure; safe both for the handler and to the environment; non-corrosive and low cost. The efficacy of the biocide chosen should have been evaluated in the field where pH and changes brought about by air-scrubbing and the concentrating effect of water loss create conditions that are different and usually less favourable than in the laboratory. Dosing schedules should be carefully calculated to take into account the capacity of the system and its water flow rate from which can be assessed the water loss due to evaporation, drift and bleed-off. The dose and date given must be recorded in a log-book. In the DHSS code, chemical control parameters are recommended as a means of monitoring the water quality and therefore the maintenance of recirculating water systems. It should, however, be noted that there is no clear correlation between any of these parameters, or indeed between the total (non-legionella) bacterial count of a water sample and the level of its contamination with Legionellae. When tests for Legionellae have been done on well-maintained systems that were regularly cleaned and treated with biocides, both the isolation rate and the level of contamination were markedly reduced in comparison to the results from poorly maintained sites. Legionellae were isolated from only one of 22 of the well-maintained systems and then only in small numbers, while in the poorly maintained group, 20 out of 27 systems harboured the organism with counts of at least 100 000/1 in 7 (Dr J. Lee, personal communication). Codes of practice to control Legionellae should therefore include testing for the organisms. The presence of small numbers (<1000/1) in a water sample does not in itself constitute a health hazard and should not create alarm. It does indicate, however, that the system is contaminated and that conditions somewhere in it are suitable for its growth. In such circumstances the occurrence of a favourable event might permit multiplication of those Legionellae to a level (>100 000/1) that could present a risk, especially if the strain was virulent.

In the light of present knowledge, man-made environments can be made safe with respect to infection by Legionellae.

References

BARTLETT, C.L.R., MACREE, A.D. & MacFARLANE, J.T. (1986). *Legionella Infections* London: Edward Arnold Ltd.

BHOPAL, R.S., FALLON, R.J., BUIST, E.C., BLACK, R.J., URQUHART, J.D. (1991). Proximity of the home to a cooling tower and risk of non-outbreak Legionnaires' disease. *British Medical Journal*, 302, 378–83.

COMMITTEE OF INQUIRY INTO THE OUTBREAK OF LEGIONNAIRES' DISEASE IN STAFFORD IN 1985. (1986). First Report . Cmnd. 9772, London: HMSO.

COMMITTEE OF INQUIRY INTO THE OUTBREAK OF LEGIONNAIRES' DISEASE IN STAFFORD IN 1985. (1987). Second Report. Cmnd. 256, London: HMSO.

GOLDBERG, D.J., WRENCH, J.G., COLLIER, P.N., EMSLIE, J.A., FALLON, R.J., FORBES, G.I., McKAY, T.M., MACPHERSON, A.C., MARKWICK, T.A. & REID, D. (1989). Lochgoilhead fever : outbreak of non-pneumonic Legionellosis due to *Legionella micdadei*, *Lancet* **i**, 316–18.

GREAT BRITAIN. DEPARTMENT OF HEALTH (1988). *The Control of Legionellae in Health Care Premises. A Code of Practice*. London: HMSO.

GREAT BRITAIN. DEPARTMENT OF HEALTH (1989). *Report of the Expert Advisory Committee on Biocides*. London: HMSO.

GREAT BRITIAN. HEALTH AND SAFETY EXECUTIVE (1991). The control of legionellosis including legionnaires' diease. HS(G)70. London HMSO.

HERWALDT, L.A., GORMAN, G.W. & McGRATH, T. (1984). A new Legionella species, *Legionella feeleii* species nova, causes Pontiac fever in an automobile plant. *Annals of Internal Medicine*, **100**, 333–8.

MAYS, G.D. ed. (1989). *Water Fittings and Materials Directory*. Ashford, Kent: Martins.

SHANDS, K.N., HO, J.L., MEYER, R.D., GORMAN, G.W., EDELSTEIN, P.H., MALLISON, G.F., FINEGOLD, S.M., & FRASER, D.W. (1985). Potable water as a source of Legionnaires' disease. *Journal of the American Medical Association*, **253**, 1412–16.

4

BACTERIA, FUNGI AND OTHER MICRO-ORGANISMS

P. SHERWOOD BURGE

Introduction

This chapter will assess the main effects of microbiological contamination of the indoor environment on health, ranging from the rare but more serious diseases to the common and milder diseases. Extrinsic allergic alveolitis, humidifier fever, asthma, rhinitis, sick building syndrome and infection will be considered. The sources of microbiological contamination of the indoor air, particularly from humidifiers, chillers, duct work, water leaks and the building occupants will be reviewed. The assessment of the indoor environment by culturing organisms or measuring antigen will be described and methods of control evaluated. Legionella is covered in a separate chapter.

Effects on health

Extrinsic allergic alveolitis

Extrinsic allergic alveolitis is a potentially severe disease of the gas exchanging parts of the lungs (alveoli) which can lead to permanent disability and occasionally to premature death. The most common causes of the disease are unrelated to indoor air quality and are found in farmers allergic to mouldy hay (*M. faeni and T. vulgaris* and sometimes *Aspergillus fumigatus*) or to those who have become allergic to birds, such as budgerigars and pigeons.

The disease has an acute and a chronic form. The acute form characteristically starts 6 to 8 hours after exposure with a 'flu'-like illness, fever, chills, breathlessness, cough and general aches and pains. An acute attack usually lasts 1 to 2 days if further exposure is avoided but may last far longer. The chronic form is more common when there is continuous low level exposure, such as when birds are kept indoors at home. Successive exposure leads to increasing disease (unlike humidifier fever). Weight loss is a prominent feature in some cases and progressive breathlessness occurs. The chronic disease is usually fairly advanced by the time the diagnosis is made. The inflammation in the alveoli (alveolitis) leads to scarring and shrinking of the lung and often to permanent disability.

Extrinsic allergic alveolitis due to contamination of the indoor air in office workers was first described by Banaszak *et al.* in 1970 when four workers in an office building became ill. The source of the alveolitis was from the humidifier which was growing *Micropolyspora faeni*. This is the same organism responsible for most cases of farmers' lung and was growing in the humidifier which contained warm water. These moulds are thermophilic, i.e. they require warm temperatures in which to grow. There have been several other outbreaks with similar causes, mostly related to home cool mist and furnace humidifiers (Pestalozzi, 1959; Fink *et al*. 1971), and others where the exact cause is uncertain (Sweet *et al*, 1971; Stricker *et al.,* 1986; Newman Taylor *et al.,* 1978; Ganier *et al.,* 1980; Friend *et al.,* 1977). Other outbreaks have been attributed to *thermoactinomyces vulgaris* (Tourville *et al.,* 1972; Michel *et al.,* 1976; van Assendelft *et al.,* 1979) and *Aspergillus fumigatus* (Jacobs *et al.,* 1989), which are also well-recognised causes of extrinsic allergic alveolitis in other circumstances, particularly in farmers lung. Other implicated organisms include *Penicillium* species (Solley & Hyatt, 1980; Bernstein *et al.,* 1983) and particularly *Penicillium chrysogenum* and *Penicillium cyclopium* (Fergusson *et al.,* 1984), *Aureobasidium pululans* (Torok, de Weck & Scherrer, 1981; Woodard *et al.,* 1988), *Bacillus subtilis* (Molina *et al.,* 1982), *Cephalosporium* species (Patterson *et al.,* 198l) and *Sphaeropsidales* species (Baur *et al.,* 1988). All of these grow preferentially in warm water. Cold water spray humidifiers may become contaminated with different organisms and at least one outbreak of allergic alveolitis was related to thermophilic actinomyces (Robertson *et al.,* 1987). In this outbreak the humidifier of a printing works became heavily contaminated with many micro-organisms. The humidifier temperature was cold at 15 °C.

One outbreak in a textile works was found to be due to a new organism *Cytophaga allerginae* (Liebert *et al.,* 1984). This organism was isolated from among 700 found in the humidifier as the one most likely to be the cause of the disease from serological testing. The antibodies were directed to an endotoxin produced by this organism, rather than the whole organism, which might explain the problems others have had in identifying specific organisms as the cause of disease. *Cytophaga allerginae* has some similarities to flavobacteria, with which it was originally confused.

In Japan there is a seasonal variety of extrinsic allergic alveolitis particularly occurring between June and October. It occurs amongst family groups and remits when the patient is removed from his home. It is more common following damp weather and particularly affects residents of older wooden dwellings. For a long time the cause was elusive; *cryptococcae* were originally postulated as the cause, but these were never isolated from the patients' houses. It is now known to be due to *Trichosporon cutaneum*, a fungus which can be isolated from the houses, and to which affected individuals have precipitating antibodies in their blood (Shimazu *et al.*, 1984; Soda *et al.*, 1986). There are some cross-reacting antibodies between *cryptococcae* and *Trichosporon cutaneum* accounting for the original confusion. There is little work on this fungus outside Japan, but it is unlikely that it is confined to these islands. Allergic alveolitis in domestic dwellings due to moulds is rare, but has been described (together with asthma), due to the dry rot fungus *Merulius lacrymans* (O'Brien *et al.*, 1978), and to mould growth in a bedroom containing *Penicillium casei* and *Aureobasidium pullulans* (Torok, de Weck & Scherrer, 1981).

Outbreaks of extrinsic alveolitis (and more commonly symptoms more suggestive of humidifier fever) have also occurred due to contamination of mains water, particularly from shallow lakes in Finland where cases occurred following bathing, showering or taking a sauna (Muittari *et al.*, 1980; Aro *et al.*, 1980). This outbreak occurred during a dry summer when the lake from which the town's water was drawn turned blue-green. More than half of the population were affected. Excess coliform organisms and *Aspergillus fumigatus* were cultured from the water, as well as *Candida guillermondii* and an *Absidia* species, but precipitating antibodies to the individual organisms were not found in the blood of affected individuals. The endotoxin level in the water was 0.2–1.0 μg/ml. The outbreak has been attributed to the effects of endotoxin without further proof (Muittari *et al.*, 1982). The outbreak could not be cured by further chlorination of the water or replacing the filterbed sand, but was cured by using an alternative source for the town's water. This outbreak was not related to drinking the water, but to inhaling aerosols during showering, etc in the home.

Householders in New Guinea have developed a form of extrinsic allergic alveolitis when living in houses with thatched roofs. The cause is thought to be in the thatch, as precipitating antibodies to the thatch were found (Blackburn & Green, 1966). Thatch can also be the home of bloodsucking triatomine bugs in South America. These drop out of the thatch onto sleeping householders and transmit *Trypanosoma cruzi*, the cause of Chagas disease.

The identification of a case of extrinsic allergic alveolitis (or of humidifier fever or occupational asthma) requires a prompt investigation of possible sources. The most likely source within a building is in the humidifier but contaminated duct-work and mould growth in damp housing are also possible sources. The levels of exposure required to produce alveolitis are relatively high compared with levels required to produce other allergic diseases.

Humidifier fever

The symptoms of humidifier fever are similar to those of extrinsic allergic alveolitis with fever, sweating, aches in the limbs, sometimes breathlessness and wheeze. Again characteristically it starts 6–8 hours after exposure but continued exposure leads to tolerance. Attacks are usually worse following a break from work such as over the weekend and commonly occur on the first day of work only, although longer attacks may occur. There are no known long-term sequelae of humidifier fever and complete recovery is always thought to occur (Pickering, 1980; MRC Symposium, 1977; Pickering *et al.*, 1976; Ekstrom, 1980). The chest X-ray is always normal in this disease (in about half of the cases of extrinsic allergic alveolitis there are nodular changes on the X-ray). It is unclear why some patients develop extrinsic allergic alveolitis and some humidifier fever. The same contaminated humidifier may give rise to both diseases. The precipitating antibodies in the blood are similarly positive in both diseases and again relate more to exposure than to the disease itself. Many of the outbreaks have occurred in the printing industry, perhaps because such factories have been humidified for a long time and tend to have older humidifiers, partly perhaps because of the paper dust in the air which forms a good substrate for microbiological growth.

Both humidifier fever and extrinsic allergic alveolitis can be reproduced in individuals by exposing them to extracts of the contaminated humidifier water (Stricker *et al.*, 1986; Newman Taylor *et al.*, 1978; Friend *et al.*, 1977; Robertson *et al.*, 1987; Edwards and Cockcroft, 1981). An example of such a challenge is shown in Figure 4.1 where the worker has kept a

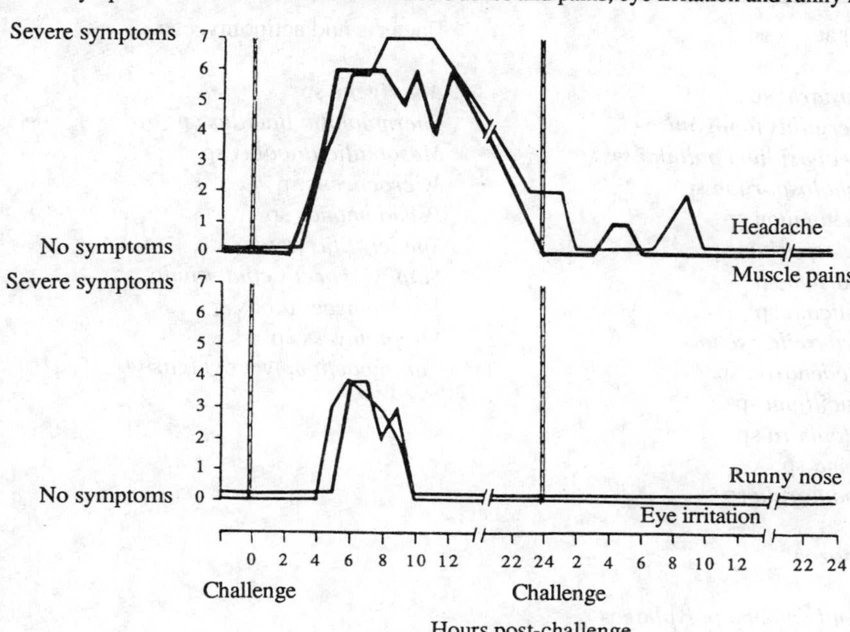

Fig. 4.1. An extract of water from the workplace humidifier was nebulised for 5 minutes each morning for two successive mornings, in a worker with humidifier fever. The record shows the results of hourly diary card assessments of symptoms. Following the first exposure headache, muscle pains, eye irritation and runny nose start after 6 hours and resolve within 24 hours. There is no reaction on the second day, showing that tolerance has developed.

symptom diary following exposure to a control solution and to antigen from a humidifier on separate days. The challenges were repeated on two consecutive days: the figure shows that the reaction was much less on the second day than on the first day, demonstrating that tolerance had developed. The identification of humidifier fever should again lead to a thorough search of the cause within the building.

The role of individual organisms in humidifier fever and extrinsic allergic alveolitis

Most outbreaks of building related extrinsic allergic alveolitis have resulted from exposure to aerosols from a mixed growth of a wide range of organ-

Table 4.1. *Organisms identified in a water spray humidifier, responsible for an outbreak of extrinsic allergic alveolitis, humidifier fever and occupational asthma*

Fungae	Bacteria and actinomycetes
Alternaria sp	*Alkaligene* sp
Aspergillus fumigatusa[a]	*Thermophilic Bacillus* sp
Aureobasidium pullulans[a]	*Mesophilic Bacillus* sp
Cephalosporium sp	*Micrococcus* sp
Chaetomium sp	*Pseudomonas* sp
Cladosporium sp[a]	*Sphaerotilus natans*
Fusarium sp	*Staphylococcus epidermidis*
Hemicola sp	*Streptomyces albus*
Mortierella zychae	*Streptomyces* sp
Oidodendron sp	*Thermoactinomyces vulgaris*[a]
Penicillium sp[a]	
Philophora sp	
Phoma sp	
Rhodotorula sp	
Protozoa	
Acanthamoeba polyphaga[a]	
Hartmanella vermiformis	
Vahlkamfia inornata	
Oikomonas terno	

[a] reported as a specific cause in some outbreaks

isms, many of which are usually not identified (Liebert *et al.*, 1984; Burge *et al.*, 1985; Austwick *et al.*, 1986). Table 4.1 lists the organisms identified in one humidifier responsible for an outbreak of extrinsic allergic alveolitis, humidifier fever and occupational asthma. No individual organism was shown to be the cause of this outbreak, but patients reacted to extracts of the humidifier water (Burge *et al.*, 1985). In some reports individual organisms have been thought to be the cause; however, in general, strict proof that these individual organisms are the cause is lacking. There seems to be cross reaction between the antigens from some, but not all, outbreaks (Austwick *et al.*, 1986). Most contaminated humidifiers contain amoebae as well as bacteria and fungae. The amoebae have been thought to be the cause of the symptoms on the grounds of antibody detection in affected individuals (Edwards, 1980); however, antibody levels do not differentiate between those with and without disease, and are likely to be another example of cross-reacting antibodies, as the amoebae ingest the bacteria also present in contaminated humidifiers (Finnegan *et al.,* 1987). Currently, opinion is divided between individual organisms, or a soluble product (perhaps endotoxin) from these organisms, as being the cause of individual outbreaks. It is possible that disease which progresses with successive exposure, such as typical extrinsic allergic alveolitis, requires sensitisation to particular and specific organisms, while disease that remits despite repeated exposure, such as humidifier fever, is less dependent on specific organisms, and may be due to toxins derived from the organisms (Rylander & Haglind, 1984).

Occupants who develop allergic alveolitis or humidifier fever also develop precipitating antibodies in the blood directed to the relevant antigens. These antibodies (called precipitins) are found in many more workers than actually have the disease and are more a measure of exposure than disease. They can, however, be used as a method of biological monitoring to see if the source of the contamination has been controlled, as the levels decline after exposure ceases (Finnegan *et al.,* 1985).

Asthma

Asthma is a disease of the airways which become inflamed and narrow resulting in variable breathlessness, chest tightness and often wheeze. Asthma is generally at its worst during the night and attacks initiated by exposure during the day may occur predominantly in the night. Asthma is common in the general population (about 5% of adults). Many asthmatics are better in air-conditioned buildings. However, the building itself may also be a source of the asthma. This can sometimes be because the particular allergen is in the indoor environment, such as an animal or bird. The most common indoor allergen in the United Kingdom is the house dust mite (*Dermatophagoides pteronyssinus*) which particularly is found in bedding and soft furnishing. It is only just too small to see with the naked eye. Its growth is favoured by moist, warm environments. It eats both moulds and skin scales. House dust mites are rare in desert climates and in northern Europe, probably because of the lack of humidity in the indoor environment. They are similarly rare at altitude in central Europe. The low levels of house dust mites in mountain resorts in Europe are probably responsible for the improvement in asthmatics sent to special schools there. Mite asthma has not been shown to be a significant problem in office buildings. Affected

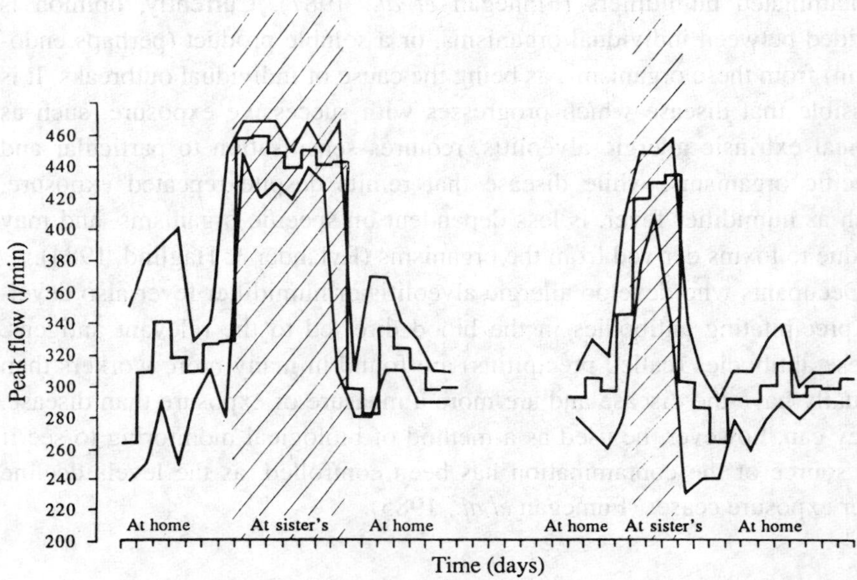

Fig. 4.2. Plot of daily maximum, mean and minimum peak expiratory flow rate in an asthmatic whose asthma was related to living in his own house. His symptoms subsided when he removed his ducted hot air heating.

individuals have been shown to have higher concentrations of mites in their houses than others (relative risk about seven fold), and come from homes that are older and more humid (Korsgaard, 1983). Korsgaard has recently shown that asthmatics with house dust mite sensitivity improve when moved into housing designed to keep the relative humidity below 40%, this being achieved with mechanical ventilation. Asthmatics without house dust mite sensitivity did not improve, suggesting that the lower number of mites rather than the lower humidity itself was the cause for the improvement (Harving *et al.*, 1988).

Asthma can occasionally be precipitated by living in a particular house. Such houses tend to be poorly insulated, damp and mouldy but it has been difficult to identify specific moulds as the cause of the asthma. An example of an asthmatic who is bad in his own house and improves in his sister's house is shown in Figure 4.2.

The precipitating factors in an asthmatic can be identified by relating the timing of attacks to exposures. Attacks may occur soon after exposure (an immediate reaction usually at its peak within about 20 minutes of exposure) or may be delayed for several hours (a late asthmatic attack commonly starting 4–8 hours after exposure and often lasting for 24 hours or more). At work the most common reactions occur several hours after exposure and are often at their worst during the night following work, often leading to delay in diagnosis as the patient looks to a cause at home rather than a cause at work. Occupational asthma can be identified when the attacks are related to an agent at work such that the worker is worse on a day at work and better on a day away from work. Once occupational asthma has developed non-specific precipitants such as exercise, infection, paint fumes, cold weather,

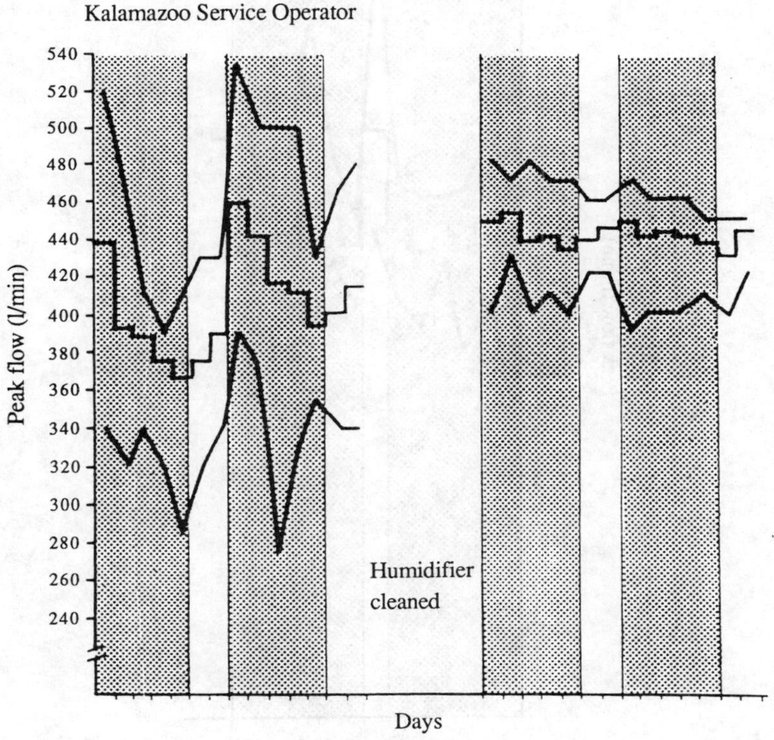

Fig. 4.3. Plot of daily maximum (top line), mean (middle line) and minimum (bottom line) peak expiratory flow rate in an asthmatic whose asthma was related to working in a building with a contaminated humidifier. The days at work have a stippled background, the days away from work a clear background. There is progressive daily deterioration during each of the first two workweeks, with worse values at the end of each working week. After the humidifier was cleaned the peak flow is much improved.

fog, etc can usually also induce the attacks. Most cases of occupational asthma due to poor indoor air quality have occurred in office buildings with either humidifiers or chillers which have become microbially contaminated (Burge *et al.,* 1985; Finnegan & Pickering, 1984). Solomon (1974) reported two cases due to contamination of home mist vaporizers with *Rhodotorula.* Other indoor sources are sometimes a problem. One outbreak of occupational asthma occurred in a bank where clerks were counting old banknotes which had become mouldy from storage in a damp room. The cause was thought

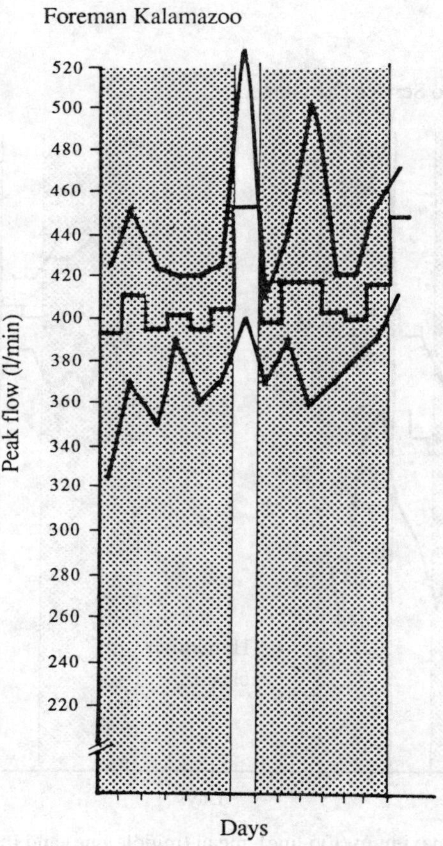

Fig. 4.4. Plot of daily maximum (top line), mean (middle line) and minimum (bottom line) peak expiratory flow rate in an asthmatic whose asthma was related to working in a building with a contaminated humidifier. The days at work have a stippled background, the days away from work a clear background. There is equivalent daily deterioration on each workday with rapid recovery on the one day off work.

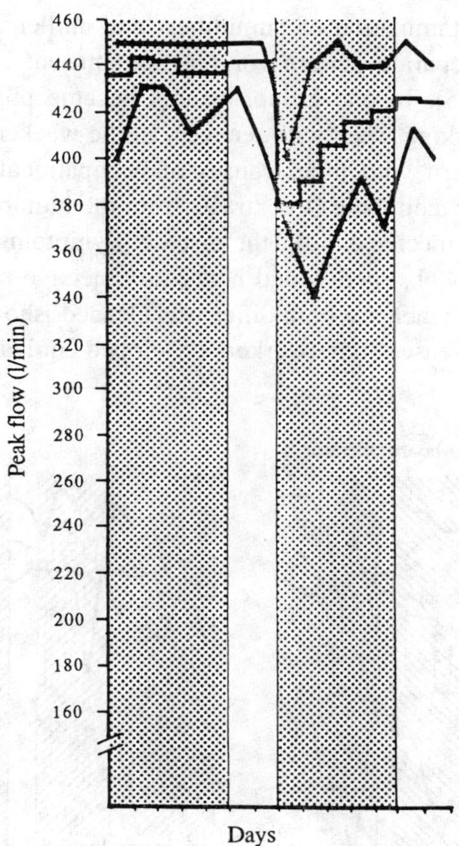

Fig. 4.5. Plot of daily maximum (top line), mean (middle line) and minimum (bottom line) peak expiratory flow rate in an asthmatic whose asthma was related to working in a building with a contaminated humidifier. The days at work have a stippled background, the days away from work a clear background. There is deterioration which is maximal on the first workday and improves with continuing exposure, showing that tolerance has developed.

to be the *Aspergillus fumigatus* growing on the notes (el Batawi *et al.* 1982). Other rare cases have arisen when material from a neighbouring factory has got into the air input of an office building (Carroll *et al.,* 1976). Neighbourhood cases have also arisen in schools and other buildings following emission of potent allergens into the air such as castor bean dust in feed mills, soya bean dust from unloading cargo vessels and aerosol produced from biotechnology plants containing yeasts and other organisms (Ekenvaal, 1983; Topping *et al.*, 1985).

It is likely that there are different mechanisms underlying occupational asthma due to contamination of humidifiers and chillers. In one outbreak, there were four separate patterns of reaction present in the work force (Burge *et al.*, 1985). In one group, asthma became progressively severe throughout the working week with recovery at the weekend (Fig. 4.3). This is a common pattern with other causes of occupational asthma and was associated with IgE antibodies to extracts from the humidifier, thereby suggesting an allergic mechanism for the disease. Symptoms and peak flow (a measure of airways obstruction and hence an objective measure of asthma severity) increased when the humidifier was cleaned, showing that this was the cause of the disease. Some workers developed equivalent daily deterio-

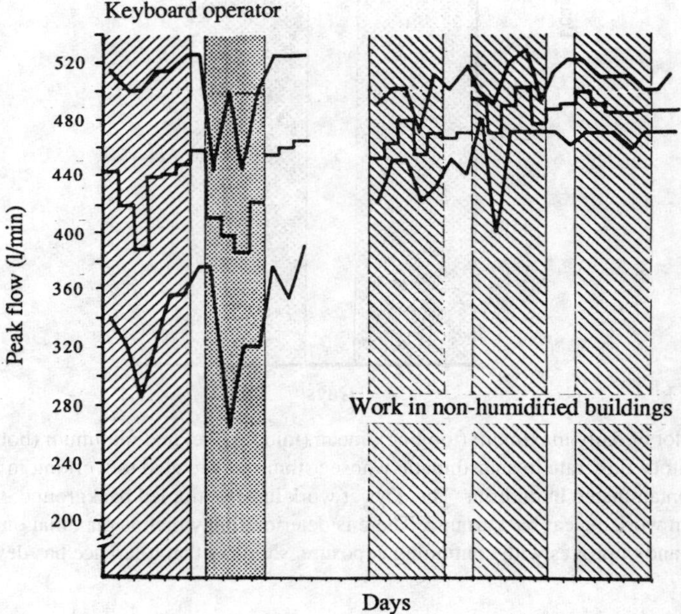

Fig. 4.6. Plot of daily maximum (top line), mean (middle line) and minimum (bottom line) peak expiratory flow rate in an asthmatic whose asthma was related to working in a building with a contaminated humidifier. The days at work have a stippled background, the days away from work a clear background. There is deterioration which is maximal midweek and improves with continuing exposure as a more severe variant of the first day pattern. During the final three workweeks he moved to a different building without a humidifier, when his asthma resolved.

ration in their asthma (Fig. 4.4) whereas others were worse on the first day or midweek (Figs. 4.5 and 4.6). These latter two groups were not associated with specific IgE to humidifier antigens and it is likely that the mechanism is different. There is some evidence that endotoxin exposure causes such reactions. This pattern of reaction is similar to that seen in cotton mills (Byssinosis) where there is a substantial amount of endotoxins in the air.

Allergic rhinitis

Allergic rhinitis is a disease similar to hay fever where the nose is inflamed and sneezing occurs. Running or blockage of the nose is also a symptom. Attacks may be temporarily related to acute exposures. Most reactions occur shortly after exposure although occasionally late reactions may develop. Asthma and rhinitis commonly go together but in some situations rhinitis occurs without asthma. The house dust mite (*Dermatophagoides pteronyssinus*) is a common cause for rhinitis occurring all the year round.

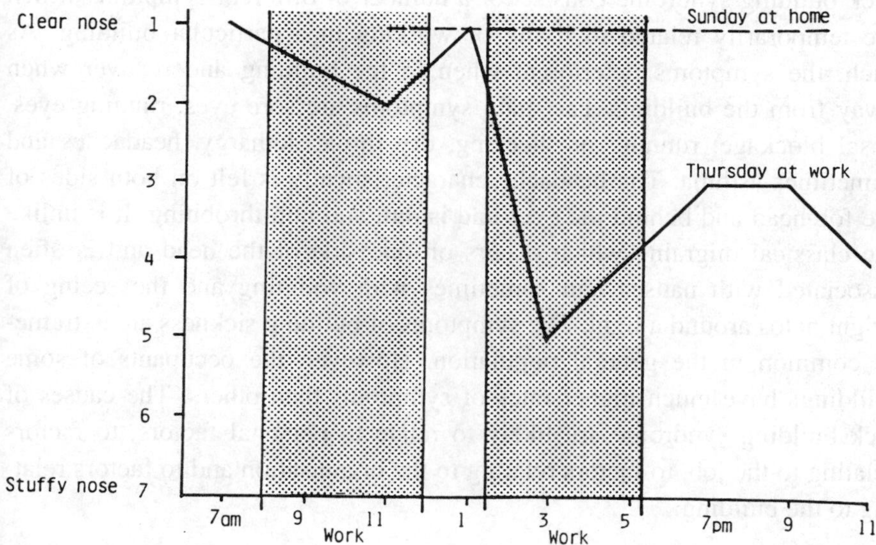

Fig. 4.7. Plot of two-hourly diary card scores for an office worker with occupational rhinitis. The periods at work have a shaded background, the periods away from work are clear. There was no rhinitis on a Sunday at home. On a Thursday at work the nose started to become stuffy in the morning, improved away from the building at lunchtime, and became much worse at work in the afternoon, with some resolution at home in the evening.

Rhinitis may also be caused by microbiological contamination of the indoor environment. Objective measures of rhinitis are difficult and are not applicable to routine use. Diary cards can, however, be used to assess the current cause of symptoms. An example is shown in Figure 4.7.

The health consequences of living in damp and mouldy houses are difficult to classify, as the major investigation has used self-completed questionnaires rather than medical diagnoses. Platt and colleagues (1989) in an important study, investigated 597 households, mainly in Scotland. About one-third of houses studied were damp or mouldy. Adults living in these houses had significantly more breathlessness, blocked noses, nausea, backache, fainting and bad nerves than occupants of dry houses. Children in the same damp houses had significantly more wheeze, runny noses, sore throat, headache and fever. The prevalence of symptoms increased with increasing levels of mould growth and dampness suggesting a causative for these factors. Confounding factors were well controlled.

Sick building syndrome

Sick building syndrome consists of a number of different symptoms which are temporarily related to living or working in a particular building. As such, the symptoms deteriorate when in the building and recover when away from the building. The main symptoms are sore eyes, running eyes, nasal blockage, running or sneezing, dry throat, lethargy, headaches and sometimes asthma. The headache characteristically is felt on both sides of the forehead and behind the eyes and is dull and non-throbbing. It is unlike the classical migraine which occurs on one side of the head and is often associated with nausea and sometimes with vomiting and the seeing of bright halos around a light. The symptoms of building sickness are extremely common in the general population. However, the occupants of some buildings have much higher rates of symptoms than others. The causes of sick building syndrome are likely to relate to personal factors, to factors relating to the job, to factors relating to the organisation and to factors relating to the building.

Personal factors

Females have more problems than males and workers lower down the hierarchy, independent of sex, have more problems than workers higher up the hierarchy. The reason why females have more problems than males is unclear. One hypothesis is that females have more body awareness and are therefore more likely to report symptoms although this has not been substantiated by psychological tests. Workers lower in the hierarchy often work in poorer quality parts of the building, are more crowded, have more boring jobs and are less able to get improvements made in the indoor environment than are managers who usually have better accommodation, are more mobile within the workplace and have more control of their environment.

Agents in the workplace

Some office materials are potential causes of occupational asthma and rhinitis and perhaps headaches as well. This particularly applies to acrylics used in correcting fluids although these are fairly rare causes of problems. Some old fashioned photocopying machines produced increased concentrations of ozone in the air which is a respiratory irritant. Some modern photocopiers use a wet process which increases solvents in the air which have been incriminated as a cause of sick building syndrome although they are not a common problem.

The rol‿ of indoor plants in the production of asthma is unclear. They can be used to add water vapour to the indoor air, but they liberate potential allergens. Weeping Fig trees are commonly used indoors; those who look after the trees, rather than office occupants, have developed occupational asthma from allergens exuded from their leaves. This might be a greater problem in less humid offices, where there is a greater water loss from the leaves (Axelsson *et al.*, 1987).

Organisational factors

Public sector buildings have more problems than private sector buildings. This may either be due to the poor quality of the buildings in the public sector, poorer maintenance, or perhaps to the management structure and career prospects, etc in the workforce. Current evidence would suggest that the main cause for the difference relates to the buildings rather than to the job itself. This is supported by a study in one building with both a public sector and a private sector occupant where the rates of symptoms were very similar in the two groups (Fig. 4.8). Conversely, in a separate public sector workforce some worked in an air-conditioned building with humidifiers and

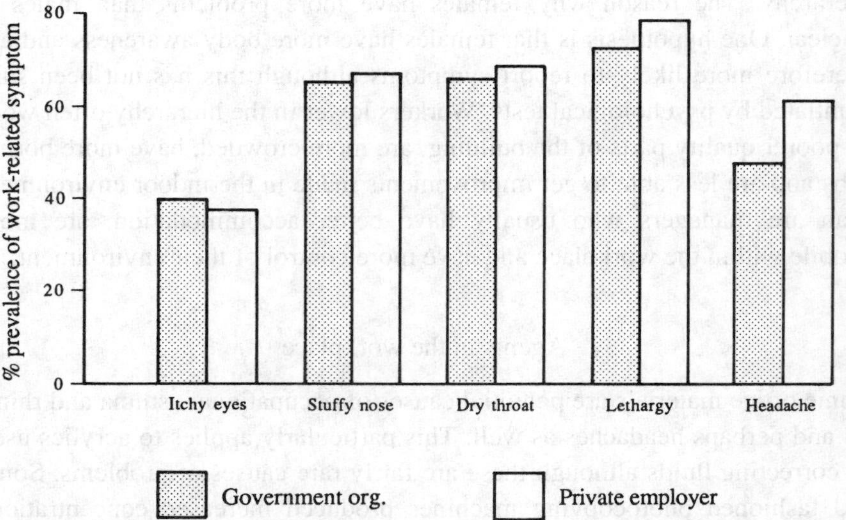

Fig. 4.8. Comparison of the prevalence of symptoms between two separate workforces, one private sector and one a government agency, both working on different floors of the same air-conditioned building. The prevalence of work-related symptoms is very similar in both groups.

chillers and some worked in a naturally ventilated building with opening windows. Both buildings were connected and workers moved their workplace from time to time from one building to another. The workforce in the air-conditioned building had two to three times as many symptoms as those in the naturally ventilated building, suggesting that the building rather than the job was the cause of the problem (Fig. 4.9) (Robertson *et al.*, 1985).

Building-related factors

The role of indoor air microbiological contamination in the aetiology of sick building syndrome is less clear than with the other diseases and syndromes that we have dealt with so far. From an epidemiological point of view, buildings with humidifiers and chillers have more problems than buildings without these and the best buildings tend to be naturally ventilated (Fig. 4.10). However, naturally ventilated buildings usually have higher levels of colony-forming units of bacteria and fungi in the air than air conditioned buildings so it is unlikely that the organisms in the indoor air *per se* are the cause of the symptoms. Again it is possible that soluble products from the organisms are responsible. Symptoms of building sickness are also

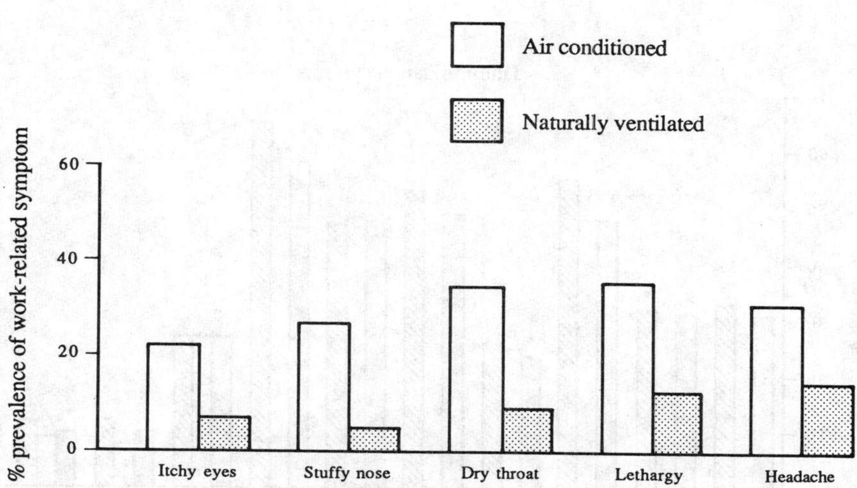

Fig. 4.9. Prevalence of work-related symptoms in one workforce split between two connected buildings, one naturally ventilated with a lower prevalence of symptoms, and one air-conditioned with a higher prevalence.

more common in buildings with more active smokers suggesting that passive cigarette smoke exposure is also a factor (Robertson *et al.*, 1985). Problem buildings also tend to have more exposed shelves and cupboards with paper dust, their floors tend to be dirtier and there tends to be more fabric in the indoor environment which could possibly absorb and desorb airborne contaminants (Valbjorn & Skov, 1987). Symptoms are also more common in large open-plan offices than in cellular offices and are more common when individual control of temperature, ventilation and lighting is inadequate (Burge *et al.*, 1987).

There are few adequate studies of the relationship between microbial contamination with viable organisms and building sickness, but current evidence does not support a direct relationship between the two. Turiel *et al.* (1983) found higher levels inside a problem air-conditioned building than outside (suggesting an indoor source). However, the maximum level found was 179 cfu/m^3 when the ventilation was in recirculation mode, which is a low level. British office buildings studied by Pickering and colleagues have shown consistently higher colony-forming unit levels in naturally ventilated buildings with less symptomatic workers compared with problem air-conditioned buildings. Average levels for all buildings were under 1000 cfu/m^3

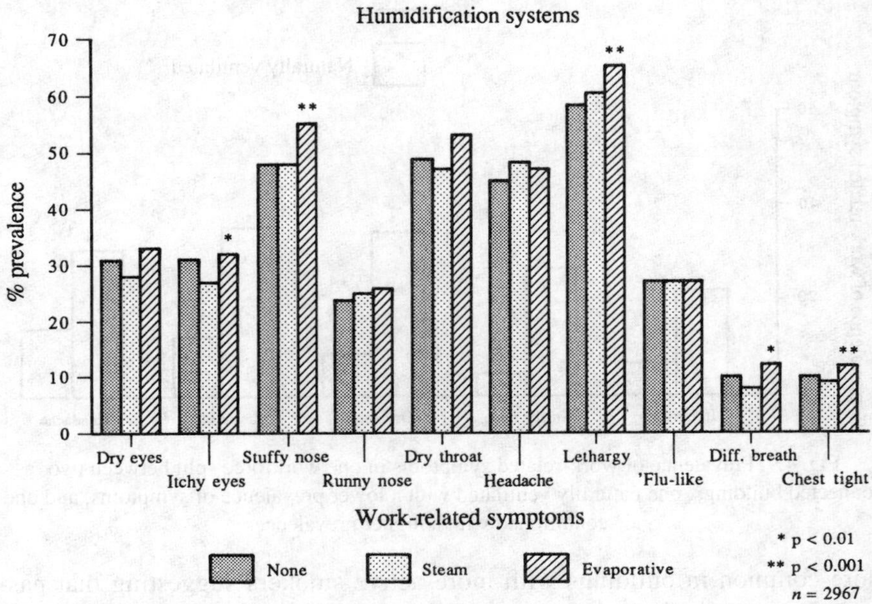

Fig. 4.10. The average number of work-related symptoms per worker (the building symptom index, maximum possible 10), in workers in buildings with different types of ventilation.

(Austwick *et al.*, 1989). They also studied a building with a clean room, the room being positively pressurised and supplied with air from high grade filters. The microbial load in the clean area (bacteria plus fungae) was 125 cfu/m^3 compared to 400 cfu/m^3 for the area supplied by the standard air-conditioning system. Total dust levels were also halved in the clean area. Despite these changes, the symptoms of sick building syndrome were, if anything, greater in the clean area (Harrison *et al.*, 1987).

The epidemiological evidence shows a correlation between the presence of air-conditioning with either chillers or humidifiers and an increased prevalence of building sickness symptoms (Fig. 4.11). The limited microbiological data show that total viable organism counts do not account for this difference. The most likely explanation is that a soluble product which requires the presence of water in the air-conditioning system is the missing link (a similar situation to humidifier fever where the humidifier is clearly the usual source but the microbiology is equally confusing). An alternative explanation is that either there is a confounding variable that is not being measured, or that other agents associated with humidifiers or chillers are the cause. Biocides are a possible link. They are not designed to be put into air-conditioning systems and subsequently inhaled (as apart from cooling towers where the biocide aerosol disperses in the outside air); their use may result in widespread exposure to these agents via the ventilating system, and are therefore a potential cause for sensitisation. Sensitisation to glutaraldehyde (Jackuck *et al.,* 1989) and isothiazolinones (Clark, 1987), both sometimes used in air-conditioning systems, have been reported.

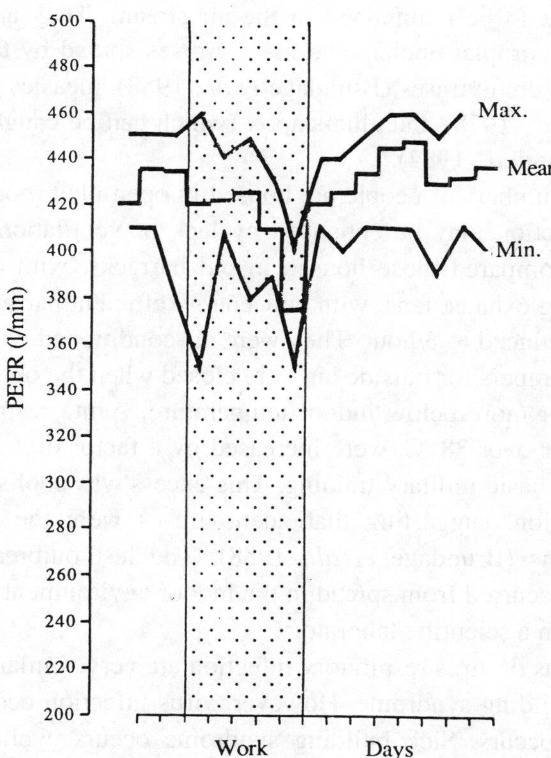

Fig. 4.11. Plot of daily maximum (top line), mean (middle line) and minimum (bottom line) peak expiratory flow rate in an asthmatic whose asthma was related to working in a building with contaminated chillers. There were no humidifiers in this building. The days at work have a stippled background, the days away from work a clear background. There is deterioration which is progressive during the workweek and improves on days away from work.

Infection

Infections may be spread via the indoor air environment and in particular by ventilating systems. The development of infection depends on the type of organism in the air and how virulent the strain is. It also depends on the inhaled dose, although there is no absolute dose which is safe or an absolute dose which is bound to cause infection. Some people are more susceptible than others. The micro-organism needs to survive in the air so that it is still living when inhaled. Survival is usually favoured by humidity and warm temperature (Hemmes *et al.,* 1960; Knight, 1980; Arundel *et al.,* 1986). The main infectious agents which spread via the indoor air are viruses. They do not multiply outside humans but need to stay alive before inhalation and therefore require to be maintained in the air stream. They are commonly present in water droplet nuclei. The main viruses spread by the indoor air are the common cold viruses (Brundage *et al.,* 1988), measles (Bloch *et al.,* 1985; Riley *et al.,* 1978) and chickenpox (which can be caught from shingles) (Gustafson *et al.,* 1982).

When large numbers of people are housed in open-plan rooms, spread of respiratory infection may be enhanced by lack of ventilation. A study of army recruits compared those housed in old barracks, with opening windows and ceiling exhaust fans, with new energy efficient barracks where air changes were reduced to 3/hour. They were air-conditioned with 95% of air recirculated. Dampers for outside air were closed when the outside temperature was 5 °C or more below indoor temperature. Acute respiratory infections, with fever over 38 °C, were increased by a factor of 1.51 during the seven weeks of basic military training. This excess was abolished by adenovirus vaccination, suggesting that adenoviruses were the cause of the excess infections (Brundage *et al.,* 1988). The last outbreak of human smallpox also occurred from spread in the indoor environment via a ventilation system from a scientific laboratory.

The symptoms of virus respiratory infection are very similar to the symptoms of sick building syndrome. However, virus infection occurs once and then recovery occurs. Sick building syndrome occurs cyclically, usually every week at work with improvement at the weekend. It is highly unlikely that sick building syndrome is caused by infection.

The main bacterium causing infection which is related to buildings is Legionella which is discussed elsewhere. This, however, is more often acquired outside the building than inside although it can be acquired from hot water systems in hospitals and hotels. Patients in hospital may be particularly susceptible to organisms with low virulence. An outbreak of pneumonia with *Nocardia asteroides* occurred in a renal transplant unit (Lovett *et al.*, 1981; Houang *et al.*, 1980). This outbreak occurred when the ground outside the hospital was being excavated for new building work. The ground was a possible source of the organism, which appeared to be spread by aerosol.

Tuberculosis is very occasionally spread by ventilating systems (Iwasaki, 1979) but it more often occurs from direct patient to patient spread via coughing. The risks of catching tuberculosis from an infectious source at work are very small. However, children are particularly susceptible and can catch it from infected teachers, swimming pool attendants, etc. Adult tuberculosis varies greatly in infectivity. A significant risk of infection arises when the tubercle bacilli can be seen on direct microscopy of the sputum (a smear positive case). One outbreak occurred in a playgroup of preschool children where an attending adult had untreated smear positive tuberculosis. Eleven out of 60 children attending the playgroup developed tuberculosis as well as 21 children whose only contact was at a Christmas party (Bosley *et al.*, 1986). One outbreak occurred in a swimming pool where the source case was a lifeguard usually seated on a high stool by the side of the pool, which had mechanical ventilation directing the airflow from end to end of the pool. 16 of 3764 children attending the pool developed tuberculosis (Rao *et al.*, 1980). Wales *et al.* (1985) described an outbreak spread from a smear-positive teacher in an open-plan primary school. There were 24 cases of tuberculosis in her class and the next door class separated by a movable partition, with a small number of tuberculin conversions in other children within the school. These outbreaks are likely to have been spread directly by droplet nuclei to susceptible children who had not been protected with BCG vaccination. Working next door to a colleague with infectious tuberculosis at work usually results in far less than a 1% risk of acquiring tuberculosis and does not warrent surveillance of work contacts on medical grounds. There has, however, been one outbreak of tuberculosis in office workers in Japan, in an air-conditioned building with no makeup fresh air, and carbon dioxide levels often over 1000 mg/m^3 (Shigematsu & Minowa, 1985).

Sources of microbiological contamination in the indoor environment

Micro-organisms usually enter a building either via the incoming water or via the incoming air. Most of these organisms are not pathogenic in themselves (i.e. they do not cause direct infection) and must multiply to much higher numbers before problems develop. Multiplication is fostered by warm damp environments. Water spray humidifiers are a particular place where such proliferation may occur. It appears that the initial microbiological growth is on the walls of the humidifier tanks, followed by colonisation of the surface. Bacterial colonies in particular are found attached to the walls and may spread for several centimetres into the water, from where they can break off and enter the airstream to be distributed around the building (Austwick *et al.,* 1986). Colonisation may also be a problem in the drip feed-type humidifiers where there is a cellulose matrix through which water drips and evaporates leading to ideal culture conditions for micro-organisms which sometimes grow within such a matrix and shed spores into the incoming air. There is a very wide range of organisms which can be found in such systems including bacteria, fungi, algae, and sometimes nematodes, Table 4.1 (Liebert, 1984; Burge *et al.,* 1985; Austwick *et al.,* 1986). If biocides are put into such systems, fewer organisms grow but the biocides then become aerosolised and spread around the building. Few of these biocides are designed for human exposure and may in themselves be responsible for some of the symptoms.

Chillers which sometimes dehumidify the air are also a possible source of microbiological contamination. Water collects in the trays under chillers when the supply air is moist and the chiller is in action. Many of these are not properly drained and many are in positions which are difficult to service. At least one outbreak of occupational asthma occurred due to contaminated chillers in a building without a humidifier (Fig. 4.11). Other possible places where micro-organisms grow are in duct work which has become damp or where water has collected. Some buildings have developed microbiological contamination following major water leaks (Patterson *et al.,* 198l; Morey, 1984; Hodgson *et al.,* 1985). In domestic dwellings, the major reason for microbiological indoor contamination is from poor ventilation, poor insulation and dampness.

Individual occupants are also a source of microbiological contamination and shed bacteria and fungi from the skin and also, when infected, by the respiratory tract.

Risk assessment

At present, the main method of assessing risk is by finding disease in the building occupants. These diseases are described above. If there is microbiological contamination then the microbiological source such as a contaminated humidifier can be used to prepare antigen to which sensitised individuals will produce precipitating antibody. A survey of precipitating antibodies in building occupants can be used as a method of risk assessment. Such studies have shown that the levels of precipitating antibodies relate both to the length of exposure and to cigarette smoking (Finnegan *et al.*, 1985).

Attempts have been made to quantify microbiological contamination in the indoor air as opposed to the source. This can be done by exposing fungal and bacterial culture media to indoor air and culturing the micro-organisms and producing a colony count (colony-forming units). Harriet Burge (no relation) has reviewed the methods used (Burge H., 1984). Separate culture media are required to grow bacteria and fungae. The type of each organism grown also depends on the culture medium and the temperature of incubation. There are no standardised methods and so results cannot usually be compared directly between studies. Such studies have shown that, in general, naturally ventilated buildings have higher numbers of colony-forming units than air-conditioned buildings and that humidified buildings have similar levels to non-humidified buildings. Occasionally there is a microbiological source downstream of the ventilation plant room, such as wall mounted chillers and humidifiers and flooded duct-work. A colony-forming unit count of 1000/m^3, collected by unspecified means, has been used as an indicator for further investigation in office buildings (Morey, 1984). This study found fungal counts averaging 10 200 cfu/m^3 in previously flooded areas, 3020 cfu/m^3 in non-flooded areas and 1290 cfu/m^3 outside for fungae grown at 28–30 °C. Similar values for thermophilic fungae grown at 42–45 °C were 3090, 650 and 220 cfu/m^3 respectively. The outbreak of allergic alveolitis reported by Bernstein *et al.*, (1983) found fungal colony-forming unit levels in an office approaching 10 000/m^3. Here the source was in heating and chilling coils distal to the air intake filtration. Similarly, when the source was a cold mist vaporizer in the room, the levels were over 1000 cfu/m^3 in 11 of 15 homes studied (Solomon, 1974). The buildings investigated by Austwick and colleagues, where workers were developing severe disease, all had airborne microbial levels below 1000 cfu/m^3. The sources were in the central plantroom humidifiers themselves (Austwick *et al.*,

1986). The levels of colony-forming units may therefore be satisfactory for identifying sources of microbiological contamination within the working space, or in the duct-work distal to the plantroom, but these levels cannot be used to detect severe microbiological contamination in the central plant-room, as water spray humidifiers precipitate particulate material, and subsequent filtration may also remove viable organisms from the airstream, leaving soluble products to enter the workplace and cause disease.

Measurement of antigen in air

Because of the problems in interpreting colony-forming units, and because some outbreaks of disease occur with low viable organism levels, techniques of antigen detection in air have been developed. Antigen quantification requires a supply of the relevant antigen and antibodies. Sludge from humidifiers, sometimes partially purified, has been used as the antigen, and the serum from affected workers as the source of antibody. Antibody could also be raised in animals if diseased workers were not available. Air samples for antigen detection currently need to be of very large volume, collected with a sampler running at about 1700m^3 in 24 hours. Antigen is then collected from the filter and assayed by competitive inhibition of the weighed antigen previously collected. Using this technique, Reed and colleagues (1983) found levels of antigen in air around 0.3 μg/m^3 in working areas and 0.4 μg/m^3 in duct-work in a factory with an outbreak of allergic alveolitis. Since antigens sometimes vary between oubreaks, a panel of relevant antigens and antibodies will be needed if these methods are to be used to monitor workplaces without identified problems.

The relevance of moulds in domestic dwellings is unclear from a strictly health point of view. Hunter *et al.* (1988) studied 15 English and 47 Scottish homes between November and early April. Thirty-seven types of filamentous fungae were isolated. The number of viable colony forming units of fungae, including yeasts, varied between 12 and 449 800/m^3. There were considerable differences both between different buildings and over time in the same building. High levels were associated with vacuuming. Only one-third of all samples had levels below 200 cfu/m^3. The most common species found were *Penicillium* species (91% of dwellings), *Cladosporium* sp (89%), *Aspergillus* sp, mainly *versicolor* (75%), *Ulocladium* sp (62%), *Geomyces pannorum* (57%), *Sistotrema brinkmannii* (51%) and yeasts (94%).

Possibilities for control

The most appropriate method for control is to remove the sources of microbial growth from the building. The main source in buildings in the UK is from humidifiers and chillers. The need for humidification in temperate climates is unclear, and is based more on the standards for comfort rather than any health effects. Although many workers in problem buildings complain that the air is too dry, there is no correlation between measured water content of the air and perceived dryness, i.e. the worker is responding to some other aspect of the indoor environment (Andersen *et al.*, 1973). There is no evidence that relative humidity levels down to 25% are associated with increasing symptoms of building sickness, although levels over 50% are associated with more housedust mite infestation and a greater viability of airborne virus. In temperate climates, the need for humidification for health reasons is unproven and probably unnecessary. If humidification is necessary, steam humidifiers are much less likely to lead to microbial growth and are preferred. Once contamination of a humidifier has occurred it can be very difficult to eradicate. Reed *et al.* (1983) treated a problem humidifier with a chlorinating agent, slimicide and by regular cleaning. Antigen levels in the workplace decreased from levels up to 15 μg/m^3 down to 0.06 μg/m^3. No fresh cases of sensitisation occurred once the antigen levels were reduced below 0.05 μg/m^3 (and probably none below 0.2 μg/m^3). Other outbreaks have not been controlled by cleaning and chlorination, and have required replacement of the water spray with steam humidification (Finnegan *et al.*, 1985). In this type of outbreak, the initial antigen levels were between 1 and 10 μg/m^3.

Chillers are a possible source of problems: control rests with proper cleaning and maintenance. This requires good design so that the chillers are drained and the whole of the drip tray is readily accessable for cleaning. Chillers sited in central plantrooms are much easier to service than chillers spread around the ceiling space, and wall cavities of peripheral offices. Mould growth in domestic dwellings can be controlled by reducing the humidity. Excessive humidity is often a problem of poor design, with inadequate ventilation, poor insulation and inappropriate heating, such as paraffin stoves.

Control can often be achieved by better maintenance. Building services management in the UK tends to be a low status job, with unclear accountability. Maintenance workers are often employed by the building owner and have no clear responsibility to building users in leased buildings. Bonuses often depend more on energy conservation than on standards of maintenance. In a recent study of sick building syndrome, three pairs of buildings with high and low prevalences of building sickness were studied. Environmental measurements of temperature, humidity, air change rates, lighting and noise did not differentiate between the good and bad buildings. The principal differences were in the standards of maintenance in the plantrooms, which, in turn, related to the documentation provided by the plant installers (Burge *et al.,* 1990). The importance of design and maintenance has been recognised in the WHO guidelines described below.

Standards and recommendations

At present, there is insufficient evidence to set standards for indoor air microbial contamination. Recent recommendations have been made by Walkinshaw (1988), based on unpublished work of Miller. Their recommendations for fungae are:

1. The presence of certain pathogens such as *Aspergillus fumigatus* and certain toxogenic fungae such as *Stachybotrus atra* should be considered unacceptable.
2. More than 50 colony-forming units per cubic metre should be a reason for concern if only one species is present.
3. Less than l50 colony-forming units per cubic metre should be considered as acceptable if there is a mixture of species.
4. Less than 300 colony-forming units per cubic metre should be considered as acceptable if the predominant species are cladosporium or other common phylloplane fungae.

The WHO working group on indoor air quality, suspended viable particles, allergens and other particulate matter, has included the following amongst its recommendations:

1. The building and its heating, ventilation and air conditioning systems should not produce biological contaminants which are introduced into the ventilation air. If biocides are unavoidable, they should be prevented from entering space which can be occupied.
2. Standards and building codes should ensure effective maintenance of ventilation systems by specifying adequate access paths, regular inspection and maintenance schedules.

3. In a building, in which occupants cannot effectively control the quality of ventilation air, an individual who is responsible for this task should be identified to the occupants.
4. The maintenance personnel of public and office buildings should be given adequate training for routine inspection and maintenance for the building systems.

The American Conference of Government Industrial Hygienists has made the following recommendations for office buildings (Morey *et al.*, 1986).

1. The sum of the total counts of fungae, bacteria (35 °C) and thermophilic *Actinomycetes* (55 °C) should not exceed 10 000 cfu/m^3.
2. If the levels of *Bacillus* spp and gram negative rods exceed 500 cfu/m^3, a building associated source is presumed.
3. If any of the following exceed 500 cfu/m^3 remedial action is needed: *Staphylococcus, Streptococcus, Corynebacterium, Acinetabacter, Pseudomonas* or *Micrococcus*.
4. Levels of *Actinomycetes* over 500/m^3 are considered high.

References

ANDERSEN, I., LUNDQVIST G.R. & PROCTOR, D.F. (1973). Human perception of humidity under four controlled conditions. *Archives of Environmental Health*, **26**, 22–7.

ARO, S., MUITTARI, A., VIRTANEN, P. (1980). Bathing fever epidemic of unknown aetiology in Finland. *International Journal of Epidemiology* **9**, 215–18.

ARUNDEL, A.V., STERLING, E.M., BIGGIN, J.H. & STERLING, T.D. (1986). Indirect health effects of relative humidity in indoor environments. *Environmental Health Perspectives*, **65**, 351–61.

VAN ASSENDELFT, A., FORSEN, R.O., KESKINEN, H. & ALANKO, K. (1979). Humidifier associated extrinsic allergic alveolitis. *Scandinavian Journal of Work Environment & Health*, **5**, 35–41.

AUSTWICK, P.K.C., DAVIES, P.S., COOK, P. & PICKERING, C.A.C. (1986). Comparative microbiological studies in humidifier fever. In *Maladies des climatiseurs et des humidificateurs*. Molina, C. ed., INSERM Symposium. **135**, 155–64.

AUSTWICK, P.K.C., LITTLE, S.A., LAWSON, L., PICKERING, C.A.C. & HARRISON, J. (1989). *Microbiology of Sick Buildings in Airbourne Deteriagens and Pathogens*, ed. Flannigan, B., pp. 122–8, Biodeterioration Society, Kew.

AXELSSON, I.G.K., JOHANSSON, S.G.O. & ZETTERSTROM, O. (1987). Occupational allergy to weeping fig in plant keepers. *Allergy*, **42**, 16–7.

BANASZAK, E.F., THIEDE, W.H. & FINK, J.N. (1970). Hypersensitivity pneumonitis due to contamination of an air conditioner. *New England Journal of Medicine*, **283**, 271–6.

el BATAWI, M.A. & el BATAWI, Y.A. (1982). Exposure to airborne fungi of workers handling old paper currency in a bank. *Proceedings of the XVI Congress of Occupational Health*, pp. 712–13, Cairo.

BAUR, X., BEHR, J., DEWAIR, M., EHRET, W., FRUHMANN, G., VOGELMEIER, C., WEISS, W. & ZINKERNAGEL, V. (1988). Humidifier lung and humidifier fever. *Lung*, **166**, 113–24.

BERNSTEIN, R.S., SORENSON, W.G., GARABRANT, D., REAUX, C. & TREITMAN, R.D. (1983). Exposures to respirable, airborne Penicillium from a contaminated ventilation system: clinical, environmental and epidemiological aspects. *American Industrial Hygiene Association Journal*, **44**, 161–9.

BLACKBURN, C.R.B. & GREEN, W. (1966). Precipitins against extracts of thatched roofs in the sera of New Guinea natives with chronic lung disease. *Lancet*, **2**, 1396–7.

BLOCH, A.B., ORENSTEIN, W.A., EWING, W.M., SPAIN, W.H., MALLISON, G.F., HERMANN, K.L. & HINMAN, A.R. (1985). Measles outbreak in a pediatric practice: airborne transmission in an office setting. *Pediatrics*, **75**, 676–83.

BOSLEY, A.R.J., GEORGE, G. & GEORGE, M. (1986). Outbreak of pulmonary tuberculosis in children. *Lancet* , **i**, 1141–3.

BRUNDAGE J.F., SCOTT, R.McN., LEDNAR, W.M., SMITH, D.W. & MILLER, R.N. (1988). Building-associated risk of febrile acute respiratory disease in army trainees. *Journal of the American Medical Association*, **259**, 2108–12.

BURGE, H.A. (1984). Indoor sources of indoor airborne microbes. In *Indoor Air and Human Health*. Gammage, R.B., Kaye, S.V., Jacobs, V.A., eds, Proceedings of the 7th Life Sciences Symposium. Chelsea, M1., Lewis Publishers Inc.

BURGE, P.S., FINNEGAN, M., HORSFIELD, N., EMERY, D., AUSTWICK, P., DAVIES, P.S. & PICKERING, C.A. (1985). Occupational asthma in a factory with a contaminated humidifier. *Thorax*, **40**, 248–54.

BURGE, P.S., HEDGE, A., WILSON, S., HARRIS BASS, J. & ROBERTSON, A.S. (1987). Sick building syndrome: a study of 4373 office workers. *Annals of Occupational Hygiene*, **31**, 493–504.

BURGE, P.S., JONES, P., HEDGE, A. & ROBERTSON, A.S. (1990). Environmental measurements in pairs of buildings with high and low prevalences of sick building syndrome. *Annals of Occupational Hygiene,* in press.

CARROLL, K.B., SECOMBE, C.J.P. & PEPYS, J. (1976). Asthma due to non-occupational exposure to toluene di-isocyanate. *Clinical Allergy*, **6**, 99–104.

CLARK, E.G. (1987). Risk of Isothiazolinones. *Journal of Social and Occupational Medicine*, **37**, 30–1.

EDWARDS, J.H. (1980). Microbial and immunological investigations and remedial action after an outbreak of humidifier fever. *British Journal of Industrial Medicine*, **37**, 55–62.

EDWARDS, J.H. & COCKCROFT, A. (1981). Inhalation challenge in humidifier fever. *Clinical Allergy*, **11**, 227–35.

EKENVALL, L., DOLLING, B., GOTHE, C.J., EBBINGHAUS, L., von STEDINGK, L.V. & WASSERMAN, J. (1983). Single cell protein as an occupational hazard. *British Journal of Industrial Medicine*, **40**, 212–15.

EKSTROM, N.A. (1980). Frequency of illness and immuno-response related to infected humidifiers in a book-bindery. *European Journal of Respiratory Disease*, 61 suppl. **107**, 151–3.

FERGUSSON, R.J., MILNE, L.J.R. & CROMPTON, G.K. (1984). Penicillium allergic alveolitis: faulty installation of central heating. *Thorax*, **39**, 294–8.

FINK, J.N., BANASZAK, E.F., THIEDE, W.H. & BARBORIAK, J.J. (1971). Interstitial pneumonitis due to hypersensitivity to an organism contaminating a heating system. *Annals of Internal Medicine*, **74**, 80–3.

FINNEGAN, M.J. & PICKERING, C.A.C. (1984). Occupational asthma and humidifier fever in air conditioned buildings. In *Indoor Air*, Berglund, B., Lindvall, T. & Sundell, J. eds, Vol. 3. pp 257–62. Stockholm: Swedish Council for Building Research.

FINNEGAN, M.J., PICKERING, C.A.C., DAVIES, P.S. & AUSTWICK, P.K. (1985). Factors affecting the development of precipitating antibodies in workers exposed to contaminated humidifiers. *Clinical Allergy*, **15**, 281–92.

FINNEGAN, M.J., PICKERING, C.A.C., DAVIES, P.S., AUSTWICK, P.K.C. & WARHURST, D.C. (1987). Amoebae and humidifier fever. *Clinical Allergy*, **17**, 235–42.

FRIEND, J.A.R., GADDIE, J., PALMER, K.N.V., PICKERING, C.A.C. & PEPYS, J. (1977). Extrinsic allergic alveolitis and contaminated cooling-water in a factory machine. *Lancet* **i**, 297–300.

GANIER, M., LIEBERMAN, P., FINK, J. & LOCKWOOD, D.G. (1980). Humidifier lung. An outbreak in office workers. *Chest*, **77**, 183–7.

GUSTAFSON, T.L., LAVELY, G.B., BRAWNER, E.R., HUTCHESON, R.H., WRIGHT, P.F. & SCHAFFNER, W. (1982). An outbreak of nosocomial airborne varicella. *Pediatrics*, **70**, 550–6.

HARRISON, J., PICKERING, C.A.C., FINNEGAN, M.J. & AUSTWICK, P.K.C. (1987). The sick building syndrome, further prevalence studies and investigation of possible causes. In *Indoor Air*, Proceedings of the 4th conference on indoor air quality and climate, Vol. **2**, 487–96, Berlin: Institute for water, soil and air hygiene.

HARVING, H., KORSGAARD, J. & DAHL, R. (1988). Mechanical ventilation in dwellings as a preventative measure in mite-asthma. NER *Allergy Proceedings*, **9**, 283 [abstract].

HEMMES, J.H., WINKLER, K.C. & KOOL, S.M. (1960). Virus survival as a seasonal factor in influenza and poliomyelitis. *Nature*, **188**, 430–1.

HEPALAININ, A. (1989). *Bacterial Aerosols in Indoor Air*, Finland: National Public Health Institute, Helsinki (NPHl A3/1989).

HODGSON, M.J., MOREY, P.R., ATTFIELD, M., FINK, J.N., RHODES, W.W., SORENSON, W. & VISVESVARA, G.S. (1985). Pulmonary disease associated with cafeteria flooding. *Archives of Environmental Health*, **40**, 96–110.

HOUANG, E.T., LOVETT, I.S., THOMPSON, F.D., HARRISON, A.R., JOEKES, A.M. & GOODFELLOW, M. (1980). Nocardia asteroides infection, a transmissible disease. *Journal of Hospital Infection*, **1**, 31–40.

HUNTER, C.A., GRANT, C., FLANNIGAN, B. & BRAVERY, A.F. (1988). Mould in buildings: the air spora of domestic dwellings. *International Biodeterioration*, **24**, 81–101.

IWASAKI, T. (1979). Results of an investigation of infantile tuberculosis in a Japan Red Cross maternity hospital. *Nihon Ishikai Zasshi*, **56**, 1140–46.

JACKUCK, S.J., BOUND, C.L., STEEL, J. & BLAIN, P.G. (1989). Occupational hazard in hospital staff exposed to 2% glutaraldehyde in an endoscopy unit. *Journal of Social and Occupational Medicine*, **39**, 69–71.

JACOBS, R.L., ANDREWS, C.P. & JACOBS, F.O. (1989). Hypersensitivity pneumonitis treated with an electrostatic dust filter. *Annals of Internal Medicine*, **110**, 115–18.

KNIGHT, V. (1980). Viruses as agents of airborne contagion. *Annals of the New York Academy of Science*, **353**, 147–56.

KORSGAARD, J. (1983). Mite asthma and residency. A case control study of the impact of exposure to housedust mites in dwellings. *American Review of Respiratory Disease*, **128**, 231–5.

LANCET (1978). Smallpox research after Birmingham. *Lancet Editorial*, **2**, 560.

LIEBERT, C.A., HOOD, M.A., DECK, F.H., BISHOP, K. & FLAHERTY, D.K. (1984). Isolation and characterization of a new Cytophaga species implicated in a work related disease. *Applied Environmental Microbiology*, **48**, 936–43.

LOVETT, I.S., HOUANG, E.T., BURGE, S., TURNER-WARWICK, M., THOMPSON, F.D., HARRISON, A.R., JOEKES, A.M. & PARKINSON, M.C. (1981). An outbreak of Nocardia asteroides infection in a renal transplant unit. *Quarterly Journal of Medicine*, **50**, 123–35.

MICHEL, F.B., AIACHE, J.M., GAYRAUD, J.P. & PIERREDON, M. (1976). Maladie des humidificateurs, a propos de 4 observations. *Revue Française Maladies Respiratories*, **4**, 537–44.

MOLINA, C., AIACHE, J.M., BEDU, M., MENAUT, P., WAHL, D., BRESTOWSKI, J. & GRALL, Y. (1982). La maladie des climatiseurs. Resultats d'une enquete de medicine du travail. *Nouvelle Presse Medicale*, **11**, 2325–9.

MOREY, P.R. (1984). Case presentations: problems caused by moisture in occupied spaces in office buildings. *Annals of the American Conference of Industrial Hygienists*, **10**, 121–7.

MOREY, P., CHATIGNY, M., OTTEN, J., FEELEY, J., BURGE, H., La FORCE, F.M. & PETERSON, K. (1986). Airborne viable micro-organisms in office environments. Sampling protocols and analytical procedures. *Applied Industrial Hygiene*, **1**, R19–23.

MRC Symposium. (1977). Humidifier fever. *Thorax*, **32**, 653–63.

MUITTARI, A., RYLANDER, R. & SALKINOJA-SALONEN, M. (1980). Endotoxin and bath water fever. *Lancet*, **ii**; 89.

MUITTARI, A., KUUSISTO, P. & SOVIJAVI, A. (1982). An epidemic of bath water fever – endotoxin alveolitis? *European Journal of Respiratory Diseases*, **63**, suppl 123,108–16.

NEWMAN TAYLOR, A., PICKERING, C.A.C., TURNER-WARWICK, M. & PEPYS, J. (1978). Respiratory allergy to a factory humidifier contaminant presenting as pyrexia of undetermined origin. *British Medical Journal*, **2**, 94–5.

O'BRIEN, I.M., BULL, J., CREAMER, B., SEPULREDA, R., HARRIES, M., BURGE, P.S. & PEPYS, J. (1978). Asthma and extrinsic allergic alveolitis due to *Merulius lacrymans. Clinical Allergy*, **8**, 535–42.

PATTERSON, R., FINK, J.N., MILES, W.B., BASICH, J.E., SCHLENTER, D.B., TINKELMAN, D.G. & ROBERTS, M. (198l). Hypersensitivity lung disease presumptively due to *Cephalosporium* in homes contaminated by sewage flooding or by humidifier water. *Journal of Allergy and Clinical Immunology*, **68**, 128–32.

PESTALOZZI, C. (1959). Febrile group diseases in a carpentry shop caused by inhalation of moisture from air humidifiers contaminated with moulds. *Schweizerische Medizinische Wochenschrift*, **89**, 710–13.

PICKERING, C.A.C. (1980). Humidifier fever. In *Advances in Allergology and Immunology*, Oehling, A., ed. Proceedings of the 10th International Conference on Allergology, pp. 357–60. Oxford: Pergamon Press.

PICKERING, C.A.C. (1982). Humidifier fever. *European Journal of Respiratory Disease*, **63** suppl. 123,104–7.

PICKERING, C.A.C., MOORE, W.K.S., LACEY, J., HOLFORD-STEVENS, V.C. & PEPYS, J. (1976). Investigation of a respiratory disease associated with an air-conditioning system. *Clinical Allergy*, **6**, 109–18.

PLATT, S., MARTIN, C.L., HUNT, S.M. & LEWIS, C.W. (1989). Damp housing, mould growth and symptomatic health state. *British Medical Journal*, **298**, 1673–8.

RAO, V.R., JOANES, R.F., KILBANE, P. & GALBRAITH, N.S. (1980). Outbreak of tuberculosis after minimal exposure to infection. *British Medical Journal*, **281**, 187–9.

REED, C.E., SWANSON, M.C., LOPEZ, M., FORD, A.M., MAJOR, J., WITMER, W.B. & VALDES, T.B. (1983). Measurement of IgG antibody and airborne antigen to control an industrial outbreak of hypersensitivity pneumonitis. *Journal of Occupational Medicine*, **25**, 207–10.

RILEY, E.C., MURPHY, G. & RILEY, R.L. (1978). Airborne spread of measles in a suburban elementary school. *American Journal of Epidemiology*, **107**, 421–32.

ROBERTSON, A.S., BURGE, P.S., HEDGE, A., SIMS, J., GILL, F.S., FINNEGAN, M., PICKERING, C.A. & DALTON, G. (1985). Comparison of health problems related to work and environmental measurements in two office buildings with different ventilation systems. *British Medical Journal*, **291**, 373–6.

ROBERTSON, A.S., BURGE, P.S., WIELAND, G.A. & CARMALT, M.H.B. (1987). Extrinsic allergic alveolitis caused by a cold water humidifier. *Thorax*, **42**, 32–7.

RYLANDER, R., HAGLIND, P. (1984). Airborne endotoxins and humidifier disease. *Clinical Allergy*, **14**, 109–12.

SHIGEMATSU, I. & MINOWA, M. (1985). Indoor infection in a modern building. *Tokai Journal of Experimental Clinical Medicine*, **10**, 407–13.

SHIMAZU, K., ANDO, M., SAKATA, T., YOSHIDA, K. & ARAKI, S. (1984). Hypersensitivity pneumonitis induced by *Trichosporon cutaneum. American Review of Respiratory Disease*, **130**, 407–11.

SODA, K., ANDO, M., SHIMAZU, K., SAKATA, T., YOSHIDA, K. & ARAKI, S. (1986). Different classes of antibody activities to *Trichosporon cutaneum* antigen in summer-type hypersensitivity pneumonitis by enzyme-linked immunosorbent assay. *American Review of Respiratory Disease*, **133**, 83–7.

SOLLEY, G.O. & HYATT, R.E. (1980). Hypersensitivity pneumonitis induced by *Penicillium* species. *Journal of Allergy and Clinical Immunology*, **65**, 65–70.

SOLOMON, W.R. (1974). Fungus aerosols arising from cold-mist vaporizers. *Journal of Allergy and Clinical Immunology*, **54**, 222–8.

STRICKER, W.E., LAYTON, J.E., HOMBURGER, H.A., KATZMANN, J.A., SWANSON, M.C., HYATT, R.E. & REED, C.E. (1986). Immunologic response to aerosols of affinity-purified antigen in hypersensitivity pneumonitis. *Journal of Allergy and Clinical Immunology*, **78**, 411–16.

SWEET, L.C., ANDERSON, J.A., CALLEIS, Q.C. & COATES, E.O. (1971). Hypersensitivity pneumonitis related to a home furnace humidifier. *Journal of Allergy and Clinical Immunology*, **48**, 171–8.

TOPPING, M.D., SCARISBRICK, D.A., LUCZYNSKA, C.M., CLARKE, E.C. & SEATON, A. (1985). Clinical and immunological reactions to *Aspergillus niger* among workers in a biotechnology plant. *British Journal of Industrial Medicine*, **42**, 312–18.

TOROK, M., de WECK, A.C. & SCHERRER, M. (1981). Allergische alveolitis intolge verschimmelung der schladzimmerwand. *Schweiz. Med. Wschr.*, **III**, 924–9.

TOURVILLE, D.R., WEISS, W.I., WERTLAKE, P.T. & LEUDEMANN, G.M. (1972). Hypersensitivity pneumonitis due to contamination of home humidifier. *Journal of Allergy and Clinical Immunology*, **49**, 245–51.

TURIEL, I., HOLLOWELL, C.D., MIKSCH, R.R., RUDY, J.V., YOUNG, R.A. & COYE, M.J. (1983). The effects of reduced ventilation on indoor air quality in an office building. *Atmospheric Environment* , **17**, 51–64.

VALBJORN, O. & SKOV, P. (1987). Influence of indoor climate on the sick building syndrome prevalence. In *Indoor Air 1987*, Proceedings of the 4th international conference on indoor air quality and climate. Vol. 2. pp. 593–7, Berlin: Institute of water, soil and air hygiene.

WALES, J.M., BUCHAN, A.R., COOKSON, J.B., JONES, D.A. & MARSHALL, B.S.M. (1985). Tuberculosis in a primary school: the Uppingham outbreak. *British Medical Journal*, **291**, 1039–40.

WALKINSHAW, D.S. (1988). Canadian indoor air quality standards, guidelines and research activities. In *Indoor and Ambient Air Quality*, Perry, R. & Kirk, P.W.W., eds. pp. 345–54. London: Selper Ltd.

WOODARD, E.D., FRIEDLANDER, B., LESHER, R.J., FONT, W., KINSEY, R. & HEARNE, F.T. (1988). Outbreak of hypersensitivity pneumonitis in an industrial setting. *Journal of the American Medical Association*, **259**, 1965–9.

WHO (1990). *Euro reports and studies III.* Indoor air quality: organic pollutants. Copenhagen: World Health Organization Regional Office for Europe.

5

THE OXIDES OF NITROGEN

J. M. SNEDDON

It is recognised increasingly that indoor air quality could be an important factor affecting respiratory function. People in Northern Europe and North America now spend more than 70% of their time indoors, either in the home, at work, or in transport, so indoor exposure constitutes the major fraction of total individual exposure to many environmental pollutants. With energy conservation strategies in industrial countries limiting indoor–outdoor air exchange, such pollutants can occur in the indoor environment at higher concentrations than outdoors. These pollutants include gases such as carbon monoxide (CO) and nitrogen dioxide (NO_2) which are produced by combustion of fossil fuels, as well as formaldehyde and organic vapours which evaporate from solvents, resins, and glues used in the construction of both buildings and furniture. The indoor atmosphere can also be contaminated with fine particles such as dust, powders, aerosols (from spray cans), fungal spores and other organisms. Water vapour itself is not usually considered an environmental pollutant, but the concentrations generated in an enclosed space, such as a kitchen using gas for cooking, can favour the growth of house dust mites, moulds, and fungi, which in many cases can lead to allergic reactions with a respiratory component. In addition, radioactive decay products from trace amounts of ^{238}U in building materials and soils are present at higher concentrations in enclosed spaces than outdoors.

This chapter is concerned with indoor, primarily domestic, exposure to the oxides of nitrogen. It has been suggested that repeated exposure to the oxides of nitrogen from gas burning appliances and/or tobacco smoke may aggravate pre-existing lung disorders, increase susceptibility to respiratory infection, particularly in children, and perhaps contribute to the development of chronic pulmonary disease. The evidence for this is reviewed critically here.

The major contributors to atmospheric contamination with the oxides of nitrogen are the forces of nature, where bacteria, volcanic activity and lightning lead to a background level of about 9.4 ng/m^3 in the global atmosphere. In terms of global generation of nitrogen oxides, man's contribution is small, although it is human activity that determines the average urban air concentration due to the combustion of fossil fuels and industrial processes. In the industrial environment, workers may be exposed to a variety of nitrogen oxides which would not only include nitric oxide (NO) and nitrogen dioxide (NO_2) but also higher oxides as well as nitrous and nitric acids. For domestic exposure, only NO and NO_2 are thought to have biological importance.

In most urban areas, annual mean NO_2 concentrations in outdoor air are in the range 0.02–0.09 mg/m^3. Reliance on such ambient outdoor levels to characterise exposure levels may be inadequate, as it is generally believed that NO_2 released from gas burning appliances and from smoking is most important with respect to respiratory health. In most rural and low pollution urban areas the ambient NO_2 concentration is about 0.01–0.02 mg/m^3. In the same areas, in homes using gas for cooking, indoor concentrations in winter average 0.07–0.08 mg/m^3 (summer 0.03–0.04 mg/m^3), whereas, in homes using electricity indoor NO_2 concentrations are similar to those outdoors at all times of the year. Cigarette smokers also experience voluntarily an intense, repeated exposure to high concentrations of gases and particles in tobacco smoke which may contain 0.272–1.231 mg/m^3 of nitric oxide.

Toxicity of nitrogen oxides

Nitrogen dioxide is a relatively insoluble gas and little is removed by the mucus secretions in the upper airways so that, after inhalation, it penetrates into the lower airways. Pathological studies in animals demonstrate that the zone between the alveolar and terminal bronchiolar airways is most sensitive to NO_2 exposure. At the ultrastructural level, morphological changes occur in pulmonary tissues at extremely low exposure levels, which could lead to impaired blood–gas exchange, though the mechanism by which this happens is unclear.

In relation to epidemiological studies linking indoor pollution by NO_2 and respiratory illness in children, the most interesting response in animals may be increased susceptibility to respiratory infection. In mice and rats, exposure to NO_2 (0.75–9.44 mg/m^3) for 2–3 months results in a higher mortality rate when the animals are infected by airborne *Klebsiella pneumonie*. After 12 months' exposure, mice were unable to clear viable bacteria from their lungs. Squirrel monkeys exposed to the relatively high concentration of 9.4 mg/m^3 for 1 month show increased susceptibility to infection with *Klebsiella pneumonie*, marked decrease in respiratory function, and a reduction in bacterial defence mechanisms which allows the infecting organism to multiply. However, the episodes of acute respiratory infection in children which are used as a clinical indicator in epidemiological studies are primarily viral in origin, and it is unclear whether NO_2 exposure potentiates viral as well as bacterial infection.

Our knowledge of the effects of NO_2 inhalation in humans comes from controlled laboratory studies and a small number of occupational accidents where exposure levels have been much higher. There is a wide variability in the acute mortality from NO_2 among different laboratory species but extrapolation of the animal data gives an estimate for the 1 hr LD_{50} in man of 319 mg/m^3. This agrees with accidental human exposure.

Smell is a sensitive indicator of exposure as the odour of NO_2 is perceptible by some persons at 0.02 mg/m^3 and for most at 0.37 mg/m^3. Throat irritation may not occur until concentrations of 212 mg/m^3 are reached and cough at 353 mg/m^3. This lack of immediate, pronounced upper airway irritation allows exposed subjects to breathe high concentrations of NO_2 for some time without distress, and onset of acute respiratory symptoms is delayed for 3–30 hours. The patient then becomes acutely ill with paroxymal coughing, wheezing, frothy blood-stained sputum, nausea, vomiting, increasing dyspnoea, restlessness and anxiety. Death from respiratory failure may occur. For the survivors, there is a period of apparent recovery followed by a relapse into a second acute illness which can be more severe than the first.

Laboratory studies in humans require the administration of NO_2 at concentrations sufficient to elicit a measured physiological response, usually quantitated as changes in baseline pulmonary functions such as specific airway resistance (SG_{aw}), flow rate, or forced expiratory volume in 1 s (FEV_1), and on the bronchoconstriction induced by inhalation of cholinergic stimulants. Such studies on the acute, short-term health effects may have limited relevance to chronic health effects resulting from long-term exposure to NO_2 levels which are below those required to elicit such acute physiological responses. In addition, the published human studies show a marked inconsistency in their findings, which may result from the use of inappropriate or insensitive methods to measure pulmonary function, and to the variable number of susceptible individuals in the study population, the so-called 'responders'. In their 1980 review, Shy and Love conclude that the published experimental studies showed no statistically significant changes in pulmonary function following exposure to NO_2 below 2.9 mg/m^3 for 2 hours. Other controlled laboratory studies have demonstrated that the most frequently observed effect following single short-term exposure to NO_2 is an increase in airway resistance. This phenomenon occurs in most individuals tested following exposure to 3.8–13.2 mg/m^3 for 1 or 2 hours, but has been inconsistently demonstrated at lower exposure levels.

In addition to NO_2, the indoor environment may contain a significant concentration of NO. In a number of common combustion processes, the amount of NO_2 initially produced is usually small compared with that of NO, with a NO/NO_2 ratio greater than 2. Nitric oxide is normally considered to be a very reactive gas, rapidly oxidised in air to NO_2 and the higher oxides. In pure air, at concentrations above 1%, the oxidation of NO is almost instantaneous; however, the rate of oxidation slows as the concentration in air decreases, so at more usual concentrations e.g. 37.6 mg/m^3 the half-life is 2.6 hours. Controlled inhalation studies in humans have demonstrated that, in the airways, the oxidation of NO proceeds at approximately the same rate as it does in pure air so that after inhalation 80–90% of the inhaled NO is rapidly transferred from the alveoli to the blood without being oxidised to NO_2. Following absorption into the blood stream the NO combines almost immediately with haemoglobin in the red blood cells to form nitrosyl-haemoglobin, which is then rapidly oxidised to yield a mixture of nitrite and nitrate. Nitric oxide is also one of the principal components of the gas phase of fresh cigarette smoke. At 600 °C, the minimum temperature for the burning tip of a cigarette during active smoking, the breakdown of NO_2 to NO and oxygen is complete. The maximum concentration produced per cigarette can range between 0.56 and 1.88 mg/m^3 NO for a poorly ventilated room.

In experimental studies, where tobacco smoke is filtered to remove particulate material, the oxidation of NO to NO_2 is very rapid. However, in whole smoke the NO appears to be stabilised, perhaps by constituents in the particulate phase, which prevents the interaction of NO with oxygen-centred free radicals. This observation could have important toxicological implications. If the gas phase can interact with the particulate phase, not only could this prolong the half-life of a contaminant gas such as NO, but it may also potentiate the toxic effects of an absorbed gas on pulmonary function by promoting deposition and retention in the respiratory tract, or by providing a surface for the catalytic transformation of the gas into a more toxic form. Such a gas/particle interaction for NO_2 has been demonstrated in mice, where short-term exposure to 47 mg/m^3 NO_2 or inhalation of fine carbon particles was without gross pathological effects in mouse lung. However, when the mice were exposed to carbon particles previously exposed to NO_2 the mice developed focal destructive lesions with loss of cells from the alveolar walls (Boren, 1964).

Mechanisms of toxicity

The mechanistic basis for NO_2 is unclear. Nitrogen dioxide is a strong oxidising agent and it is evident that NO_2 can oxidise important biological constituents such as polyunsaturated lipids in cell membranes as well as sulphydryl groups which form part of the cellular defence mechanisms against oxidative damage. However, the evidence regarding lipid peroxidation as a primary mechanism of NO_2-induced lung damage is open to more than one interpretation. The possibility that the measured alteration in lipid peroxidation is a consequence of cell death makes it impossible to decide whether it is the cause of NO_2 toxicity or a consequence of cellular damage induced by other mechanisms. The recent studies of Sagai and Ichinose (1987) do provide evidence that lipid peroxidation may be a primary mechanism of NO_2-induced pulmonary damage. In these studies, rats were exposed to increasing concentrations of NO_2 for periods of 9 to 27 months. In the early stages of exposure the activities of the antioxidative enzymes which protect cells from oxidative stress increased, but then decreased during the period of chronic exposure. During this latter period, there was a clear dose–response relationship between NO_2 exposure and lipid peroxidation as reflected in increased exhalation of the gas ethane, and this varied inversely with the activities of the protective antioxidative enzymes. This study demonstrated that chronic exposure to relatively high concentrations of NO_2 leads to a reduction in cellular defence mechanisms which may result in progressive damage to cell membranes (lipid peroxidation) leading to cell injury and death. In this case individuals with low cellular reserves of antioxidants (ascorbic acid, glutathione, etc) may be at greater risk for the adverse health effects of oxidant gases such as NO_2 or O_3. This hypothesis has been tested experimentally in man by investigating the effects of increasing body ascorbic acid levels by pretreatment with vitamin C. The administration of vitamin C does not affect baseline pulmonary function, or apparently protect against histamine-induced bronchoconstriction. However, it reduces or abolishes methacholine-induced bronchoconstriction in normals, and methacholine or exercise induced bronchoconstriction in asthmatics. Vitamin C administration is also reported to abolish the potentiation of methacholine-induced bronchoconstriction induced by inhalation of 3.73 mg/m^3 NO_2 for 60 min. (Mohsenin, 1987).

Measurement of exposure

Epidemiological studies on respiratory symptoms caused by indoor pollutants require quantification of the concentration of the contaminant in the atmosphere and an appropriate marker of individual exposure. The atmospheric concentration of nitrogen oxides can be determined by several chemical methods, such as chemoluminescence or the Greiss reaction, with various degrees of sophistication and sensitivity. For individual NO_2 exposure personal monitors to sample the atmosphere have been used, but evaluating the consequences of individual exposure to NO_2 is more difficult.

Clinical assessment and tests of pulmonary function have proved useful in studying the effects of active smoking, but are less discriminating when applied to the study of ETS or other environmental pollutants. For indoor contaminants, such as NO_2, which are present in concentrations that do not elicit measurable physiological changes, suitable biochemical markers of individual exposure are needed.

Studies on humans who were active smokers have demonstrated a dose–response effect between the number of cigarettes smoked and urinary hydroxyproline excretion, and one study (Kasuga, 1985) has attempted to use this method as a marker for NO_2 exposure. In this study, hydroxyproline excretion was measured in children exposed to varying concentrations of NO_2 derived from automobile exhaust, smoking among members of the same family, and the type of heating used in the home. All homes used gas or kerosene cooking stoves. An analysis of the data demonstrated a significant relationship between hydroxyproline excretion in non-smoking children and parents and the number of cigarettes smoked by the smoking parent. There was an additional effect due to automobile exhaust independent of the effect of ETS. The relationship between hydroxyproline excretion and personal NO_2 exposure in the indoor environment was assessed by estimating NO_2 exposure with personal exposure meters. Indoor NO_2 levels were found to be two to three times that of the outdoor concentration in winter but almost identical in summer. There was a small but significant seasonal variation in urinary hydroxyproline excretion but the winter increase was small compared to the effect of ETS exposure. Personal NO_2 exposure was also unrelated to cigarette smoke exposure in both active smokers and persons exposed to ETS, which perhaps reflects the virtual absence of NO_2 from fresh cigarette smoke.

Thus the hydroxyproline technique is not only limited by this apparent insensitivity to NO_2, but also by a number of confounding factors. Diet, pregnancy and specific diseases such as hormone disorders or any other disease that may alter collagen metabolism impose obvious limitations on the technique. Urinary hydroxyproline levels are high in children and rapidly decline with age to reach a constant level at about age 20, so values for children have to be closely age matched. Future research should attempt to define a more appropriate biochemical marker.

Indoor contamination by NO_2 and respiratory disease in children

A number of large-scale epidemiological studies have suggested that there is an increase in respiratory illness in children exposed to NO_2 generated in the home (for review see Morrow, 1984). In the Chattanooga School Children Study (Shy *et al.*, 1970) it was reported that children residing in areas of high ambient outdoor NO_2 concentrations had the highest incidence of respiratory illness. However, the ambient air contained other pollutants which may have been the causative agents, or their presence, in addition to NO_2, was required to elevate illness rates. In the Columbian study (Keller *et al.*, 1979) a 12 month monitoring of 441 families divided into those using gas for cooking and those using electricity was undertaken. Continuous chemoluminescence methods were employed over 3-day periods in 46 homes in order to measure instantaneous variation in indoor NO_2 and NO concentrations in relation to cooking times. In gas cooking homes NO_2 levels ranged from 1.8–19 mg/m^3, and NO from 1.8–46 mg/m^3, three to four times higher than in electric cooking homes, but there were no significant differences in reported respiratory illness between the gas and electricity cooking homes. Nor was there a greater incidence of respiratory illness in gas cooking homes with NO_2 above the median than in households below the median. Other studies have suggested that there may be a correlation between indoor NO_2 exposure and respiratory health in children (see Morrow, 1984, for references) although many studies are compromised by failure to account adequately for confounding factors. The problem is illustrated in two excellent studies.

Speizer *et al.* (1980) originally reported that NO_2 produced from gas cookers had a negative effect on respiratory function in children under 2 years of age. One of the confounders in this study was the socio-economic status of the parents and various criteria were applied to the study population to reduce bias introduced by this factor. In the initial study the criteria adopted to define socio-economic status were not sufficiently discriminating and in a second larger study (Ware *et al.,* 1984) socio-economic status was defined by the number of years of parental schooling. Analysis of the data from this larger group, using the new criteria, only partially supported the original observations that gas cooking affected pulmonary function in children. Thus, while respiratory symptoms were slightly higher in children living in gas cooking homes, this was not statistically significant after controlling for socio-economic status, place of residence and sex.

A similar secular drift in the association between indoor NO_2 levels and respiratory illness was found by Melia and her colleagues. In 1977 they studied 5738 children aged 6-11 years from 28 randomly selected areas of Scotland and England (Melia *et al.,* 1977). Children from homes where gas was used for cooking were reported to have more respiratory symptoms than children from homes where electricity was used for cooking ($P < 0.01$ for boys; $P < 0.001$ for girls). Asthmatic children were excluded from this study and this gas effect appeared to be independent of the effects of age, father's social class, latitude, population density, family size or outdoor levels of particulates and SO_2. Four years later a different group of children of the same age, from the same areas were studied (Melia *et al.,* 1979). Data on family smoking habits were collected and after allowing for this and other variables there still appeared to be an excess of children with respiratory symptoms in gas cooking homes, although the association was weaker. In addition, the relative risk for respiratory illness in children from gas cooking homes varied considerably between age-groups and between boys and girls, and there were groups where the risk was either negligible or greater in homes using electricity. In a further study, 8086 children aged 6–11 from poor social classes were studied during the winter in the town of Middlesbrough (Florey *et al.,* 1979). Weekly outdoor NO_2 measurements were made and compared with NO_2 concentrations in kitchens and 25% of bedrooms. The gas cooker was the main factor associated with high levels of NO_2 indoors and these levels were linked to the recent respiratory illness history of both parents and children, supported by direct measurement of respiratory function in the children. The analysis of the data took account of a wide range of interfering variables but no association was found between illness, respiratory functions, and kitchen levels of NO_2. There was a weak association between bedroom NO_2 levels and illness, but no association with respiratory function. In a further, smaller study, comparable winter indoor NO_2 levels were found in homes where gas was used for cooking but in this study there was no association between respiratory illness and bedroom levels of NO_2 (Melia *et al.*, 1982*a,b*). There is thus conflicting evidence of the association between NO_2 exposure in the home and respiratory illness in children, and the lack of consistency in the findings must throw doubt upon the primary hypothesis.

The other major source of indoor nitrogen oxides is tobacco smoke, which, if fresh, contains predominantly NO, but following ageing, NO_2. More than 30 epidemiological studies have been published on respiratory symptoms or disease in children in relation to the smoking habits of their household. Almost all cross-sectional studies which compare children of smoking parents with the children of non-smoking parents report an increased prevalance of respiratory symptoms in children of smoking parents. However such studies can be confounded by under-reporting of active smoking on the part of school age children, recall bias leading to over-reporting of symptoms by parents, and the incidence of respiratory infection in the parents. Such symptoms are more frequent in smoking families, and children have a higher prevalance of respiratory symptoms if they live in a household with adults with the same symptoms, regardless of the family smoking habits. When the relationship between adult and child respiratory illness is taken into consideration, the association between tobacco smoke exposure and childhood respiratory symptoms becomes significantly weaker.

Individuals with chronic lung disease, such as asthmatics, are more sensitive than normals to the acute inhalation of near ambient concentrations of SO_2. However, whether such individuals are more sensitive to the acute, irritating effects of other common air pollutants such as tobacco smoke, O_3 or NO_2 is unclear. In epidemiological studies on the health effects of environmental NO_2, children with pre-existing respiratory illness are usually excluded, but a study of asthma epidemics in Barcelona during periods of high atmospheric pollution suggests that the relatively high levels of NO_2 in the city atmosphere did not correlate with an increase in symptoms or hospital admissions, suggesting factors other than NO_2 exacerbate asthmatic attacks (Anto *et al.,* 1986).

A number of laboratory studies investigating the effects on asthmatics of inhalation of near ambient NO_2 concentrations are confounded by the lack of a normal control group. Orehek *et al.* (1976), investigating carbachol-induced bronchoconstriction in 20 asthmatics before and after 60 min exposure to 0.18 mg/m^3, found 13 subjects had small but significant increases in baseline SG_{aw}, and significant potentiation of methacholine bronchoconstriction in asthmatics following exposure to 0.37 mg/m^3 NO_2 for 60 min. In a more recent study inhalation of 0.54 mg/m^3 NO_2 for 30 min by 15 asthmatics, who responded to cold air provocation by airway constriction, produced no significant change in baseline pulmonary function, but a significant reduction in FEV_1, and potentiation of bronchoconstriction induced by cold air provocation and exercise. Both these studies suggest that asthmatic subjects might be sensitive to NO_2 levels below those required to influence pulmonary function in healthy individuals.

In contrast, the inhalation of 0.18 mg/m^3 NO_2 for 60 min in 15 normal and 15 asthmatic subjects was found by Hazucha *et al.* (1983) to produce only a slight, insignificant increase in baseline airway resistance. Applying Orehek's criteria, only 3 of the 15 asthmatics studied would be classified as responders as opposed to his 13 out of 20, and, perhaps because of this, and the more stringent criteria used for entry into the study, NO_2 exposure did not potentiate methacholine-induced bronchial reactivity in either normals or asthmatics.

In a comparison of ten healthy adolescents and ten asthmatics drawn from a population previously found to be extremely sensitive to SO_2, 60 min exposure to 0.23 mg/m^3 O_3 or 0.23 mg/m^3 NO_2 at best produced no significant changes in baseline respiratory function in either group. Following exposure to either 0.23 mg/m^3 and 0.33 mg/m^3 O_3 or NO_2 combined with exercise, O_3 produced a small but significant change in FEV_1 and RT in both asthmatics and normals. No consistent changes were found in either group after NO_2 exposure. The authors also attempted a post-exposure methacholine challenge to assess non-specific bronchial reactivity after exposure to air, O_3 or NO_2, but this was not possible as the normal subjects were found to be 2.5–5 times more sensitive to methacholine after sham exposure to air, suggesting inhalation via a mouthpiece during moderate exercise affected bronchial reactivity to bronchoconstrictors (Koenig *et al.,* 1987).

In another study, the effects of exposure to 0.19 mg/m^3 NO_2 for 60 min in 20 normal and 20 asthmatic subjects were assessed. In addition to standard tests of respiratory function this study assessed bronchial reactivity by constructing a dose–response curve to inhaled carbachol and calculating a cumulative dose of carcachol which caused a 35% decrease in SG_{aw}. Subjects who showed sensitivity to the effects of the provocation dose of carbachol were classified as responders (16/20 of the normals and 13/20 of the asthmatics). Exposure to NO_2 did not alter baseline airway function or influence carbachol sensitivity in the non-responder group, but significantly increased the sensitivity to carbachol in the responders in both normal and asthmatic groups. Thus the results from controlled laboratory studies suggest that asthmatics are not more sensitive to NO_2 than normals.

Summary and conclusions

It is frequently stated that atmospheric pollution with the oxides of nitrogen has a negative impact upon respiratory health. Almost all of the research that has been published has been concerned with NO_2 and not the other oxides of nitrogen and still any association between ambient NO_2 concentrations and potential impact on respiratory health is not clear. It is not possible to dismiss the oxides of nitrogen completely as a factor in respiratory illness, but a review of the current data would indicate that any adverse effect is likely to be small and difficult to detect.

Many studies reporting an association between NO_2 exposure and respiratory health have methodological flaws or fail to give sufficent weight to confounding factors. Genetic background, climate, presence of influenza virus or other respiratory infections, frequency of cigarette smoking, exposure to environmental tobacco smoke and socio-economic status are frequently not taken into account and often confuse the apparent relationship between respiratory symptoms and indoor NO_2 levels. Both NO_2 and respiratory illness might be related to a third variable such as cold weather. For example, cold weather might increase the incidence of upper respiratory tract infections and also increase the incidence and frequency of use of open gas and kerosene heaters, which combined with closed windows leads to the winter increase in indoor NO_2 concentrations. A correlation would then exist between respiratory infections and indoor air pollution and might lead to the erroneous assumption that these two variables are causally related.

However, the problem of indoor air pollution and its impact on health will not go away. The levels of many indoor pollutants in the home have never been measured systematically, but indirect evidence would suggest that the levels may be on the increase. For example, among the major causes of indoor pollution in the Japanese home are the increased use of economic but non-ventilated kerosene stoves, instant gas water heaters, and gas cooking tables (Tominaga & Ono, 1985). In addition, the widespread use of air conditioners and effectively sealed windows and doors for reasons of fuel economy appear to aggravate the indoor air pollution by reducing ventilation and exchange between indoor and outdoor atmospheres.

Future needs

(a) A simple lightweight and cheap instrument to continuously monitor exposure to ambient levels of NO_2.

(b) An investigation into the possible importance of the large concentration of oxides of nitrogen in environmental tobacco smoke. We lack any data on the reactions of NO in polluted air and on the transfer of NO from the gas phase to the particulate phase in tobacco smoke which could be important with respect to the retention of NO in the respiratory tract.

(c) A study of the way in which other particles in the environment may absorb or adsorb trace amounts of NO_2 and so serve as a carrier mechanism keeping local NO_2 levels elevated in the alveoli.

(d) Control of the levels of pollutants in the indoor environment, identification of the sources of pollution and development of measures to control or eliminate the pollutant.

(e) Research is still needed to assess the influence of NO_2 on the respiratory health of infants, the elderly and social groups where the intake of vitamin C may be low, particularly in winter.

Acknowledgements

Dr T.M. Bearpark is thanked for many helpful discussions and assisting in the preparation of the manuscript.

References

ANTO, J.M., SUNGER, J. & PLASENCIA, A. (1986). Nitrogen dioxide and asthma attacks. *Lancet*, **ii**, 1096–7.

BOREN, H.G. (1964). Carbon as a carrier mechanism for irritant gases. *Archives of Environmental Health*, **8**, 119–24.

FLOREY, C.duV., MELIA, R.J., CHINN, S., GOLDSTEIN, B.D., BROOKS, A.G., JOHN, H.H., CRAIGHEAD, I.B. & WEBSTER, X. (1979). The relation between respiratory illness in primary school children and the use of gas for cooking III. *International Journal of Epidemiology*, **8**, 347–53.

HAZUCHA, M.J., GINSBERG, J.F., McDONNELL, W.F., HAAK, E.D., PIMMELL, R.L., SALAAM, S.A., HOUSE, D.E. & BROMBERG, P.A. (1983). Effects of 0.1 ppm nitrogen dioxide on airways of normal and asthmatic subjects. *Journal of Applied Physiology,* **54**, 730–739

KASUGA, H. (1985). A review of hydroxyproline as a biochemical marker on health effects of smoking and air pollution with nitrogen dioxide. *Tokai Journal of Experimental and Clinical Medicine*, **10**, 439–44.

KELLER, M.D., LANESE, R.R., MITCHELL, R.I. & COTE, R.W. (1979). Respiratory illness in households using gas and electricity for cooking. 1. Survey of incidence. *Environmental Research,* **19**, 495–503.

KOENIG, J.Q., COVERT, D.S., MARSHALL, S.G., VAN BELLE, G. & PIERSON, W.E. (1987). The effects of ozone and nitrogen dioxide on pulmonary function in healthy and in asthmatic adolescents. *American Review of Respiratory Diseases,* **136**, 1152–7.

MELIA, R.J., FLOREY, C.DuV., ALTMAN, D.C. & SWAN, A.V. (1977). Association between gas cooking and respiratory diseases in children. *British Medical Journal*. **2**, 149–52.

MELIA, R.J., FLOREY, C.DuV. & CHINN, S. (1979). The relation between respiratory illness in primary schoolchildren and the use of gas for cooking. 1. Results from a national survey. *International Journal of Epidemiology*, **8**, 333–8.

MELIA, R.J.W., FLOREY, C.duV., MORRIS, R.W., GOLDSTEIN, B.D., CLARK, D. & JOHN, H.H. (1982a). Childhood respiratory illness and the home enviroment I. Relation between nitrogen dioxide, temperature and relative humidity. *International Journal of Epidemiology,* **11**, 155–63.

MELIA, R.J.W., FLOREY, C. DuV., MORRIS, R.W., GOLDSTEIN, B.D., JOHN, H.H., CLARK, D., CRAIGHEAD, I.B. & MACKINLAY, J.C. (1982b). Childhood respiratory illness and the home environment II. Association between respiratory illness and nitrogen dioxide, temperature and relative humidity. *International Journal of Epidemiology*, **11,** 164–9.

MOHSENIN, V. (1987). The effect of vitamin C on NO_2 induced airway hyperresponsiveness in normal subjects. A randomized double-blind experiment. *American Review of Respiratory Diseases*, **136**, 1408–11.

MORROW, P.E. (1984). Toxicological data on NO_X: an overview. *Journal of Toxicology and Environmental Health*, **13**, 205–27.

OREHEK, J., MASSARI, J.P., GAYRARD, P., GRIMAUD, C. & CHARPIN, J. (1976). Effect of short term low level nitrogen dioxide exposure on bronchial sensitivity of asthmatic patients. *Journal of Clinical Investigation,* **57**, 301–7.

SAGAI, M. & ICHINOSE, T. (1987). Lipid peroxidation and antioxidative protection mechanism in rat lungs upon acute and chronic exposure to nitrogen dioxide. *Environmental Health Perspectives*, **73**, 179–89.

SHY, C.M., CREASON, J.P., PEARLMAN, M.E., McCLAIN, K.E., BENSON, F.B. & YOUNG, M.M. (1970). The Chattanooga School Children Study: Effects of community exposure to nitrogen dioxide II. Incidence of acute respiratory illness. *Journal of the Air Pollution Control Association*, **20**, 582–8.

SHY, C.M. & LOVE, G.J. (1980). Recent evidence on the human health effects of nitrogen dioxide. In: *Nitrogen Oxides and Their Effects on Health*, Lee, S.D., ed, pp. 291–305. Ann Arbor, Michigan: Butterworth.

SPEIZER, F.E., FERRIS, B.G., BISHOP, Y.M.M. & SPENGLER, J.D. (1980). Respiratory disease rates and pulmonary function in children associated with NO_2 exposure. *American Review of Respiratory Diseases,* **121**, 3–10.

TOMINAGA, S. & ONO, M. (1985). A plan of the comprehensive study on indoor pollution and its health effects by the Aichi Prefecture, Japan. *Tokai Journal of Experimental and Clinical Medicine,* **10**, 391–4.

WARE, J.H., DOCKERY, D.W., SPIRO, A., SPEIZER, F.E. & FERRIS, B.G. (1984). Passive smoking, gas cooking and respiratory health of children living in 6 cities. *American Review of Respiratory Diseases*, **129**, 366–74.

6

MINERAL FIBRES

BY R. C. BROWN, J. A. HOSKINS and A. POOLE

Introduction

Fibrousness is a desirable property in a material and one which has been exploited throughout man's history. The fact of capillary attraction through a bundle of fibres plus thermal stability made natural mineral fibres (asbestos) a good choice for the wicks of oil-lamps from prehistoric times to the present. Other uses in antiquity for some types of asbestos included reinforcement for pottery while the American Indians and some Europeans used cloth woven from these minerals for its 'magical' fire-resistant properties.

The industrial exploitation of asbestos began about 100 years ago with the discovery of large deposits in the New World. The uses of these minerals then expanded dramatically until the health hazards associated with their use became well known. The materials were used alone for their insulating properties, or in a composite, where they added strength, as in cement, or increased friction, as in brake shoes. These and other uses together with the market they represent have led to the development of many man-made mineral fibres (MMMF). The development of these synthetic materials has been given a great impetus in recent years by the need for more efficient insulation and by the health risk posed by many of the natural materials.

During the mining and subsequent processing of asbestos and the manufacture and processing of MMMF, fibres are released into the atmosphere. As a result of the health risks associated with the inhalation of fibres, legislation has been enacted to ensure that environmental dust levels are monitored and kept within 'acceptable' levels. These subjects are outside the scope of this chapter except in so far as release from mining or manufacturing sites will affect ambient fibre levels. The prime concern of this chapter is with fibre release from fibre-containing products in use. This happens for example, when brake shoes wear and they produce dust, and when thermal lagging is disturbed during breaking or demolition. The weakest link in the legislative chain of control is at the stage of destruction of the fibre-containing material.

The presence of mineral fibres in ambient air reflects both these releases and a background of natural fibrous material. In this chapter we will review the composition of mineral fibres, their uses and the properties responsible for their potential pathogenicity. The levels which have been found in the ambient atmosphere will be discussed and consideration will be given to whether these pose a risk to human health.

What are mineral fibres?

Asbestos

Asbestos is a collective name given to a group of highly fibrous minerals that are readily separated into long, thin, strong fibres which have high tensile strength; are heat resistant, chemically inert and possess high electrical resistance. These desirable properties have found many industrial uses.

There are six minerals called asbestos of special interest in commerce. These come from two mineral groups: serpentine and amphibole. Only one serpentine mineral is called asbestos, that is chrysotile, or white asbestos. The other minerals are all amphiboles: amosite (brown asbestos); crocidolite (blue asbestos); anthophyllite; tremolite; and actinolite. (See Table 6.1).

Asbestos minerals are silicates with the silicate anion, SiO_4^{4-} being the common structural element. The silicate anion in simple orthosilicates has a net negative charge evenly distributed over the four oxygen atoms while in the silicate minerals the charge is concentrated on particular oxygen atoms. Silicate minerals consist of chains, sheets, or three-dimensional arrays of silicate tetrahedra in which oxygen is either covalently bound to two silicon atoms (leaving no charge) or to one silicon atom (producing a net negative charge). This arrangement gives, in effect, very large polysilicate anions. It is important to note that each oxygen is bonded to silicon only and not to other oxygen atoms. The common representation of silicates as tetrahedra apparently bonded between apices merely represents the relative arrangement of the constituent atoms in space.

The fact that the silicate tetrahedra can bond together in several ways produces a convenient method for classifying silicate minerals. When two

Table 6.1. *Characteristics of some fibrous minerals*

Mineral	Class	Theoretical formula	Colour	Thermal stability [a]	Density
Chrysotile	Serpentine	$Mg_3(Si_2O_5)(OH)$	White to pale green	450–700	2.55
Crocidolite	Amphibole	$Na_2Fe(II)_3Fe(III)_2(Si_8O_{22})(OH)_2$	Blue	400–600	3.3–3.5
Amosite	Amphibole	$(Fe,Mg)_7(Si_8O_{22})(OH)_2$	Light grey to pale brown	600–800	3.4–3.5
Anthophyllite	Amphibole	$(Mg,Fe(II))_7(Si_8O_{22})OH)_2$	White to grey or pale brown	600–850	3.3–3.5
Tremolite	Amphibole	$Ca_2Mg_5(Si_8O_{22})(OH)_2$	White to grey	950–1040	2.9–3.1
Actinolite	Amphibole	$Ca_2(Mg,Fe(II))_5(Si_8O_{22})(OH)_2$	Pale to dark grey	620–960	3.0–3.2
Erionite	Zeolite	$(Na_2K_2CaMg)_{4.5}(Al_9Si_{27}O_{72}).27H_2O$	Grey		
Attapulgite	Clay	$Mg_5Si_8O_{20}(OH)_2(H_2O)_4.4H_2O$	Grey		
Sepiolite	Clay	$Mg_8Si_{12}O_{30}(OH)_4(H_2O)_4.8H_2O$			
Wollastonite		$CaSiO_3$	Grey		

[a] Temperature range (in °C) over which dehydroxylation or dehydrogenation (condensation) occurs with consequent breakdown of the crystal lattice and loss of strength.

oxygen atoms from silicate tetrahedra are shared, the simple silicate chains found in minerals called pyroxenes result. Double silicate chains are the backbone of amphiboles in which some tetrahedra share two oxygens, others three. The negative charge is balanced by hydroxylated metal ions in an octahedral configuration which bridge the silicate chains. Most commonly the metals are magnesium, iron, calcium, sodium and aluminium. The consequence of this structure is a crystal which can be cleaved readily in directions parallel to the silicate chains, but not in directions which cut the chains. This is the reason for the fibrous nature of such minerals. In contrast, the structure of chrysotile, a fibrous sheet silicate, is totally different and is described below.

Chrysotile

This is the principal asbestos of commerce and is the magnesium analogue of the aluminosilicate mineral kaolinite. It is chemically similar to talc and is a sheet silicate in which there are interleaving layers of magnesium oxide–hydroxide octahedra (known as brucite). The silicate and brucite layers are bonded through apical oxygen atoms. The particular properties of chrysotile arise because, unlike other sheet silicates such as mica which are flat composites, the layers are rolled up into concentric tubes or scrolls with their long axes parallel to the fibre. To form a flat composite the constituent layers would have to match correctly in size. However, because the repeat distance of the silicate tetrahedra is less than that of the brucite octahedra, the sheets of chrysotile roll up with the magnesium layer on the outside. Electron micrographs show that the individual fibrils are 0.01–0.05 μm in diameter with an aspect ratio of 100:1 or greater. The size of the scrolls is limited by an increasing radius of curvature with increasing number of turns and consequent increase in strain. The strain-free layer of chrysotile has a radius of curvature of about 8.8 nm, therefore the diameter of the tube should not be much less than 17 nm; the measured average diameter of the scrolls is about 25 nm.

Chrysotile fibrils are long, flexible and curved, and they tend to form bundles that are often curvilinear with splayed ends. They occur naturally in lengths varying from 1–20 mm, occasionally as long as 100 mm. Comminution of the fibres may produce separated unit fibrils since the binding between them is weak; there may also be inter-fibril amorphous magnesium silicate material.

The chemical composition of chrysotile is uniform compared with that of the amphiboles. However, traces of other metals can occur either within the tetrahedral silicate layer (e.g. Al for Si) or within the octahedral brucite layer (e.g. Ni or Fe). The chemical properties of chrysotile arise largely because the external surface of the fibrils is magnesium hydroxide and therefore chemically they behave as if they were this compound. Chrysotile reacts with acid to leave a silica residue. The rate of solution is appreciable. Chrysotile is almost completely destroyed within 1 h at 95 °C by 1 M hydrochloric acid. Chrysotile, however, is more resistant to alkali than the amphiboles.

Amphiboles

Amphiboles are double chains of silicate tetrahedra, cross-linked with bridging cations; a structure less complicated than that of chrysotile. Amphiboles always contain hydroxyl groups attached to the metal ions which hold the parallel silicate chains together. The structure is such as to allow considerable latitude in cation replacement which is reflected in varying chemical and physical properties and which distinguishes the various minerals in the amphibole group. Only rarely does the composition of a field sample coincide with the assigned theoretical or idealised formula. However, these idealised compositions are useful for identifying the various fibres.

Comminution of amphiboles produces breakage (both parting and cleavage) along defined crystallographic planes. This is important biologically with regard to resultant particle number, surface area and general respirability (see below) and perhaps with regard to the chemistry of the exposed fibre surface.

The amphiboles are all much more resistant to acid than chrysotile since vulnerable metal ions in the crystal are protected from attack by the silicate chains and are only slowly lost through diffusion. The order of resistance to acid attack (tremolite > anthophyllite > crocidolite > actinolite > amosite) is largely a function of magnesium content.

Unlike chrysotile the amphiboles are negatively charged at physiological pH.

Other mineral fibres

Many minerals, other than asbestos, exist which, on comminution, can produce fibrous particles. These may enter the environment either through human activities or through natural erosion processes. There are also a number of artificially produced mineral fibres which are generally grouped together under the collective title 'man-made mineral fibres' (MMMF). However, the terminology used to classify such fibres is not always consistent or clear. The non-asbestos mineral fibres include several which have been proposed (natural) or produced (man-made) as substitutes for asbestos. Some MMMF have been prepared to have properties which are unique or else in some way superior to existing materials. One distinction between natural fibres and MMMF is that generally naturally occurring fibres are crystalline in structure, while most MMMF are amorphous when manufactured. Such vitreous fibres can devitrify in use. For example, aluminosilicate ceramic fibres which are amorphous when produced can partially convert on heating to a composite of crystalline materials such as cristobalite (a form of silica) and mullite (aluminium silicate).

Other natural mineral fibres

There are other naturally occurring fibrous silicates including certain zeolites, attapulgite (fibrous palygorskite), sepiolite and wollastonite. Naturally occurring glass filaments can be produced during explosive volcanic eruptions. The zeolites are crystalline aluminosilicates, that is, silicate minerals in which some of the silicon atoms have been replaced by aluminium and the net increase in negative charge is balanced by various readily exchangeable alkali and alkaline earth counter ions. Of the several fibrous zeolites which are known, one – erionite – has been found to have some of the same adverse effects on health as the asbestos minerals. Erionite fibres are similar in diameter to asbestos though they are, on average, shorter. Certain clay minerals, such as attapulgite and sepiolite, occur in forms similar to both chrysotile and amphibole fibrils. Under the electron microscope they may appear to have a hollow tube structure, or look like an amphibile fibril.

A few other natural fibrous minerals may occur which are not silicates. For example, fibres of dawsonite – sodium aluminium carbonate – may occur in bauxite deposits.

Any of these natural fibres may contaminate the environment in which they occur or are shipped and used.

MMMF

The term man-made mineral fibres is self-explanatory and a variety of types are produced. Generally MMMF are classified according to their origins: glass wool, rock wool, slag wool, etc. In this chapter, the term MMMF will be used to describe the amorphous glassy fibres made from molten slags, natural rocks (such as basalt, diabase and olivine) and borosilicate or calcium aluminium silicate glasses. These materials are also known by the collective term of mineral wools. All these MMMF are amorphous silicates which can be classified into four broad groups: continuous filament glass fibre made by extrusion and winding processes, insulation wool (including rock/slag wool, and glass wool), refractory (including ceramic fibre) and special purpose fibres. The ceramic fibres used to line industrial furnaces are made from aluminium silicate minerals such as kaolin. The non-continuous fibres are made by dropping molten material onto spinning discs, by air jet impingement on molten material or by similar processes in which slurries are fired during and/or after fibrisation. The non-continuous materials contain a wide range of fibre sizes and are often contaminated by small glassy balls called shot.

The dimensions of MMMF are of the same order as those of natural fibres: continuous filament glass: 6–15 μm; various insulation wools: rock wool, slag wool, glass wool: 2–9 μm; refractory fibres: ceramic and others: 1.2–3.5 μm; special purpose fibres: flame attenuated glass fibres: 0.1–3 μm.

In addition to these bulk materials there are other, usually crystalline, synthetic mineral fibres, which include fibres or 'whiskers' produced from alumina, silica and zirconia. Also there are non-oxide MMMF which include graphite, silicon carbide and nitride and boron nitride.

A number of other fibres have been produced with important industrial properties; some such as the plastic Kevlar and the ceramic potassium octatitanate have been shown to produce significant levels of disease in animal experiments. This may indicate that they could pose a health risk. There are also fibrous forms of calcium silicate and calcium sulphate. This latter material occurs in plaster products and is manufactured as a filler and reinforcing fibre. Fibres from plaster are perhaps the commonest such particles found in man's indoor environment.

There is a large and rapidly growing market in synthetic zeolites chemically related to erionite. These materials are usually cuboidal rather than fibrous in structure but their similar properties to erionite suggest they should be treated with caution. They are used either as ion-exchangers or molecular sieves and a typical use is in the newer, phosphate-free, washing powders.

Uses of fibrous materials

Perhaps the oldest industrial use of chrysotile asbestos is in the manufacture of textiles. These asbestos textiles have been used for heat, electrical and acoustic insulation and, in the form of yarn or rope, as packing material in pumps, glands and pistons. Chrysotile is also used as a reinforcement and friction enhancer in various products such as brake shoes and brake and clutch pads. Today, there is considerable pressure to stop the use of chrysotile and replace it by newer and, hopefully, safer materials.

The construction of any domestic or public building may include fibrous materials as thermal and acoustic insulation or fire proofing. These may be used as a low density product, generally a MMMF wool, or as a higher density product, often containing asbestos used alone or in a cement. This type of product is found as thermal lagging of pipes and boilers or sprayed or trowelled onto walls, roofs or structural members as fire protection.

In addition, fibres can be employed in a number of structural elements. Chrysotile and the amphiboles are used as reinforcing for cement products for cladding and roofing such as asbestos cement slates, troughed or corrugated sheeting, soffit boards or fire proofing. Asbestos cement rainwater pipes and gutters are used extensively and even large water mains may be made of this material.

The types of asbestos used vary with the date of manufacture and the country of origin. Thus, while materials made in the recent past in the UK contain exclusively chrysotile, other types of asbestos have been used elsewhere, particularly in some continental products. Chrysotile, but sometimes amosite, is found in materials for surface treatment for thermal and acoustic insulation and amosite containing asbestos–cement products are common in fume ducts and wall panelling in some older buildings. Some textured paints may contain 2–4% chrysotile asbestos as originally formulated; the dried film may contain a much greater proportion of fibre.

Crocidolite has been used more or less in a pure form as fire insulation on steel frames of buildings as well as in thermal insulation in heating systems. Amphibole asbestos has been used as a filler or stabiliser in, and as the backing of, vinyl floor sheeting and tiles. It has also been used with bitumen in roofing felts.

The choice of asbestos type is not just dictated by whim, though whim has certainly played a part in such choices. Blue, brown and white asbestos are different materials with different properties and therefore different uses.

Only chrysotile can be woven into textiles and the durability and chemical resistance of the fibre are important properties determining choice. When asbestos is used as the reinforcing medium in cement products the final use of the product dictates the fibre type. Chrysotile is readily attacked by acid and so acid-resistant cement is made using brown or blue asbestos.

Some building materials which are apparently non-fibrous may contain asbestos; vermiculite used for insulation can sometimes contain tremolite, an amphibole asbestos, as an impurity. Fibrous forms of calcium silicate have been used as an asbestos substitute in boards, and such boards have been used for roofing, as soffit and verge boards and ceiling and wall linings. Calcium sulphate in plaster or plaster board is frequently fibrous and, as mentioned above, provides the commonest source of fibres in indoor air.

Rock and glass wools are perhaps most commonly used as loft and cavity wall insulation. These wools may also be compressed into bats or boards for use in partitions or into various shapes for use as pipe insulation. Compressed to a higher density they are used in wall and ceiling tiles. Various MMMF are used as reinforcing agents in solid plastics and cement and as the base in bitument felts. More recently MMMF have been used as an asbestos substitute in vinyl flooring. Glass-reinforced plastic (GRP) is used for cold-water storage tanks, as roofing, and for other more specialised uses. One type of alkali-resistant glass is used to reinforce cement products.

Diseases associated with exposure to mineral fibres

The pathogenicity of mineral fibres is best seen in populations of asbestos exposed persons. The effects of such exposures can include the excessive deposition of collagen (fibrousis) at various thoracic sites, tumours of the pleura or peritoneum (mesothelioma) and lung cancer. Attempts have been made to link asbestos exposure to tumours at other sites but these are problematic and poorly justified. Many of these conditions can be reproduced in animal experiments and the potential pathogenicity of newer fibres is best examined by such methods since too few humans will have been exposed for any epidemiological investigation.

In general, it seems that a greater exposure is necessary for the development of fibrosis than for neoplasms. The risk of an individual developing lung cancer increases in a way linearly related to cumulative exposure. This is in contrast to the dose relationship for the development of mesothelioma which is proportional to fibre concentration and exponentially related to time since first exposure (Peto *et al.*, 1982).

Types of fibre exhibiting pathogenicity

Clearly, the first requirement for any mineral fibre to cause disease is that it should gain access to the body. While fibres can be ingested or penetrate the skin, the most important route is the entry of airborne fibres via the lungs.

Only a subset of fibres will have properties such that they can be inhaled and deposited in the respiratory tract. This subset will have falling speeds equivalent to a sphere of unit density of less than 10 μm (usually called 'aerodynamic diameter'). Particles with aerodynamic diameters between 5 and 10 μm will be deposited in the large airways and those of less than 5 μm will reach the terminal bronchioles and alveoli. Particles capable of inhalation and reaching the terminal parts of the respiratory tract are called 'respirable'. Particles with aerodynamic diameters of less than about 0.1 μm will not be deposited.

The aerodynamic diameter and hence respirability of fibres is largely determined by their density and cross-sectional area (proportional to actual diameter) with length playing only a small role. It is important to note, however, that fibre morphology is so different from that of the ideal sphere that long fibres will impact on airway bifurcations and other areas of turbulence in a way difficult to predict from any simple theory. The actual diameter of the largest asbestos fibres (specific gravity approximately 3) found in samples of human lung is about 3.5 μm. Larger fibres of less dense materials would be respirable.

Asbestos exposed to abrasion will split into fibrils of smaller diameter with little effect on fibre length, any dust generated will therefore be more and more respirable with even the ultimate fibrils of chrysotile large enough to be deposited in the lung. Unlike asbestos fibres, MMMF do not split longitudinally but may break transversely into shorter segments. The product will become dustier as the particles generated become shorter though the diameters will not change and such particles will not become more respirable. It is important to note that the insulating properties of MMMF are partially determined by fibre diameter, finer fibres having lower thermal conductivity. This, and the fact that a lower mass is needed for equivalent insulating properties, has provided an incentive to produce MMMF of smaller and smaller fibre diameter with the greater possibility that respirable dust will be generated.

The interaction of deposited fibres with the cells of the lung requires that the fibres remain present in the lung for some finite time, the length of which is unclear. It is conventional wisdom to state that, to be pathogenic, fibrous minerals must be both respirable and durable. Indeed, unlike most toxic materials, mineral fibres are extremely durable and this is a prerequisite for their use in industry. However, there are considerable differences in durability even amongst the commercial types of asbestos as described above.

These variations in durability are reflected in the persistence of the fibres in the biological environments encountered within the body. In a series of animal experiments Wagner *et al.* (1974) found that in rats exposed to either chrysotile or amphiboles by inhalation there was a steady increase in the amount of retained amphibole whereas a steady state level of chrysotile was rapidly reached. A possible explanation for the differences in accumulation is that chrysotile is being cleared from the lung while amphiboles are not; however the mechanism of clearance is unknown. This low retention of

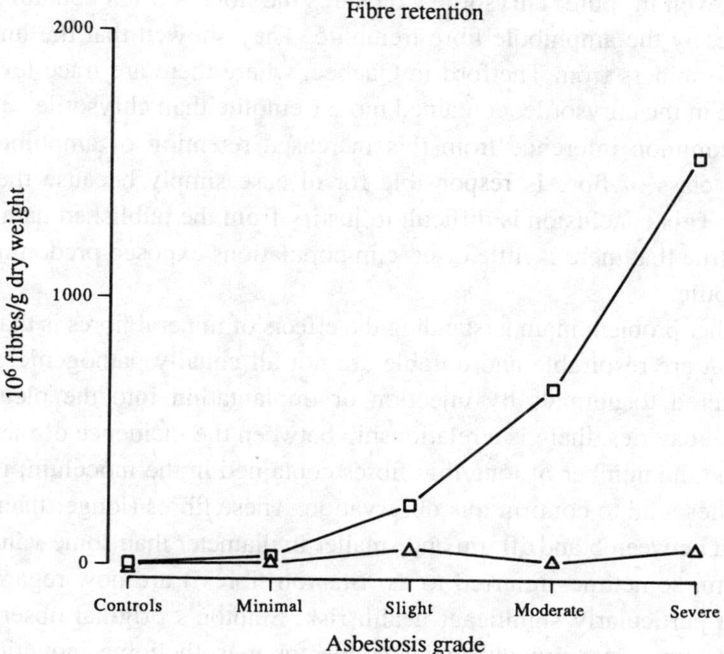

Fig.6.1. The fibre content of lung samples taken from cases of asbestosis in East London. The data are derived from those presented in Table 4 of Wagner *et al.* (1989). The squares are the amphibole content (amosite + crocidolite) and the triangles the chrysotile content in millions of fibres/g lung dry weight. The number of cases are: controls: 56, minimal: 3, slight: 11, moderate: 12 and severe: 9.

chrysoltile can also be seen in an analysis of the dust recovered from human lungs. While the commonest type of asbestos in urban air samples is chrysotile (hardly surprising given that this mineral has accounted for over 99% of the world trade in asbestos), in studies of fibre in human lungs from control populations, little chrysotile is found. If the lungs from people with mesothelioma are examined, few chrysotile fibres are found but amphibole concentrations are high, reflecting the patients' exposure to this class of material. It can be shown from the data of Wagner *et al.* (1989) on the fibre content of lungs from factory workers with asbestosis that the content of amphibole increased with the grade of fibrosis but the chrysotile content was approximately constant (Fig. 6.1). Chrysotile, crocidolite, and amosite were used in the factory in which these workers were employed and it is safe to assume that the chrysotile was deposited but not retained.

A further example of the differences in retention of chrysotile and amphiboles was described by Sebastien *et al.* (1989) who took advantage of the fact that even in 'pure' chrysotile exposures the fibre is often contaminated, at source, by the amphibole fibre tremolite. They showed that the lungs of chrysotile miners from Thetford in Quebec, where there are trace levels of tremolite in the chrysotile, contained more tremolite than chrysotile.

One common inference from this increased retention of amphiboles is that this class of fibre is responsible for disease simply because they *are* retained. This conclusion is difficult to justify from the published data alone but it is true that there is little disease in populations exposed predominantly to chrysotile.

A further problem in understanding the effects of mineral fibres is that even fibres that are respirable and durable are not all equally pathogenic. When administered to animals by injection or implantation into the pleural or peritoneal cavities, there is a relationship between the incidence of mesothelioma and the number of long/thin fibres contained in the inoculum: inhalation studies tend to confirm this observation. These fibres (longer than some threshold between 5 and 10 μm and smaller in diameter than some value less than 1 μm, sometimes referred to as 'Stanton fibres') are now regarded as posing a particularly significant health risk. Stanton's original observation that these fibres are probably responsible for mesothelioma induction has been confirmed in several laboratories and similar fibres have been found to be uniquely active in various *in vitro* systems (Brown *et al.*, 1978).

For some time it was believed that all equally sized fibres are equally active but it is now clear that this is not the case. The discovery of a very

Fig.6.2. Bright field light micrographs of some mineral fibres. Magnification of all pictures 1 mm = 4 μm except number 8 where 1 mm = 8 μm. In many cases biologically active fibres are below the resolution of the light microscope.

1	2	3
4	5	6
7	8	9

1. Respirable sample of 2 prepared using a Hexlet sampler, the larger diameter fibres in 2 have been removed by horizontal elutriation so that all the particles in this sample would be capable of reaching the alveolar regions of the lung.
2. Milled sample of micro glass fibres (Johns Manville code 108).
3. Dust sample from the village of Karain in central Anatolia Turkey, one erionite fibre is visible. In this village pleural mesothelioma is the commonest cause of death.

4.–6. Three samples of tremolite asbestos, sample 6 is not biologically active in animal or *in vitro* experiments; sample 5 is cytotoxic *in vitro* but not carcinogenic *in vivo* and sample 4 is both cytotoxic and carcinogenic. The main difference between these three samples is the distribution of fibre sizes. The full details of composition, size distribution and biological activities are given in *Wagner et al.* (1982).

7. A fibrous sample of gypsum (calcium sulphate) sold as Franklin fibre and advertised as completely non-hazardous.
8. A modern sample of ceramic wool with a mean diameter of 0.21 μm.
9. A sample of glass wool insulation installed in a house in 1929. Considerable devitrification has occurred and small fragments can be seen, the large fibre has crystalline areas at several sites visible as transverse marks or 'growths' on the surface.

Source: Wagner, J.C., Chamberlain, M., Brown, R.C., Berry, G., Pooley, F.D., Davies, R. and Griffiths, D.M. (1982). Biological effects of tremolite. *Br. J. Cancer*, **45**, 352–60.

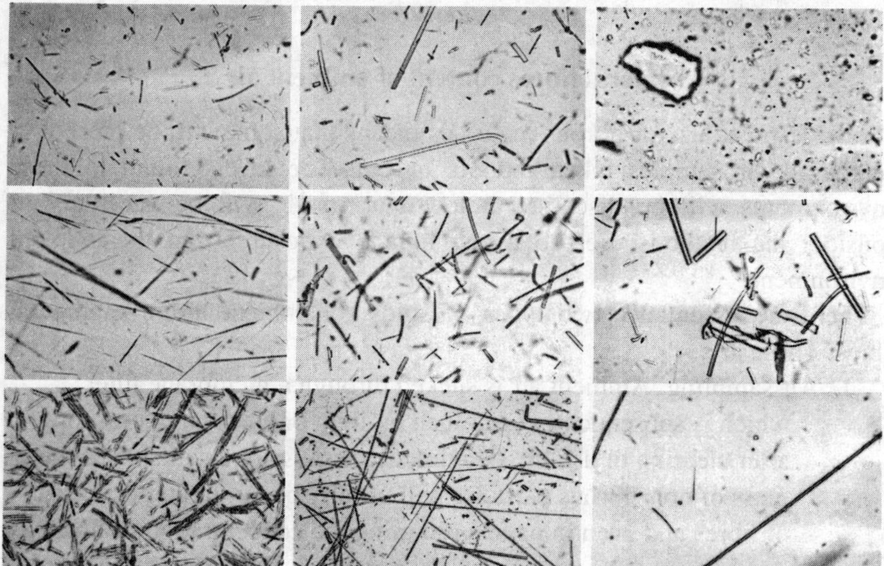

high incidence of mesothelioma in the village of Karain in Central Turkey has led to the discovery that erionite (a naturally occurring fibrous zeolite found in that village) is more capable of producing mesotheliomas, even on an equal fibre number basis, than any other mineral. Thus, in addition to size and durability, there is some effect of fibre composition.

While something is known about the properties of fibres responsible for causing mesothelioma, less is known for the other diseases associated with fibre exposure. There have been few experimental studies on the quantitative relationships between fibre type and fibrosis. There is no firm evidence as to the properties or doses which will determine whether or not fibres will produce fibrosis or parenchymal tumours.

In summary, the factors that seem to be important in fibre pathogenicity are:

1. To produce any disease, a fibre must be respirable and probably durable.
2. A long/thin subset of respirable fibres can cause mesotheliomas.
3. The composition of the long, thin fibres does play a role in determining their activity in producing mesotheliomas.
4. The properties of fibres determining their activity in producing other diseases is inadequately studied though some authorities have suggested that these are similar to those responsible for mesothelioma induction. We suggest that there is little justification for this assumption.

Mineral fibres content of ambient air

The widespread use of fibrous materials means that some will be present in nearly all air samples taken indoors or out. The levels found in these environments will be discussed below but first it will be necessary to consider the methods used to assess human exposure from the ambient environment.

There are essentially two types of study to ambient airborne mineral fibres. These are:

1. Air sampling. Ambient air is sucked through a membrane filter which is subsequently examined by light or electron microscopy after clearing or ashing. Gravimetric assays are unsuitable as the mass of non-fibrous particles will usually be far greater than that of fibres and even in papers reporting the concentration of fibres as

a mass/unit volume this concentration has usually been derived by microscopy.

2. The examination of fibres in autopsy lung samples. The lung samples are ashed or dissolved before collecting the residue and examining it as above.

Clearly, these two approaches do not give the same type of information: air samples give the concentration at the time of sampling while the retained lung dose reflects historical levels of exposure complicated by clearance and by losses during preparation.

Concentrations in air as determined by air sampling

In what follows, concentrations have been converted to fibres/litre of air (f/l) whatever the original units used.

Any comparison of data from different workers is not simple. In most published reports a fibre has been defined as an elongated particle with an aspect ratio (length/diameter) of more than 3. Also most of these studies have only counted fibres longer than 5 μm. However, these criteria are not universal. For example, there is in West Germany a draft guideline for the measurement of the fibre content of ambient air (VDI 3492) which involves the measurement of all fibres more than 2.5 μm long. The filters used to collect the particulate phase of ambient air, the methods of clearing or ashing this filter and whether the dust is resuspended or not before final examination all affect the results obtained. To complicate matters, light, electron and scanning electron microscopy have all been used to count the fibres. The variable definitions of a fibre, the different resolutions of the microscopic techniques and the variations in collection make all summaries of the available data suspect. Ideally, comparison between studies should only be made where the same methods have been used.

Outdoor air

Any sample of ambient air collected indoors or outdoors will probably contain some fibrous particles including chrysotile, other types of asbestos, calcium sulphate, man made mineral fibres (MMMF) and fibres of unknown origin. Most of these fibres derive from man's activities but there is a background derived from natural sources. This is best illustrated in a study of the fibre content of Antarctic ice in which nearly 10^6 f/l have been found in

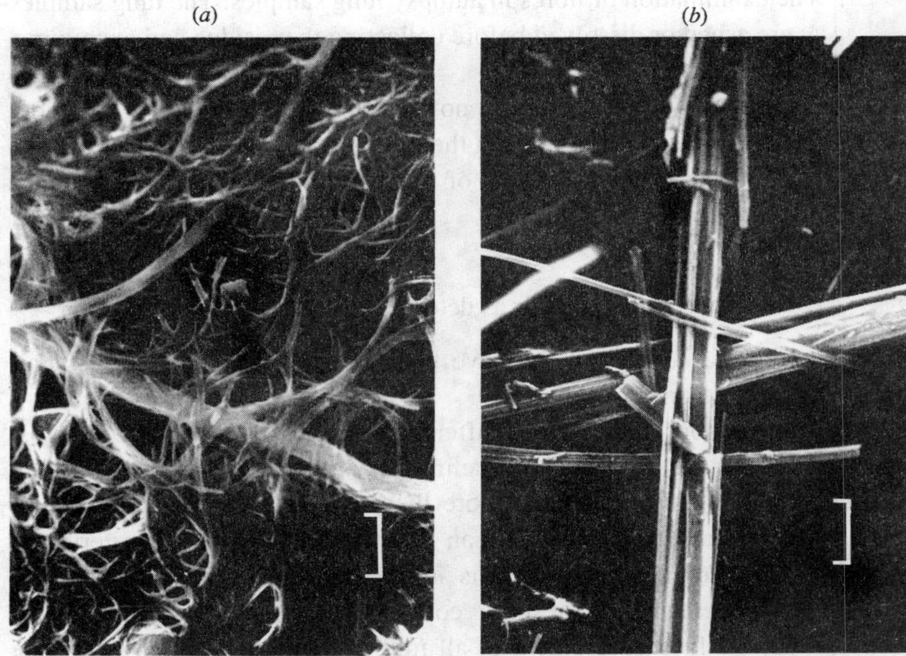

Fig.6.3. (a) Candian chrysotile, note range of sizes and curly nature of this fibre. Magnification: bar = 10 μm.
(b) Crocidolite asbestos, note the range of sizes of the straight fibres and the longitudinal splitting. Magnification: bar = 2.5 μm.

10 000 year-old ice. In comparison, just over 10^7 f/l were found in snow falling in Japan in 1984 (Kohyama, 1989).

In urban outdoor air there are several sources of mineral fibres, including motor vehicle brakes, weathered or friable building materials, releases from industrial operations, or increasingly, in the case of asbestos, from clearance and disposal operations. The evidence for the release of asbestos from brakes is best illustrated by the increased levels found in areas of high braking activity. For example, Bruckman (1978) found increased levels in the vicinity of toll malls on US roads. Spurny (1989) has demonstrated that fibres are released into the air in the vicinity of weathered asbestos cement products. Air near buildings containing such weathered products could contain between 0.2 and 1.2 f/l from this source. The contribution from industry has been the subject of many studies. In one of the most recent, Marconi *et al.* (1989) studied the airborne fibre concentrations in the vicinity of an asbestos cement plant. Peak levels of asbestos of up to 19 f/l were found 0.4 km from the plant. Nicholson *et al.* (1980) measured levels of asbestos of up to 100 times background in the vicinity of asbestos fire proofing operations.

The World Health Organisation has concluded (WHO, 1986) that, allowing for the great variations in fibre counts, and counting techniques, levels of asbestos in outdoor ambient air are usually less than 1 f/l in rural areas and up to 10 f/l in towns. Other inorganic fibres can be found at ten times these levels. Though there have been very few studies on the levels of MMMF in outdoor air, it seems that there are far fewer of these than asbestos fibres though occasionally samples with 2–3 f/l have been reported (Balzer, 1976).

Asbestos in indoor air

Unlike most of the studies on the asbestos content of outdoor air, most studies on the indoor environment have attempted to correlate air content with the presence of fibres in the structure or more recently to determine the effects of asbestos removal programmes. Burdett *et al.* (1989) have published a summary of asbestos measurements made in the UK in which they concentrated on buildings with asbestos containing warm air heaters and buildings with trowelled or sprayed asbestos insulation. The asbestos level in the air of only 9 of the 39 sites studied exceeded the limit of quantification. At one site, where an asbestos-containing heater had been dismantled and the asbestos damaged and left exposed, the level reached 2 f/l. A level of 12 f/l was found at one site with sprayed or trowelled asbestos but here there was 'a large area of damaged sprayed asbestos'.

The levels of fibres in indoor air even in buildings containing exposed asbestos are of the same order as those in outdoor air. In some cases, asbestos-containing buildings have had measured levels inside considerably lower than those on the streets outside.

The levels found during normal occupation of a building are not reflected in those generated during removal of asbestos insulation. In the paper by Burdett *et al.* (1989) referred to above, the levels during removal operations were reviewed to show the efficiency of containment during this process. It was found that, even outside properly constructed enclosures the levels were higher than before removal operations started and that these levels remained raised in the building after completion of the work.

The same authors when discussing the removal of asbestos from schools point out that children are at school for 15 000 hours. If during that time the average asbestos level is assumed to be 0.5 f/l this would give a cumulative exposure which may be achieved in a few days by proximity to removal operations. Taken together with the fact that the fibre level remains high after removal is completed, the wisdom of the widespread clearance of asbestos must be doubted. While the risks to normal occupants of asbestos containing buildings is very low, maintenance staff and cleaners may be exposed far more and the risks to these operatives may require remedial action even at the cost of increasing the exposure to other occupants.

Man-made fibres in indoor air

There have been very few studies on the levels of MMMF in indoor air. Most of the published Figures show peak concentrations of about 1 f/l with most samples containing very many fewer, usually less than 0.1 f/l. For example Gaudichet *et al.* (1989) reported a peak value of 6.2 f/l while his median value from 79 determinations was only 0.003 f/l.

As glass fibre ages it becomes more dusty through devitrification and breakage. For this reason it seems that old MMMF insulation is likely to produce higher levels of dust than more recently installed material (Jaffrey, 1989). Unlike asbestos there is generally less contamination with MMMF outdoors than there is in inside air and this may reflect the use of asbestos in brakes and the increased weathering of outdoor building materials. If MMMF become common constituents of these materials then this might be expected to change.

Fibres in lung samples

The use of the lung as a long-term sampler for the atmospheric content of durable particles is very attractive. Using autopsy material Gibbs *et al.* (1989) have found that control groups with only ambient exposure to asbestos may have up to 10^6 amphibole fibres/g lung dry weight. However, the confusion caused by the selective clearance of some particles will affect the results con-siderably. As mentioned above, amphiboles are well retained in the lung and chrysotile is cleared readily. Other confounding variables include the smoking history of the subject since there is greater retention of fibre in the lungs of smokers. It is now possible to lavage the bronchiolar–alveolar region of the lung in living subjects and in some cases this can provide qualitative evidence for exposure to certain particulates including mineral fibres.

Very few glass fibres have been found in lung samples but this may arise from one or more of several causes. There could be low levels of exposure, these fibres could be cleared during life, or they could be destroyed during sample preparation involving alkaline dissolution of the lung tissue. As there are so many variables involved in these techniques, it seems that little solid information on environmental levels of fibre can be obtained from these measurements alone.

Do ambient levels of mineral fibres pose a health risk?

The ubiquitous nature of mineral fibres and their durability results in a level of these materials of about 10^6f/g dry weight of lung even in non-occupationally exposed people. Yet there is a clear cut difference in the incidence of mesothelioma between occupationally exposed persons and the general public. This suggests that there may be a threshold exposure for any effect to occur, or less controversially, that there is a requirement for a fairly heavy exposure before a risk becomes apparent. However, it seems that only moderate doses of fibre can cause mesothelioma. There are numerous reports of para-occupational exposure to asbestos causing this tumour in the families of asbestos workers. One of the most recent (Li *et al.*, 1989) showed that the wife, daughter and brother-in-law of an asbestos worker all contracted mesothelioma. In this case the wife laundered the husband's clothes and also sacks that he brought home. The sacks, that had contained moulded asbestos insulation, were made into nappies for the worker's children. The brother-in-law had lived in the house as a teenager. The fact that only large exposures cause fibrosis is illustrated by this family as the father died of asbestosis complicated by cirrhosis of the liver. This type of effect caused by exposures which, while unquantified, are assumed to have been fairly low has led to the risk of mesothelioma from ambient exposure to be estimated as about one or two excess deaths per 10 000 (Enterline, 1982).

The dose of fibre required to cause lung cancer is more uncertain. Many pathologists have only been willing to ascribe lung cancer to asbestos exposure if there were also asbestosis and this may have overestimated the asbestos dose needed to cause this type of tumour. In the paper by Enterline he predicted an excess mortality of two per million for deaths from lung cancer.

Both the estimates for mesothelioma and lung cancer exceed the one in one million which some people have suggested as the limit for an 'acceptable risk'. In general, it must be stated that there is insufficient evidence to give much support to this or other quantitative risk estimates for the general environmental effects of most types of fibre (see Gardner & Saracci, 1989). The situation is even more complex because of the lack of information about the interaction of fibres with other environmental carcinogens. Much more experimental and epidemiological work is necessary before these problems can be properly addressed.

Conclusions

There is, at present, little evidence that ambient levels of mineral fibres pose a real risk to human health. The near hysterical rush to clear asbestos from buildings, together with the ageing of MMMF, may cause more human exposure to mineral fibres and not less. Vigilance will be necessary to ensure that these unnecessary exposures do not become a real hazard.

The development of new insulating materials with potentially hazardous properties must also be monitored by toxicological examination of the materials themselves and by environmental hygiene measures to quantify and control human exposure. Finally, the possibility that some fibres are themselves harmless but might interact with other carcinogens cannot be ignored. This possibility requires a great deal more research. A proper assessment of the risks due to fibre exposure requires knowledge of the pathogenic mechanisms responsible for the various fibre-associated diseases, an area which contains more speculation than certainty. Brown *et al.* (1990) give a detailed review of this area.

References

BALZER, J.L. (1976). Environmental data: airborne concentrations found in various operations. In *Occupational Exposure to Fibrous Glass*. Proceedings of a Symposium, US Dept of Health and Welfare. HEW Publication no (NIOSH) 76–151, pp. 83–9.

BROWN, R.C., CHAMBERLAIN, M., GRIFFITHS, D.M. & TIMBRELL, V. (1978). The effect of fibre size on the *in vitro* biological activity of three types of amphibole asbestos. *International Journal of Cancer*, **22**, 721–7.

BROWN, R.C., HOSKINS, J.A., MILLER, K. & MOSSMAN, B.T. (1990). Pathogenetic mechanisms of asbestos and other mineral fibres. *Molecular Aspects of Medicine*, **11**, 325–49.

BRUCKMAN, L. (1978). *A Study of Airborne Asbestos Fibres in Connecticut.* US National Bureau of Standards Special Publication 506. Washington DC US Department of Commerce National Bureau of Standards, pp. 179–90.

BURDETT, G.J., JAFFREY, S.A.M.T. & ROOD, A.P. (1989). Airborne asbestos fibre levels in buildings: a summary of UK measurements. In *Non-occupational Exposure to Mineral fibres.* Bignon J., Peto, J. & Saracci, R., eds, pp. 277–90. Lyon IARC Scientific Publication no. 90.

ENTERLINE, P.E. (1982). Health effects on the general population (cancer caused by air pollution in the United States). In *Proceedings of the World Symposium on Asbestos, Quebec.* Canadian Asbestos Information Centre, Montreal.

GARDNER, M.J. & SARACCI, R. (1989). Effects on health of non-occupational exposure to airborne mineral fibres. In *Non-occupational Exposure to Mineral Fibres.* Bignon, J., Peto, J., & Saracci, R., eds, pp. 357–97. Lyon IARC Scientific Publication no. 90.

GAUDICHET, A., PETIT, G., BILLON-GALLAND, M.A. & DUFOUR, G. (1989). Levels of atmospheric pollution by man-made mineral fibres in buildings. In *Non-occupational Exposure to Mineral Fibres.* Bignon, J., Peto, J. & Saracci, R., eds, pp. 291–8, Lyon IARC Scientific Publication no. 90.

GIBBS, A.R., JONES, J.S.P., POOLEY, F.D., GRIFFITHS, D.M. & WAGNER, J.C. (1989). Non-occupational malignant mesotheliomas. In *Non-occupational Exposure to Mineral Fibres.* Bignon, J., Peto, J. & Saracci, R., eds, pp. 219–28. Lyon IARC Scientific Publication no. 90.

JAFFREY, J.A.M.T. (1989). Laboratory studies of fibre release from new and old man-made mineral fibre blankets. *London, Health & Safety Executive Report* number I/RL/DI/89/1.

KOHYAMA, N. (1989). Airborne asbestos levels in non-occupational environments in Japan. In *Non-occupational Exposure to Mineral fibres.* Bignon, J., Peto, J. & Saracci, R., eds, pp. 262–76. Lyon IARC Scientific Publication no. 90.

LI, F.P., DREYFUS, M.G. & ANTMAN, K.H. (1989). Asbestos contaminated nappies and familial mesothelioma. *Lancet*, **i,** 909–10.

MARCONI, A., CECCHETTI, G. & BARBIERI, M. (1989). Airborne mineral fibre concentrations in an urban area near an asbestos cement plant. In *Non-occupational Exposure to Mineral fibres.* Bignon, J., Peto, J. & Saracci, R. eds, pp. 336–46. Lyon IARC Scientific Publication no 90.

NICHOLSON, W.J., ROHL, A.N., WEISMAN, I. & SELIKOFF, I.J. (1980). Environmental asbestos concentrations in the United States. In *Biological Effects of Mineral Fibres.* Wagner, J.C., ed., pp. 823–8. Lyon IARC Scientific Publication no. 30.

PETO, J., SEIDMAN, H. & SELIKOFF, I.J. (1982). Mesothelioma

mortality in asbestos workers: Implications for models of carcinogenesis and risk assessment. *British Journal of Cancer*, **45**, 124–32.

SEBASTIEN, P., McDONALD, A.D., CASE, B. & HARLEY, R. (1989). Respiratory cancer in chroysotile textile and mining industries: exposure inferences from lung analysis. *British Journal of Industrial Medicine*, **46**, 180–7.

SPURNY, K.R. (1989). On the emission of fibrous particles from weathered and corroded asbestos–cement products. *Environmental Research*, **48**, 110–16.

WAGNER, J.C., BERRY, G., SKIDMORE, J.W. & TIMBRELL, V. (1974). The effects of the inhalation of asbestos in rats. *British Journal of Cancer*, **29**, 252–69.

WAGNER, J.C., CHAMBERLAIN, M., BROWN, R.C., BERRY, G., POOLEY, F.D., DAVIES, R. & GRIFFITHS, D.M. (1982). Biological effects of tremolite. *British Journal of Cancer*, **45**, 352–60.

WAGNER, J.C., NEWHOUSE, M.L., CORRIN, B., ROSSITER, C.E. & GRIFFITHS, D.M. (1989). Correlation between fibre content and disease in East London factory workers. In *Non-occupational Exposure to Mineral Fibres*. Bignon, J., Peto, J. & Saracci, R., eds, pp. 444–8. Lyon IARC Scientific Publication no. 90.

WHO (1986). Asbestos and other natural mineral fibres. *International Programme on Chemical Safety Environmental Health Criteria 53*. Geneva WHO.

7

RADON

T. LINDVALL

Exposure levels

Small amounts of naturally radioactive nuclides are present in soil, rock and building materials as well as in living organisms. The term 'nuclide' means an atom which is specified as to number of protons and neutrons, and as to content of nuclear energy. There are several different types of radiation. Gamma and alpha radiation are the most important. Gamma radiation is an electromagnetic phenomenon similar, for example, to heat or light radiation but much more energetic and penetrative. Alpha radiation particles are emitted nuclei of helium. These penetrate poorly. Beta radiation is the emission of electrons through decay. It penetrates more easily than alpha but less than gamma radiation.

The most important contributor to the total dose from natural radioactivity is radon and its shortlived decay products present in indoor air. The uranium decay chain is the source of radon. Uranium-238 and thorium-232 are radioactive with very long half-lives, 4.5×10^9 and 1.4×10^{10} years respectively. These elements are slowly transformed into stable lead-206 and lead-208 respectively. The transformation takes place by consecutive decays in a chain of several unstable intermediate nuclides. Radon is produced by decay of its parent nuclide radium. Radon is a noble gas and thus free to move by diffusion as well as by convection. One isotope of radon (radon-220 of the thorium decay chain) has a half-life of only 55 seconds. Isotopes are varieties of an atom with the same number of protons (same atomic number) but with different number of neutrons (different atomic weight). Due to the short half-life of radon-220, only a very small fraction of this isotope in the indoor air emanates from building materials or infiltrates from the ground subjacent to the building. In the uranium decay chain one intermediate nuclide is radon-222 with a half-life of 3.8 days. This is long

Uranium-238, half-life 4.5×10^9 years, alpha radiation
↓
Radium-226, 1600 years, alpha radiation
↓
Radon-222, 3.8 days, alpha radiation
↓
Polonium-218, 3.1 min, alpha radiation
↓
Lead-214, 26.8 min, beta and gamma radiation
↓
Bismuth-214, 19.9 min, beta and gamma radiation
↓
Polonium-214, 164 microseconds, alpha radiation
↓
Lead-210, 22.3 years, beta radiation
↓
Lead-206, stable ⟶

Fig. 7.1. The decay chain of radon and its short-lived daughters. Half-life times and major radioactive emissions.

enough for a partial release from building materials and soil and to become mixed with air through convective transport from the soil into a building. It may also be present in water, including ground water.

A decay of one atom of radon is followed within a short time by the consecutive decays of the short-lived radon daughters, polonium-218 (with a half-life of about 3 min), lead-214 a (27 min), bismuth-214 a (20 min), and polonium-214 a (2×10^{-4} s). (Fig. 7.1).

Radon is commonly measured either by measuring the radon gas or the radon daughters. In Sweden, guideline values are given as yearly averages of the radon daughter concentration. Both approaches of measurement are acceptable but, if radon gas is measured, it has to be recalculated into radon daughter concentration. Common measurement methods for radon gas include track registering film, thermoluminiscence dosimeters or semiconductive detectors, all placed in chambers preceded by a filter which captures the alpha particles. For measurement of radon daughters it is common for air samples to be drawn through filters and the alpha radiation of the filters to be measured by an alpha detector. Radon gas measurements tend to vary less and give more reliable averages for the space tested or for an extended period of time. By use of an open track registering film the sum of alpha radiation from radon and its daughters can be measured. This method is often useful although the results may be difficult to interpret.

In air, the nuclides originating from radon are present in varying proportions. As a simplification, the concept of potential alpha energy concentration of the equilibrium equivalent concentration of radon (EEC, or sometimes EER) is used for any combination of these nuclides having the same potential of emitting energy as alpha particles under complete decay to lead-210 as if they were all present in the same activity concentration (Bq/m^3). Thus 1 Bq/m^3 of radon daughters (EEC) refers to any combination of the individual radon daughters having the same potential of emitting alpha particles as one Bq/m^3 of each of the nuclides. In occupational health, the traditional unit of radon exposure is the working level (WL) which is defined as any combination of short-lived radon daughters per litre of air that will result in the emission of 1.3×10^5 MeV of alpha energy through their decay to lead-210. This is equivalent to 100 pCi of radon per litre of air in secular equilibrium with its progeny, or, in SI units, to 3700 Bq/m^3 EEC. Traditionally, the cumulative exposure to radon daughters is recorded in working level months (WLM), i.e. the exposure during 1 month of work (170 hours) to 1 WL (1 WLM = 170 WLh = 72 Bq per year/m^3.

A fraction of the radon daughters can occur as free ions or small neutral clusters of atoms in the air. In normal conditions, however, the majority are attached to particles in the air. These are usually referred to as the unattached and the attached fractions respectively. However, the airborne radon daughters may exist in a size continuum from small clusters of atoms up to the size of the indoor aerosol (Raes, Janssens & Vanmarcke, 1985).

Experiments have shown that moderate airborne concentrations of radon daughters indoors can increase considerably in the presence of increased numbers of particles such as those due to cigarette smoke (Bergman, Edling & Axelson, 1986). The newly formed, electrically charged radon daughters may tend to attach to the smoke particles rather than to the walls and furniture. A recent modelling study (Moghissi & Seiler, 1989) supports the view that the presence of such particles provides a mechanism for radon progeny to be attached to them and to remain in indoor air for a prolonged time. It is not clear whether or not the increased airborne concentration of radon daughters attached to particles increases the radiation risk to man.

The combined effect of all natural radioactivity contributes more to the total radiation burden than do all artificial sources of radiation including medical diagnostics, nuclear energy and fall out from nuclear weapon testing. In some areas (e.g. those in Finland, Norway, Sweden and Switzerland) radon in dwellings is not only the single most important contributor to radiation dose burden, but is even twice as important as the sum of all other sources, nuclear energy and medical diagnostics included.

In most countries, however, radon in indoor air contributes less than 50% of the total dose burden. For example, in the United Kingdom, radon decay products are the most important contributing factor responsible for 33% of the total dose burden (O'Riordan, 1983). According to the World Health Organization, Regional Office for Europe (WHO, 1987), the approximate arithmetic means of radon daughter concentrations are Finland, Norway and Sweden 50 Bq/m^3 EEC, the United States 27, Canada and the Federal Republic of Germany 25, and in the United Kingdom 10. The United Nations Scientific Committee on the Effects of Atomic Radiation (UNSCEAR, 1982) has estimated that the average indoor concentration of radon daughters is 15 Bq/m^3 in the temperate regions of the world.

Through radiation protection activities the total dose burden from artificial sources of radiation has been reduced or remained constant. However, in some countries, doses from radon in indoor air have increased considerably during recent decades. Measurements of radon in Sweden 30 years ago indicate an average concentration of radon daughters of 13 Bq/m^3. Recent radon measurements in randomly sampled dwellings indicate that the corresponding figure today is 53 $\pm$ 16 Bq/m^3 (Swedjemark, 1985), equivalent to a four to five-fold increase in the potential alpha energy concentration in 30 years. But this may be an overestimate because the measurements did not cover the same areas. Possible reasons for this increase are changes in building design and materials.

Radon enters a building from soil, ground fill or the capillary breathing layer, by emission from building materials of a fraction of the radon which is produced by decay of radium, with drinking or ground water, and with the supply air or infiltrating ambient air.

The contribution from building materials to the concentration of radon in the indoor air is inversely related to the ventilation rate. In flats with no contact with the ground, emission from building materials is typically the major source of indoor radon. Significant contributions to the indoor concentration of radon can be released from drinking water if this is taken from a well drilled in certain types of rock, for example, granite.

In detached houses and in flats at ground level, infiltration of soil gas can act as a carrier for radon from subjacent soil, fill or the capillary breathing layer. This is the most important and problematic of the different sources because it can result in extremely high indoor concentrations of radon. Infiltration of soil radon can occur from soil with quite 'normal' radium activity if the soil is permeable, e.g. gravel.

The concentration of radon in the interstices between particles in soil or in cracks of a rock is always orders of magnitude higher than can be accepted in a dwelling. In Sweden, concentrations of 20 000 Bq/m^3 of radon in the soil gas are regarded as 'normal' (Lindmark & Rosen, 1985). The concentration of radon in the soil gas is not constant. For example, in wintertime, frozen soil and snow act as a tight cover, resulting in high concentrations. Heavy rainfalls can seal the surface temporarily resulting in a peak in concentration.

It is generally accepted that radon enters a building from the subjacent soil, being carried by a convective flow of soil gas infiltrating through cracks, holes or other openings between the soil and the interior of the building. Transport by diffusion is nearly always insignificant. Since the concentration of radon in infiltrating soil gas is very high, even a small failure in the total tightness of a building interface with the soil will result in a significant increase in indoor radon concentration.

The emission rate of radon from building materials is strongly influenced by the moisture content of the material. Very dry material has a low emission rate. This is also the case when it is extremely wet.

There are some examples of building materials with unusually high radium activities and correspondingly high emissions of radon. The most well known are mill tailings used as concrete ballast, for example, in Grand Junction, Colorado (US Department of Energy, 1979), and aerated concrete based on alum shale which was widely used in Sweden until 1975. Other examples are granite stones in some parts of the UK (Cliff, 1978), phosphate slag in Alabama and byproduct gypsum from production of phosphoric acid.

Generally, the emission of radon from building materials can be regarded as constant over time. The contribution from this emission to the indoor radon concentration is inversely proportional to the amount of air passing through the building as intentional ventilation and as infiltration through leakages. A doubling of the air exchange rate will thus result in a 50% reduction in the radon concentration emanating from building materials, provided that there are no stagnant air movement zones.

In large-scale surveys of indoor radon concentration the radon and radon decay product concentrations seem to be well represented by a log-normal distribution. In Sweden, the combination of infiltration of radon from the soil and the use of aerated concrete based on alum shale has caused the elevated mean indoor concentration of radon daughters of 53 ± 16 Bq/m^3 (Swedish Radon Commission, 1983). As an effect of the log-normal concentration distribution, the main contribution to the collective effective dose equivalent to the whole population is not from the small fraction of houses with elevated radon concentrations. In Sweden, the 1% of the building stock with more than 400 Bq/m^3 has been estimated to cause 16% of the collective exposure, while 90% of the dwellings with less than 100 Bq/m^3 determine almost 50% of the collective exposure.

In Finland, the national mean indoor radon concentration is estimated as 90 Bq/m^3. In a problem area in Southern Finland, the mean for 754 measured houses was 370 Bq/m^3 while in the most extreme subarea the mean radon concentration was 1200 (Castrén *et al.*, 1985).

In Norway, the mean indoor radon concentration has been estimated as 80 Bq/m^3. About 3% of the dwellings in Norway are believed to have radon daughter concentrations above 200 Bq/m^3 and, for about 1% of the dwellings, the concentrations of radon occur in areas with alum shale in the ground. In one municipality with such ground, 38% of the measured dwellings had more than 400 Bq/m^3 of radon (Stranden, Kolstad & Lind, 1984).

In the UK, about 800 measurements have been made in areas with igneous geology and uranium mineralisations. The results indicate the existence of small areas where the mean radon concentration is of an order of magnitude above the national average. A few houses have been identified with approximately 2 000 Bq/m^3 radon daughter EEC (Cliff, 1984).

In the USA, high indoor concentrations have been reported from a part of Eastern Pennsylvania and from Maine. In six electrically heated homes in Pennsylvania, the concentration of radon in the living area ranged from 160 to 1800 Bq/m^3 (George, Duncan & Franklin, 1984). In 122 Pittsburgh area houses measurements on the first floor gave an average radon concentration of 90 Bq/m^3. The highest measured concentration was 900 Bq/m^3 (Cohen, 1985).

Uptake

The radioactive nuclides in the decay chains of uranium and thorium emit different types of radiation, of which gamma and alpha radiation are the most important. Gamma radiation is penetrative and, when emitted by nuclides in the building materials, it creates an almost uniform dose to the whole body of the occupants. Alpha radiation penetrates a few centimetres through air and only 40 to 70 μm through tissues. Alpha radiation can only reach living cells through the decay of radioactive nuclides within or on the body. The health hazard derived from the resultant effective dose equivalent is generally assumed to be a factor of 20 times higher than that from the same energy absorbed as gamma radiation in tissues. The basic reason for this is that alpha radiation has a high linear energy transfer, giving an intense impact on the small volume of tissue affected. When gamma radiation penetrates tissues the energy is distributed along its entire path of penetration.

Exposure to radiation is generally assumed to result in a linear increase in the probability of developing cancer.

Different types of radiation and doses received by different parts of the body can be compared as to their biological effect only after transformation into a common dimension, the effective dose equivalent. The internationally adopted weighting factors for the transformation for different types of radiation and various organs are continuously revised in the light of new data evaluated by the International Commission on Radiological Protection (ICRP).

The amount of radiation energy absorbed in a tissue is expressed in the SI unit Gray (Gy, joule/kg; 1 Gy = 100 rad). The absorbed energy is transformed to a dose equivalent (SI unit Sievert, Sv; 1 Sv = 100 rem) to that tissue by multiplication with a quality factor specific to the type of radiation. For gamma radiation, this factor is 1 and for alpha radiation it is 20 as mentioned before.

The dose equivalents for a specific organ are transferred to effective dose equivalents (unit Sv) by internationally adopted weighting factors for each organ. After these transformations it is possible to consistently compare health hazards from alpha radiation to the lungs with gamma radiation to the whole body. It is then generally assumed that the dose equivalent 1 Sv to the lungs corresponds to 0.12 Sv effective dose equivalent or to a uniform dose of 0.12 Sv to the entire body.

According to the ICRP (1984), it is commonly assumed that the effective dose equivalents can be indicators of the probability of developing cancer induced by radiation.

Effects on health

Present estimates of the relationship between exposure to radon daughters in indoor air and the added risk of developing lung cancer are derived from dosimetric modelling or by analogies or from extrapolations from epidemiological studies on the excess prevalence of lung cancer among miners. The dosimetric models calculate the distribution and movement of radon daughters in the body. From this the doses received by different tissues are calculated and, subsequently, the effective dose equivalent modelled. Such modelling inevitably involves several assumptions and a considerable uncertainty. Several epidemiological studies have been reported, resulting in a fair consensus on the relationship between increased lung cancer incidence and exposure to radon daughters in mines (Archer, Gillam & Wagoner, 1976; Axelson *et al.,* 1981; Edling, 1983; Kunz *et al.,* 1979; Sevc, Kunz & Placek, 1976; UNSCEAR, 1982).

When applying results from miners to the general population, some differences can be compensated for, i.e. breathing rate and duration of exposure. Other differences can be compensated for only with assumptions which result in considerable uncertainty. For example, miners are healthy adult males whereas in dwellings the whole population is exposed. The higher breathing rate during work in mines can result in a distribution of the radon daughters in the respiratory tract different from the distribution after the slower breathing rate in dwellings. The exposures in mines are characterised by high concentration of short duration limited to 40 hours a week for 40 years; in dwellings, the exposure is extended over longer times but at lower concentrations.

Epidemiologists have been investigating the health effects of indoor radon for some years. Published results of a few pilot studies indicate a statistically significant association between bronchial cancer and estimated exposure to radon in dwellings (Axelson *et al.,* 1981; Pershagen, Damber & Falk, 1984).

During recent years, several scientific bodies have evaluated the results from dosimetric modelling and epidemiological studies and thus attempted to reach consensus on an acceptable relationship between exposure to radon daughters in dwellings, expressed as a conversion factor between exposure, concentration and the resultant effective dose equivalent. UNSCEAR (1982) gave as a reference conversion factor 0.061 mSv/year per Bq/m^3, assuming that 80% of the time was spent indoors. This corresponds to 0.08

mSv/year per Bq/m^3, assuming 100% indoor occupancy. The range of uncertainty was expressed as 0.04-0.1. The OECD (1982) assumed a conversion factor of 0.065 mSv/year per Bq/m^3 of radon EEC in buildings with low ventilation rates, and of 0.095 mSv/year per Bq/m^3 in buildings with average ventilation rates. The National Radiological Protection Board in the UK has adopted 0.05 mSv/year per Bq/m^3 (5 mSv/WLM) as a reference conversion factor for under 75% occupancy (O'Riordan, 1983). In the US the National Commission on Radiation Protection and Measurement estimated the lifetime excess risk for domestic exposure as equivalent to a conversion coefficient of 0.13 mSv/year per Bq/m^3 (12 mSv/WLM) (James 1984).

If the conversion factor of UNSCEAR (1982) is used in combination with the risk estimate of 0.02 fatal cancers per person Sv given by the ICRP (1984) then one Bq/m^3 of radon daughters is assumed to result in 0.061 mSv/year. This is a risk of $0.061 \times 10^{-3} \times 0.02 = 1.2 \times 10^{-6}$ fatal cancers per year, or in the lifetime risk over 70 years of 9×10^{-5} fatal cancers. If this risk estimate is correct, it would mean that, in Sweden, 400 out of the present 2700 cases of lung cancer per year are related to inhalation of radon daughters in indoor air. This risk estimate is based on the so-called absolute risk concept. This concept considers only the excess radiation risk as a function of the cumulative dose or exposure. However, analysis of the increase with time of the lung cancer death rates in exposed populations suggests that a relative risk model may apply. According to this model the incidence of lung cancer in exposed populations, after an initial latency period, follows that of non-exposed populations, but at a higher level by a factor constant in time after exposure.

The World Health Organization (WHO, 1987) has evaluated radon in indoor air assuming a mean indoor radon daughter concentration of 15 Bq/m^3 EEC averaged over the whole population in the temperate region and an occupancy factor of 0.8 for indoor environments. WHO concludes that 'an annual incidence of 10–40 cases of lung cancer per million persons should be expected that is attributable to radon daughter exposure'. The relative risk model leads to the conclusion that under these conditions about 5–15% of the observed lung cancer frequency or of the lifetime risk may be attributable to indoor radon daughters. This relative risk is nearly equal for males and females, and for smokers and non-smokers.

For chronic exposures at about 300–500 Bq/m^3, the lung cancer death rate might be twice as high as it would be in the same population with no excess exposure to radon. According to the relative risk model exposure to radon in absolute terms would be much more dangerous for smokers or people with otherwise enhanced risk of developing lung cancer (Jacobi, 1984; Jacobi & Paretzke, 1985).

Possibilities for control

Radon from the soil

The soil subjacent to buildings is probably the most important contributor to indoor radon. No other source can result in such extremely high concentrations as radon from the soil.

Only insignificant amounts of radon can enter a building by diffusion through building materials. Infiltrating soil gas acts as a carrier for radon from the subsoil into buildings. Thus there are three essential prerequisites for infiltration of radon from the soil:

- the subsoil must be permeable, allowing a convective flow of air,
- there must be some leakage between the soil and the interior of the building, and
- there must be a driving force, i.e. a pressure gradient drawing air from the soil into the building.

Infiltration of soil gas in new basement constructions must be prevented by designing as tight a construction as possible. Normal concrete with no cracks and holes is tight enough, and a concrete slab of conventional quality presents sufficient resistance to radon.

Often the site of the lack of tightness of a building to the ground is inaccessible or occurs over a large area. Attempts have been made to seal all the interior surfaces of concrete floors and basement foundation walls with several layers of epoxy sealants. Although epoxy sealants provide an impervious layer the effect is not normally permanent. The sealant tends to crack with cracks in the concrete slab (US Department of Energy, 1979). This phenomenon could probably be counteracted by using a permanently elastic surface sealant such as rubberised asphalt. In a furnished basement, the cost of restoring panelling and floor makes the method rather expensive.

The most commonly applied way of coping with infiltration of soil gas has been to eliminate the pressure gradient between the soil and the building. This can be achieved by balancing the negative pressure caused by the stack effect by pressurising the building. This method, however, has drawbacks. It would have the desired effect on infiltration of soil gas but would simultaneously push warm humid indoor air into the cold parts of the walls resulting in water vapour condensation and possibly mould growth. Instead, it is recommended that the pressure gradient over the slab be changed by sucking air from the soil under the slab relative to the interior of the building. The technique has been successfully applied in Canada, in the remedial action programme in Grand Junction, Colorado, and in Sweden.

In Sweden, 39 installations for creating a negative sub-slab pressure have been evaluated. the average reduction in radon concentration in indoor air was 88% or 2600 Bq/m^3. A limited flow of gas is extracted, just enough to prevent infiltration through any cracks. Sometimes it is sufficient to apply suction in one point only, but if the capillary breathing layer is sectioned by footings, suction in every section is recommended. It is essential to apply the suction to a small cavity under the slab. With no such cavity the pressure drop will often be too high close to the point of suction, preventing horizontal spreading of negative pressure under the slab (Ericson, Schmied & Clavensjö, 1984).

In certain conditions, it is also possible to prevent infiltration of soil radon by blowing air down under the slab. Such an installation will be effective if the concentration of radon under the slab is reduced to a low value by this ventilation. Optimal alternatives must be selected with due consideration to building physics and local conditions.

Increased indoor ventilation is usually not an effective remedial action against infiltration of soil gas. If soil gas infiltrates the building, the resultant indoor air concentration may become extremely high. The source of the emission then has to be reduced, since no realistic ventilation rate will suffice to dilute the radon to an acceptable concentration. Mechanical exhaust air ventilation will increase the negative pressure in the building and possibly increase the infiltration of radon from the soil. Whenever mechanical ventilation is applied in a house with suspected infiltration of radon from the soil the system should be of the balanced supply-exhaust type. Balancing the system to a neutral pressure and regular maintenance will be essential.

Radon from building materials

Building materials can be regarded, by and large, as a source of indoor radon that does not vary with time. Common building materials have such low emission rates, less than 20 Bq/m^2 h, that retrofitting with a mechanical ventilation system will not be cost-effective. The most well-known building material with excessive emission of radon is probably a type of Swedish aerated concrete with alum shale. The typical emission from this material is 50–200 Bq/m^{2}h. In such houses weather stripping can result in a rise of the radon concentration up to 1000 Bq/m^3. Then retrofitting with a mechanical balanced ventilation system is the best remedial action. Adequate air exchange rates cannot be achieved simply by increasing the general lack of tightness of the building envelope. In cold climates, the result would be an uncomfortable indoor thermal climate and excessive heating costs.

The effect of a retrofitted balanced ventilation system on indoor radon concentrations often deviates from what is calculated on the assumption of complete air mixing and dilution of a constant source for radon. Improperly designed or located air inlets and outlets may cause the ventilation air to pass straight from air inlet to air outlet, not mixing with the room air. This can often result in a surprisingly small reduction in radon concentration.

If the ventilation systems are equipped with efficient air-to-air heat exchangers or the heat in the exhaust air is recovered with a heat pump, the ventilation systems neither significantly increase nor decrease the energy demand in the building. With retrofitted mechanical and balanced ventilation systems, houses built from aerated concrete based on alum shale have been be remedied efficiently (Ericson & Schmied, 1985).

Air cleaning

When air passes through a mechanical filter or an electrostatic precipitator, particlebound radon daughters, as well as the unattached fraction of the radon daughters, are removed efficiently. These filters, however, do not remove any radon gas from the air and thus the production of radon daughters through the decay of radon is not affected. The radioactive half-lifes of the radon daughters are fairly short and limit the duration of the effect of filtration. A major reduction in concentration of radon daughters therefore requires more than 1 turnover per hour.

Filtration removes radon daughters from the air by two more mechanisms over and above the removal in the filter itself. The radon daughters produced by radon decay in filtered air, which has a low concentration of particulates, will have less chance to attach to particles, leaving a large fraction of the radon daughters in the unattached form. Unattached radon daughters are more mobile and more likely to plate out on walls and other surfaces.

The forced circulation of air which is a side-effect of filtration also increases the plate out on surfaces.

The overall reduction in concentration of radon daughters by air cleaning is partly outweighed by the fact that the unattached fraction of radon daughters is increased. The unattached fraction is assumed to be more harmful and to give a higher absorbed radiation dose per inhaled unit of activity. Radiation doses from inhalation of radon and radon daughters are calculated by models incorporating several assumptions. All the models show some reduction of absorbed radiation dose during filter use.

As the probably more harmful unattached fraction is increased in concentration by filtration, it has been argued that filtration might even increase the absorbed radiation dose. This seems not to be the case, although the reduction in radiation dose is less than is indicated by the overall reduction in activity concentration of radon daughters.

It seems that air filtration would result in only a small reduction in absorbed radiation dose. The technique is not sufficient as the sole remedial action against excessive concentrations of radon daughters except in combination with other more efficient actions. In cases of moderately enhanced concentrations of radon daughters, filtration can result in an acceptable air quality.

Risk management

The International Commission on Radiological Protection (ICRP, 1984) has given general advice on fundamental principles in radiation protection to natural sources of radiation. Radon in dwellings, as the most important exposure to natural radiation, has been used for illustration of the general recommendations.

The traditional approach in radiation protection has been to set very stringent limits on doses from artificial sources of radiation. Natural radiation, including radon in indoor air, has traditionally not been covered by any regulations.

The dose limit for occupational exposure has been set at 50 mSv in one year. This limit is applicable for a single year, but yearly doses of 50 mSv would not be acceptable over a long period of the working life of an individual. With the limit of 50 mSv/year per year for 40 years of working life, the individual would accumulate 2 Sv. Using the assumption of ICRP (1984) that each person Sv corresponds statistically to 0.02 fatalities induced by radiation, this is equivalent to a 4% individual risk. It is regarded as the upper limit of individual risk that the individual generally will accept. With no regard to the costs involved, each individual is guaranteed not to receive higher doses from occupational exposure.

For the general public the ICRP (1984) has recommended that the effective dose equivalent from exposure to artificial sources of radiation in any year be limited to 5 mSv. For repeated exposures the Commission has stated that 5 mSv/year is somewhat high and that it would be prudent to restrict the dose further to 1 mSv/year for lifelong exposure. This would limit the accumulated dose from artificial sources to each member of the public to 70 mSv, which statistically would correspond to an individual risk of 0.14%.

The setting of upper limits of accepted doses is one of three principles for radiation protection. Another principle is that a practice which causes exposure to artificial sources of radiation should be justified by providing a positive net benefit to mankind. The third principle discussed by ICRP (1984) is the rule of ALARA. ALARA stands for As Low As Reasonably Achievable, economic and social factors being taken into account. The meaning of this principle is optimisation of the use of resources allocated to further reduction of the doses below the upper limit. The underlying philosophy is that every increment in dose is equally urgent to reduce. Thus, it would be contrary to this rule to take action at a very high cost per unit of dose reduction if at the same time more cost-effective actions are neglected.

The ALARA principle is not intended to protect any single identified individual but to be used as a guideline for what is, and is not, a reasonable allocation of resources.

The annual 'whole body' effective dose equivalent from natural sources of radiation is estimated to be approximately 2 mSv as a worldwide average (UNSCEAR, 1982). Some distinguish between purely unaffected natural radiation and natural radiation enhanced by some action of man, Technologically Enhanced Natural Radiation (TENR). The ICRP finds it difficult to make such a distinction. The ICRP stated (1984): 'The Commission had already drawn attention to the difficulty of distinguishing between normal and enhanced levels of exposure. It has now concluded that this distinction is unhelpful and bases its new advice on a different approach in which the emphasis is on the extent to which the exposure to the source is controllable.'

The main parameter of interest is the long-term average concentration of radon daughters. As the concentration of radon in indoor air ranges over more than three powers of ten, a moderate uncertainty in the characterisation of a specific house must be accepted. For cases where fairly simple remedial actions can be taken, the ICRP suggests (1984) that an action level for equilibrium equivalent radon concentration in the region of 200 Bq/m^3 (annual effective dose equivalent of about 12 mSv) might be considered. For major remedial action, a value several times higher might be more appropriate, the Commission says. The World Health Organisation (WHO, 1987) recommends that simple remedial measures should be considered for buildings with radon daughter concentrations of more than 100 Bq/m^3 EEC. Remedial actions in buildings with concentrations higher than 400 Bq/m^3 EEC as an annual average should be considered without delay.

In Sweden, regulations in force stipulate that an average concentration of 400 Bq/m^3 and more of radon daughters in the inhabited space shall be regarded as a sanitary nuisance (Swedjemark, 1986). In theory, this makes remedial action mandatory. This action level was introduced as a temporary limit, giving priority to the few buildings with very high concentration.

A practical and cost-effective strategy is suggested by Ericson, Lindvall and Mansson (1986). They recommend:

- Measurements indicating less than approximately 150 Bq/m^3 of radon daughters suggest only simple and low-cost remedial actions (sealing of entry routes, ventilation maintenance).
- At concentrations 150–500 Bq/m^3 remedial actions are recommended but the decision on the specific actions be taken by the house owner following guidance from the authorities.
- Above 500 Bq/m^3 immediate simple remedial actions are justified. Repeated measurements are to be made for verification of the remedial actions.

A key problem is the design of screening programmes for identifying buildings eligible for remedial actions. The ICRP recommends (1984) that 'competent national authorities establish levels to separate exposures that require investigation from those that do not'. This procedure is meant to separate, in the least costly way, the majority of buildings with low radon concentration from the few with an elevated concentration of radon daughters above or near the action level. The ICRP recommends that not every building be subject to measurements but that characteristics such as type of building material, local geology, and ventilation principle be used in attempts to separate the small fraction of the building stock where the vast majority of buildings with radon daughter concentration above the action level can be found.

For new buildings the ICRP (1984) states that there are much better possibilities than in old buildings to achieve a low indoor radon concentration. It is recommended that 'the exposure of the most highly exposed individuals should be limited by the application of an upper bound of individual dose in the optimisation assessment'. This upper bound should be established taking into consideration the local possibilities of minimising the infiltration of soil gas and the possibility of avoiding the use of building materials with elevated activity of radium. The ICRP believes that a reasonable upper bound for the equilibrium equivalent radon concentration is of the order of 100 Bq/m^3. The WHO (1987) recommends that building codes should include sections to ensure that radon daughter levels do not exceed 100 Bq/m^3 EEC in new buildings, and that appropriate practices should be prescribed. In Sweden, it is prescribed in the Building Code that in new buildings the long-term average concentration of radon EEC must not exceed 70 Bq/m^3.

References

This chapter is largely based on the contribution of the Swedish members to the International Energy Agency, Annex IX 'Minimum ventilation rates'; for a more elaborative discussion on the subject see: ERICSON, S.O., LINDVALL, T. & MANSSON, L.G. (1986). *Indoor Ionizing Radiation: Technical Solutions and Remedial Strategy*. Stockholm: Swedish Council for Building Research, D7.

ARCHER, V.E., GILLAM, J.D. & WAGONER, J.K. (1976). Respiratory disease mortality among uranium miners. *Annals of the New York Academy of Sciences*, **271**, 280–93

AXELSON, O., EDLING, C., KLING, H., ANDERSSON, L. & RIGNÉR, A. (1981). Lung cancer and radon in dwellings. *Lancet*, 995–6.

BERGMAN, H., EDLING, C. & AXELSON, O. (1986). Indoor radon daughter concentrations and passive smoking. *Environment International*, **12**, 17–19.

CASTRÉN, O., VOUTILAINEN, A., WINGVIST, K. & MÄKELAINEN, I. (1985). 'Studies of high indoor radon areas in Finland', *The Science of the Total Environment*, **45**, 311–18.

CLIFF, K.D. (1978). Assessment of airborne radon daughter concentrations in dwellings in Great Britain, *Physics in Medicine and Biology*, **23**, 696–711.

CLIFF, K.D. (February 1984). The incidence and origin of radon and its decay products in buildings. *National Radiological Protection Board, NRPB-R159*. London: HMSO.

COHEN, B.L. (1985). Survey of one-year average Rn levels in Pittsburg area homes. *Health Physics*, 1053–9.

EDLING, C. (1983). Lung cancer and radon daughter exposure in mines and dwellings. Linköping: Linköping University, Medical Dissertations, Sweden, no. 157.

ERICSON, S.O., LINDVALL, T. & MANSSON, L.G. (1986). *Indoor Ionizing Radiation: Technical Solutions and Remedial Strategy*. Stockholm: Swedish Council for Building Research. Report D7.

ERICSON, S.O. & SCHMIED, H. (1985). Installation of supply/exhaust ventilation as a remedial action against radon from soil and/or building materials *Science of the Total Environment*, **45**, 499–505.

ERICSON, S.O., SCHMIED, H. & CLAVENSJÖ, B. (1984). Modified technology in new constructions, and cost effective remedial action in existing structures, to prevent infiltration of soil gas carrying radon. In *Indoor Air* volume 5: Buildings, ventilation and thermal climate. Berglund, B., Lindvall T. & Sundell, J. eds, Stockholm: Swedish Council for Building Research, D20, 153–8.

GEORGE, A.C., DUNCAN, M. & FRANKLIN, H. (1984). Measurements of radon in residential buildings in Maryland and Pennsylvania, USA. *Radiation Protection Dosimetry*, **7**, 291–4.

INTERNATIONAL COMMISSION ON RADIOLOGICAL PROTECTION, ICRP (1984). *Principles for Limiting Exposure of the Public to Natural Sources of Radiation*. Oxford: Pergamon Press. (ICRP Publication No. 39.)

JACOBI, W. (1984). Expected lung cancer risk from radon daughter exposure in dwellings. In *Indoor Air* volume 1: Recent advances in the health sciences and technology. Berglund B., Lindvall T. & Sundell, J. eds. Stockholm: Swedish Council for Building Research, D16,31–42.

JACOBI, W. & PARETZKE, H.G. (1985). Risk assessment for indoor exposure to radon daughters. *Science of the Total Environment*, **45**, 551–62.

JAMES, A.C. (1984). Dosimetric approaches to risk assessment for indoor exposure to radon daughters. *Radiation Protection Dosimetry*, **7**, 535–66.

KUNZ, E., SEVC, J., PLACEK, V. & HORACEK, L. (1979). Lung cancer in man in relation to different time distribution of radiation exposure, *Health Physics*, **36**, 699–706.

LINDMARK, A. & ROSEN, B. (1985). Radon in soil gas – echalation tests and *in situ* measurements, *Science of the Total Environment*, **45**, 397–404.

MOGHISSI, A.A. & SEILER, M.A. (1989). Enhancement of exposure to radon progeny as a consequence of passive smoking. *Environment International*, **15**, 261–4.

OECD, Nuclear Energy Agency (1982). Dosimetry aspects of exposure to radon and thoron daughters, report by a group of experts, Paris, OECD.

O'RIORDAN, M.C. (February 1983). *Human Exposure to Radon Decay Products Inside Dwellings in the United Kingdom*. National Radiological Protection Board, NRPB-R152. London, HMSO.

PERSHAGEN, G., DAMBER, L. & FALK, R. (1984). Exposure to radon in dwellings and lung cancer: a pilot study. In: *Indoor Air* volume 2: Radon, passive smoking, particulates and housing epidemiology. Berglund B., Lindvall T. & Sundell, J. eds, Stockholm: Swedish Council for Building Research, D17, 73–8.

RAES, F., JANSSENS, A. & VANMARCKE, H. (1985). A closer look at the behaviour of radioactive decay products in air, *Science of the Total Environment*, **45**, 205–18.

SEVC, J., KUNZ, E. & PLACEK, V. (1976). Lung cancer in uranium miners and long-term exposure to radon daughter products. *Health Physics,* **30**, 433–7.

STRANDEN, E., KOLSTAD, A.K. & LIND, B. (1984). Radon in dwellings (Radon i boliger), *Arbeidsdokument* 1984:5, National Institute of Radiation Hygiene, Norway (in Norwegian).

SWEDISH RADON COMMISSION (1983). *Final Report 'Radon i bostäder'*. Stockholm: Liber, SOU 1983:6 (in Swedish).

SWEDJEMARK, G.A. (1985). Radon and its decay products in housing estimation of the radon daughters exposure to the Swedish population and methods for evaluation of the uncertainties in annual averages. Doctoral Dissertation, Department of Radiation Physics, University of Stockholm, Stockholm.

SWEDJEMARK, G.A. (1986). *Limitation Schemes to Decrease the Radon Daughters in Indoor Air*. Stockholm: National Institute of Radiation Protection, SSI-rapport 86–01.

SWEDJEMARK, G.A. & MJÖNES, L. (1984). Exposure of the Swedish population to radon daughters. In: *Indoor Air* vol. 2: Radon, passive smoking, particulates and housing epidemiology. Berglund B., Lindvall T. & Sundell, J. eds., Stockholm: Swedish Council for Building Research, D17, 37–44.

UNITED NATIONS SCIENTIFIC COMMITTEE ON THE EFFECTS OF ATOMIC RADIATION, UNSCEAR (1982). *Ionizing Radiation: Sources and Biological Effects.* New York: United Nations.

US DEPARTMENT OF ENERGY (February 1979). *Progress Report on the Grand Junction Uranium Mill Tailings Remedial Action Program,* DOE/EV-0033.

WORLD HEALTH ORGANISATION (1987). *Air Quality Guidelines for Europe*. Copenhagen: WHO Regional Publications, European Series No. 23.

8

FORMALDEHYDE

G. R. BETTON

Introduction

Formaldehyde is a low molecular weight aldehyde (CH_2O, mol. wt.=30) existing as a pungent colourless gas with a boiling point of −19°C. It is readily soluble in water and the 37% aqueous solution known as formalin is the conventional medium for commercial distribution and manufacturing processes. Methanol is commonly added at a 10–15% concentration to formalin solution as a stabiliser to prevent polymerisation to paraformaldehyde solid.

Formaldehyde was used as a preservative or fixative of biological material both for biomedical research and for embalming purposes. It was known for many years before its chemical cross-linking reactivity was understood. Once the ability of aldehydes such as formaldehyde to serve as a basic chemical building block in polymer chemistry was appreciated, it became a major chemical raw material with many different applications.

One application has been the *in situ* generation of urea–formaldehyde foam insulation (UFFI) which provided an economic means of thermally insulating existing, as well as new, homes. Errors in production of UFFI, however, generated free formaldehyde which leaked into newly insulated homes and created considerable public awareness of the vapour's irritant and allergenic properties in the domestic environment. These and the later positive findings in animal carcinogenicity bioassays of formaldehyde prompted regulatory agencies and environmental protection groups to take a great interest in formaldehyde's health effects and in its regulation and control. The review by Samet *et al.* (1988) provides epidemiological data on such health effects.

Exposure levels

The magnitude of formaldehyde production in the USA alone exceeds 4 million tonnes per annum, approximately half of which is used in the production of polymer urea– and phenol–formaldehyde resins and in the formation of urea–formaldehyde foam insulation. The annual production in the Federal Republic of Germany was reported to be 0.5 million tonnes (BGA, 1984).

Background concentrations of formaldehyde range from 0.1 $\mu g/m^3$ in oceanic regions to 1.0 $\mu g/m^3$ in uncontaminated regions of the land mass as a result of decomposition of organic matter. In industrial regions, concentrations of 0.04 mg/m^3 have been recorded.

Indoor formaldehyde concentrations of 0.1–0.8 mg/m^3 in houses with UFFI and 0.5 mg/m^3 in mobile homes are commonly found with levels in exceptional cases exceeding 1 mg/m^3 with attendant overt clinical symptoms in occupants. Mobile homes whose occupants have complained of formaldehyde-related symptoms have recorded levels up to 3.7 mg/m^3 (Bernstein *et al.*, 1984).

High indoor concentrations in the domestic environment have been attributed to low-grade particle board products and plywoods in the construction of houses and furniture. The absence of surface sealant or laminate permits formaldehyde vapour to leach out over considerable periods of time. Reduced ventilation rates exacerbate local concentration effects (Molhave *et al.*, 1983).

The use of UFFI in houses and mobile homes has been associated with some of the more severe domestic exposure incidence. High ventilation rates or removal of foam insulation have been necessary because of the long half-life of 2–3 years for formaldehyde emission in such cases. Exposure levels to formaldehyde vapour are dependent on the following variables:

- available free formaldehyde in the source material
- temperature
- humidity
- ventilation rate

For these reasons, the more significant recorded exposure levels have been in manufacturing processes using formalin solution and in mobile homes insulated with urea–formaldehyde foam (UFFI) where high temperature and humidity and low ventilation rates have produced acutely irritant concentrations of formaldehyde vapour. Conventional houses typically have 0.5–1.0 air changes per hour but the process of 'weatherisation' to improve energy conservation can reduce this ventilation figure to 0.3 air changes/hour. Studies with other markers of indoor air pollutants have shown a 40% reduction in ventilation rates following 'weatherisation' (Lamb *et al.*, 1985).

Sources of combustion have been shown that generate formaldehyde, e.g. cigarettes (0.38 mg/pack) and gas burning stoves (15–25 mg/h, National Research Council, 1981).

Dosimetry presents a number of problems in comparing between studies, as has the wide range of measured concentrations using standardised techniques. Personal badge monitors have a lower limit of sensitivity of 0.1 mg/m^3 and a high degree of specificity but collection efficiency under-estimates values obtained by an active sampling liquid impinger system by approximately 20%. Formaldehyde also reacts with ammonia, amines and amides to form condensation products. It is also reactive with hydrochloric acid and water to form *bis*(chloromethyl)ether (BCME):

$$2HCHO + 2HCl \rightarrow ClCH_2OCH_2Cl + H_2O$$

This latter reaction is significant as BCME is a potent carcinogen.

Uptake

Formaldehyde uptake, generally, is a result of exposure to the vapour phase released by formalin solution or from polymers. Accidental ingestion of formaldehyde solution is unlikely because of the pungent odour of formalin, and cases of suicidal swallowing have been rapidly fatal.

Exposure to formaldehyde gas is generally limited by the pungent and irritant nature of the molecule with significant symptoms being experienced above 1 mg/m^3 (1.5 mg/m^3). Some tolerance to the irritant properties can develop with chronic exposure.

Formaldehyde gas is highly water soluble and the moist mucous membranes of the nasal cavity efficiently absorb all inhaled gas prior to reaching the trachea and lungs, unless bypassed by mouth breathing. The nasal mucosa is covered by a mucus film moved in defined pathways by the action of cilia. The fluid motion of this layer intended for the removal of particulate material is impaired by exposure to formaldehyde gas which increases viscosity, possibly by cross-linking glycosoaminoglycans of the mucus layer. Penetration through the mucous layer into the nasal epithelium is very rapid. Glutathione and NAD+ plus the enzymes, aldehyde dehydrogenase and catalase, with hydrogen peroxide can rapidly oxidise formaldehyde to formic acid. Subsequent metabolism proceeds in a wide range of tissues via one of two major pathways:

- peroxide-mediated oxidation to carbon dioxide using the enzyme catalase
- formation of 10-formyltetrahydrofolate from tetrahydrofolate and subsequent entry into the 1-carbon pool as methionine, serine, thymidine, etc with subsequent appearance in amino-acids, DNA and RNA, with tissues such as bone marrow being 14 °C radiolabelled within minutes of gaseous exposure.

In addition, formate is excreted in significant amounts in urine of man and other primates whereas in the rat oxidation to carbon dioxide is the principal route of elimination (Griesemer, 1982).

Formaldehyde has been shown experimentally to form methylol groups with amine bearing side chains of some amino-acids and with nucleotides with subsequent cross-linking condensation reactions between protein and single-stranded DNA chains. In view of the very short half-life of formaldehyde in the circulation (1.5 mins) and rapid metabolism, the *in vivo* biological relevance of these reactions apart from local action on the nasal mucosa is uncertain.

Effects on health

Acute exposure

Inhalation

Formaldehyde has a high toxicity as a vapour. In animals, exposure to >15 mg/m^3 is lethal with marked pulmonary, nasal and ocular mucous membrane effects. Microscopic lung pathology has been observed in hamsters exposed to 3 mg/m^3 for 4 hours.

In man, irritant effects to nose, throat, lung and eyes have been seen following exposure of 0.5 mg/m^3 and above and these effects are acute and severe at 10 mg/m^3. Pathological changes of the nasal mucosa have been reported in workers exposed to formaldehyde in the workplace with a time weighted average (TWA) of 0.07–0.74 mg/m^3 with peak concentrations of 3.3 mg/m^3. Concentrations of >2 mg/m^3 were linked to diminished lung

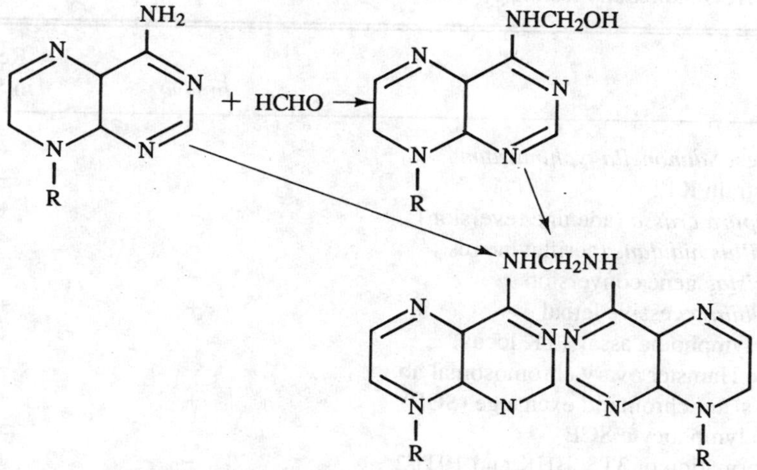

Fig. 8.1. Formation of DNA cross links by formaldehyde.

function as measured by forced expiratory volume (FEV). The ability of formaldehyde to act as an allergen (*vide infra*) has led to propositions that respiratory asthmatic hypersensitivity reactions could be initiated in sensitive individuals. Asthmatic subjects have been associated with exaggerated sensitivity to formaldehyde vapour but this has generally been attributed to the acute irritant action of the molecule on reactive airways rather than a true specific allergic response.

Skin toxicity and allergenicity

Aqueous 1% solutions of formaldehyde did not produce acute dermal irritation in rabbits. However, 2% aqueous solutions of formaldehyde have been shown to be potent allergens with concentrations as low as 0.04% being allergenic on repeated dose regimens in guinea pigs.

In man, cutaneous sensitisation and allergic contact dermatitis have been well documented in individuals chronically exposed to formaldehyde-containing materials. A 2% solution, however, does not produce acute skin irritant effects in man.

Ocular toxicity

Formalin (26%) solution has been shown to be severely irritant, producing permanent corneal injury in rabbits, guinea pigs and man. Splashes of a 4% formaldehyde solution are highly irritant in man but no permanent effects have been reported.

Table 8.1. *Genotoxicity testing of formaldehyde*

Test	*In vivo*	Result in S100
Ames test *Salmonella typhimurium*	–	+ve
E.coli strain K12	–	±ve
Neurospora crassa (adenine reversion)	–	±ve
Aspergillus nidulans (acriflavine res.)	–	±ve
S.cerevisiae gene conversion	–	+ve
Drosophila recessive lethal	+	+ve
Mouse lymphoma assay (TK locus)	–	+ve
Chinese Hamster ovary chromosomal ab	–	–ve
Mouse sister chromatid exchange (SCE)	+	+ve
Human lymphocyte SCE	–	±ve
Transformation of 3T3, BHK and 10Tl/2	–	+ve

+ve : positive ±ve : equivocal positive

Potential carcinogenicity

Being a highly reactive aldehyde, formaldehyde has been shown to produce methylene cross-links between methylol groups by a condensation reaction between nucleotides of DNA in a single strand configuration as shown in Figure 8.1.

These and other biochemical reactions have led to the expectation of genotoxic potential of formaldehyde and hence potential for carcinogenicity. Balanced against this high level of reactivity is the high rate of covalent binding to other molecules and rapid rate of metabolism with resultant short half-life of the parent molecule limiting the likely degree of DNA exposure except in sites of initial contact. Results of short-term genotoxicity testing have been somewhat varied between test systems as listed in Table 8.1.

The *in vivo* evaluation of the carcinogenic potential of formaldehyde vapour was conducted relatively late in the product's life cycle despite its widescale usage and demonstration of local sarcoma production following subcutaneous injection of rats. The latter response may represent an indirect response to local irritancy and the more relevant inhalation lifespan studies in rats and mice sponsored by the Chemical Industry Institute of Toxicology

Table 8.2. *Rodent oncogenicity tests on formaldehyde inhalation*

Species ref.	Dose (mg/m^3)	Sex	Nasal tumours[a]	
Rat (up to 30 m)	0.0	M	0/118	Kerns *et*
Fischer 344		F	0/114	*al.* (1983)
	2.0	M	0/118	
		F	0/118	
	5.6	M	1/119	
		F	1/116	
	14.3	M	51/117	
		F	52/115	
Rat (up to 20 m)	0.0	M	0/20	Albert *et*
Sprague Dawley	14.2	M	10/20	*al.* (1982)
10.0 HCl+ 14.3 (BCME)		M	12/20	
Mouse (up to 27 m)	0.0	M	0/120	Kerns *et*
B6C3Fl		F	0/120	*al.* (1983)
	2.0	M	0/120	
		F	0/120	
	5.6	M	0/120	
		F	0/120	
	14.3	M	2/120	
		F	0/120	

[a] squamous cell carcinoma, incidental tumours excluded

produced the first unequivocal evidence of formaldehyde being a nasal carcinogen in rodents (Kerns *et al.*, 1983). Groups of 120 male and 120 female Fischer F344 rats and C57Bl/6×$C3HF_1$ mice were exposed to 0, 2.0, 5.6 and 14.3 mg/m^3 formaldehyde gas for 6 h/day, 5 days a week for 24 months followed by a 6-month period without treatment. Interim sacrifices were also carried out. Results are presented in Table 8.2.

These findings therefore demonstrate the carcinogenic potential of formaldehyde on the nasal mucosa when administered by inhalation. No observed effect levels (NOELs) were therefore 2.0 mg/m^3 for the rat and 5.6 mg/m^3 for the mouse. The study of Albert *et al.* (1982) in which combination of formaldehyde with hydrochloric acid vapour was also used was inconclusive as to whether the BCME formed (<l μg/m^3) made any additional contribution to the carcinogenic response.

Possibilities for control

The great diversity of applications of formaldehyde in manufacturing, construction, medicine and the domestic environment, to cite but a few examples, requires an equal diversity of solutions for the restriction or elimination of human exposure. Monitoring of industrial processes and follow-up of personnel has permitted modification of processes, plant, machinery and operating practices so as to bring human exposure within recommended guidelines. The pungent smell of formaldehyde at concentrations of 1 mg/m^3 or less provides ready detection of transient peaks of high concentration to supplement the use of personal badge monitors and 8 hour time weighted average data from atmospheric monitoring instruments.

Proper extraction of formaldehyde vapour from reactor vessels, formalin solutions and recently treated materials or resins has significantly reduced worker exposure.

The high level of reactivity of formaldehyde with many chemical substances facilitates the scrubbing of contaminated air. Exposure of people in the domestic environment as a result of formaldehyde emission from urea–formaldehyde foam insulation or particle-board has proved a more intractable problem resulting in the removal of defective insulation emitting high levels of formaldehyde in some cases and, in some states, the banning of its use in domestic buildings and mobile homes. The inter-relationship between temperature, humidity and ventilation rate versus atmospheric formaldehyde levels has permitted costings to be made in the equation between heating costs and the increased ventilation required to reduce concentrations to acceptable levels. The emission of formaldehyde by particle board construction materials has also been quantified (Molhave *et al.*, 1983).

Risk assessment

Laboratory studies in a range of species have demonstrated formaldehyde to be a mutagen, carcinogen, irritant and allergen in one or more species. Studies in man have demonstrated comparable irritant and allergenic properties following exposure by a number of routes. Despite extensive surveys of industrial workers exposed for many years to high concentrations of formaldehyde vapour in the workplace (some at >2 mg/m^3), there is no epidemiological evidence for induction of neoplasia at any site (Acheson *et al.*, 1984).

Despite the narrow or non-existent margin of safety between the peak exposure levels recorded in the human workplace and home and the no-effect levels for carcinogenicity in the rat (2 mg/m^3) and mouse (5.6 mg/m^3), the extensive human experience with this molecule would indicate inter-species differences in carcinogenic potential do exist. Numerous epidemiological surveys, reviewed by Nelson *et al.* (1986), have reported an increased risk of lung, brain, lymphatic or haemopoietic neoplasia in exposed subjects in isolated cases but the lack of a consistent overall effect between surveys casts doubt on these findings.

There are a number of theoretical and practical explanations for this apparent discrepancy. The extrapolation of tumour incidence data from a NOEL figure from a bioassay with limited numbers of exposed animals to provide a measure of a true 'no effect level' depends on the mechanism of carcinogenicity. Formaldehyde is an irritant and causes a reactive hyperplasia in rats and mice at the 15 mg/m^3 concentration used in the CIIT oncogenicity studies (Chang *et al.*, 1983). Genotoxicity studies indicate only a weak activity in a range of test systems including the Ames test and carcinogenicity therefore may be a multistage process. Calculated virtually safe dose levels at the 1×10^{-5} incidence rate are >l mg/m^3 but the linear model for a genotoxic initiator carcinogen extrapolates to a 0.0066 mg/m^3 level (Gibson, 1983). Human epidemiology data would indicate the latter scenario not to be the case. A cautionary note should be given on the nasal cavity target for tumours in the rodent bioassay since these species are obligate nose-breathers. Since the anterior nasal cavity removes formaldhyde from inhaled air with great efficiency, lung exposure is effectively nil in these models. Man and primates on the other hand, are facultative mouth breathers, especially when exposed to irritant gases, and lung exposure and potential carcinogenesis might be expected. However, surveys of exposed workers have failed to show an increased incidence in lung cancer rates.

Safety questions relating to respiratory function call for further study as a significant correlation between formaldehyde exposure levels and impaired lung function parameters as well as irritant effects on nose, eyes, throat and chest exists (Horvath *et al.*, 1988). Effects were found to be reversible but the observation of such effects at, or below, legally permissible levels calls for better definition of a no-significant-effect concentration.

Regulatory aspects

Formaldehyde is a key chemical intermediate in many commercial processes and its elimination is neither technically nor economically feasible in many cases. The sensitivity of detection of formaldehyde vapour by its pungent odour is in the order of 0.1–0.5 mg/m^3 but chronic exposure can result in loss of sensitivity of detection and the onset of ocular, nose and throat irritation in some individuals over this range (National Research Council, 1981). Regulatory attitudes have become increasingly concerned about the narrow margins of safety between industrial and domestic exposure and the no-significant-effect levels for irritancy and carcinogenicity obtained from animal experimentation.

The United Kingdom control limit of 2 mg/m^3 and the Federal Republic of Germany MAK (Maximale Arbeitsplatz Konzentrationen) of 0.5 mg/m^3 apply to workplace exposures and Sweden has specified domestic maxima of 0.1 mg/m^3 for new homes and 0.7 mg/m^3 for old homes.

The U.S. Occupational Safety and Health Administration (OSHA) reduced the legal occupational maximal permissible concentration standard from 5 mg/m^3 with a 3 mg/m^3 8-hour time weighted average (TWA) to a 2 mg/m^3 maximum with a 1 mg/m^3 TWA (OSHA, 1985). The American Conference of Governmental Hygienists (ACGIH, 1985) recommended similar limits on exposure. The National Institute for Occupational Safety and Health (NIOSH, 1981) stated that 'Although humans and animals may differ in their susceptibility to specific chemical compounds, any substance that produces cancer in experimental animals should be considered a cancer risk to humans.'

The Report of the US Federal Panel on Formaldehyde (Griesemer, 1982) concluded that 'formaldehyde should be presumed to pose a carcinogenic risk to humans'. The European Chemical Industry Ecology and Toxicology Centre (ECETOC) (1982) placed reliance on the apparent link between irritant effects and hyperplasia and the subsequent development of nasal squamous cell carcinoma in rodents by stating that 'Where exposure is so low that metaplasia resulting from irritation does not occur, it is unlikely that tumours will develop.' This approach is flawed by the absence of concentration threshold data for histological changes to the nasal (and pulmonary) epithelium. Recent findings by Woutersen *et al.* (1989) indicate that the induction of nasal tumours by formaldehyde in the rat involves a role of preceding tissue injury.

The absence of epidemiological evidence for an increased incidence of nasal or other neoplasms in human subjects exposed to formaldehyde remains the main supporting evidence for current safety standards. A number of European nations have, however, set more stringent limits of 0.1 mg/m^3.

The finding of positive, albeit often weak, mutagenic potential in a variety of *in vitro* and *in vivo* test systems, combined with the rather inadequate level of teratogenic potential (Fielder, 1981) calls for further research into the potential effects on reproduction.

The chemical risk classification for formaldehyde in the European Community Directive 83/467 is a 'possible risk of irreversible effect' (R40).

Overview

Formaldehyde is a vital commodity in a wide range of processes and products. Economically and technically, its banning would create a large number of insuperable problems.

Equally, experimental data on irritancy, allergenicity, genotoxicity and carcinogenicity place formaldehyde high on the list of chemicals posing a significant safety hazard. In particular, the maximum permissible exposure levels of 2.0–0.1 mg/m^3 depending on country are exceedingly close to the demonstrated rodent carcinogenic dose of 5.6 mg/m^3 and no observed effect dose of 2 mg/m^3. Great reliance has been placed on the epidemiological evidence for the lack of a carcinogenic effect in man. Such evidence should continue to be collected to determine the validity of such inter-species differences and it is essential that human exposure to formaldehyde vapour or solution is progressively reduced by replacement with safe alternatives or by effective containment and ventilation procedures.

References

ACHESON, E.D., BARNES, H.R., GARDNER, M.J., OSMOND, C., PANNETT, B. & TAYLOR, C.P. (1984). Formaldehyde in the British chemical industry. An occupational cohort study. *Lancet*, **i**, 611–616.

ALBERT, R.E., SELLAKUMAR, A.R., LASKIN, S., KUSCHNER, M., NELSON, N. & SNYDER, C.A. (1982). Gaseous formaldehyde and hydrogen chloride induction of nasal cancer in the rat. *Journal of the National Cancer Institute*, **68**, 597–603.

American Conference of Governmental Industrial Hygienists. *Threshold Limit Values and Biological Indices for 1985–86* Cincinnati, ACGIH.

BERNSTEIN, R.S., STAYNER, L.T., ELLIOTT, L.J., KIMBROUGH, R., FALK, H. & BLADE, L. (1984). Inhalation exposure to formaldehyde: an overview of its toxicology, epidemiology, monitoring and control. *American Industrial Hygiene Association*, **45**, 778–85.

BUNDESGESUNDHEITAMTES (BGA) (1984). *Formaldehyd Schriftenreihe des Bundesministers für Jugend, Familie und Gesundheit*. Band 148 Verlag W Kohlhammer, Stuttgart.

CHANG, J.F.C., GROSS, E.A., SWENBERG, J.A. & BARROW, C.S. (1983). Nasal cavity deposition, histopathology and cell proliferation after single or repeated formaldehyde exposures in B6C3Fl mice and F–344 rats. *Toxicology and Applied Pharmacology*, **68**, 161–76.

ECETOC (1982). *Formaldehyde Toxicology*, Technical Report No. 6. Brussels, European Chemical Industry Ecology & Toxicology Centre.

FIELDER, R.J. (1981). *Toxicity Review* 2: Formaldehyde, London: HMSO.

GIBSON, J.E. (1983). Risk assessment using a combination of testing and research results. In *Formaldehyde Toxicity*, Gibson J.E., ed., pp. 295–302. Washington: Hemisphere Publishing.

GRIESEMER, R.A. (1982). Report of the Federal Panel on Formaldehyde. *Environmental Health Perspectives*, **43**, 139–68.

HORVATH, E.P., ANDERSON, H., HANRAHAN, L., PIERCE, W.E. & WENDLICK, J.D. (1988). Effects of formaldehyde on mucous membranes and lungs. Study of an industrial population. *Journal of the American Medical Association*, **259**, 701–7.

KERNS, W.D., PAVKOV, K.L., DONOFRIO, D.J., GRALLA, E.J. & SWENBERG, J.A. (1983). Carcinogenicity of formaldehyde in rats and mice after long-term inhalation exposure. *Cancer Research*, **43**, 4382–92.

LAMB, B., WESTBERG, H., BRYANT, P., DEAN, J. & MULLINS, S. (1985). Air infiltration rates in pre- and post-weatherised houses. *Journal of the Air Pollution Control Society*, **35**, 545–51.

MOLHAVE, L., BISGAARD, P. & DUEHOLM, S. (1983). A mathematical model of indoor air pollution due to formaldehyde from urea–formaldehyde glued particle boards. *Atmospheric Environment*, **17**, 2105–8.

NATIONAL RESEARCH COUNCIL (1981). *Formaldehyde and Other Aldehydes*. Washington DC: National Academy Press.

NELSON, N., LEVINE, R.J., ALBERT, R.E., BLAIR, A.E., GRIESEMER, R.A., LANDRIGAN, P.J., STAYNER, L.T. & SWENBERG, J.A. (1986). Contribution of formaldehyde to respiratory cancer. *Environmental Health Perspectives*, **70**, 23–35.

NIOSH (1981). Formaldehyde: evidence of carcinogenicity. *National Institute for Occupational Safety and Health Bulletin* 34, US Department of Health & Human Services. Pub. no. 81–111.

OSHA (1985). Occupational exposure to formaldehyde; notice of proposed rule-making. *Federal Register*, **50**, 50412–99.

SAMET, J.M., MARBURY, M.C. & SPENGLER, J.D. (1988). Health effects and sources of indoor air pollution. Part II. *American Review of Respiratory Diseases*, **137**, 221–42.

UBA, G., PACHOREK, G., BERNSTEIN, J., GARABRANT, D.H., BALMES, J.R., WRIGHT, W.E. & AMAR, R.B. (1989). Prospective study of respiratory effects of formaldehyde among healthy and asthmatic medical students. *American Journal of Industrial Medicine*, **15**, 91–101.

WOUTERSEN, R.A., van GARDEREN-HOETMER, A., BRUIJNTJES, J.P., ZWART, A. & FERON, V.J. (1989). Nasal tumours in rats after severe injury to the nasal mucosa and prolonged exposure to 10 ppm formaldehyde. *Journal of Applied Toxicology*, **9**, 39–46.

9

HAZARDS FROM SOLVENTS, PESTICIDES AND PCBS

D. E. RAY

Introduction

This is a varied chapter bringing together a collection of chemically unrelated substances. The sources of these substances are equally diverse, ranging from pesticides deliberately used by occupants, through compounds such as solvents often unintentionally added to the air, to trace contaminants from furnishings or building materials of which the occupants may be quite unaware. Two obvious considerations are that these substances must be firstly volatile, and secondly have either an indoor source or preferential access to the indoor environment (as from a basement or a ventilation system). In industrialised countries, especially in colder climates, room ventilation rates are generally decreasing, and most of the population spend the majority of their time inside buildings (Sterling, 1985; Lewtas, 1989). There has also been concern, especially over solvent and pesticide toxicity, heightened by deaths from solvent abuse which were estimated at 60 between 1970 and 1980 in the UK (King *et al.*, 1981), and occasional, but regular, reports of suicide by pesticide ingestion. Such concern is in many ways unwarranted, as in both of these instances the exposure levels were enormously high, and as Paracelsus (von Hohenheim) recognised as early as the sixteenth century, '[only] the right dose differentiates a poison and a remedy'. In this context it is salutary to remember that the normal atmospheric oxygen content is fully 30% of the long term lethal level for pulmonary oxygen poisoning (Clark & Lambertsen, 1971), yet this rather small safety margin is fully compatible with life!

Many of the substances described in this chapter have considerable toxic *potential* however, and this coupled with the vagaries of domestic usage means that they need to be used intelligently.

Table 9.1. *Data for some commonly used solvents*

Solvents	Multiple for mg/m^3 to ppm[a]	TWA/TLV[b] 8h (mg/m^3)	Domestic air level[c]	Blood: air ratio [e] (mg/m^3)	Pulmonary retention [g] (%)	Fraction metabolised [g] (%)
Aromatic hydrocarbons:						
Benzene	0.313	32	0.01–0.05	7	40–50	50–70
Ethyl benzene	0.230	434	0.005–0.04			
Styrene	0.235	213	0.0027	59	64	95
Toluene	0.265	377	0.01–0.61	16	50	80
o-, *m*-, *p*-xylene	0.230	434	0.01–0.14	34	64	95
Aliphatic hydrocarbons:						
n-hexane	0.284	176		1.25 [f]	28	70–90
Methyl n-ethyl ketone	0.339	590			50	70–80
Methyl n-butyl ketone	0.244	20			75–92	40
Chlorinated hydrocarbons:						
Dichloromethane	0.288	174	5 [d]	8	31–51	70–95
Carbon tetrachloride	0.159	31	0.014			50
Chloroform	0.205	49	0.0087	11		
Tetrachloroethylene	0.147	339	0.004–0.198	14	50	2
1,1,1-trichloroethane	0.183	1910	0.015–0.29	4	18–25	5
Trichloroethylene	0.186	269	0.002–0.068	9	35–60	10–20
Other						
Petrol	0.337	890				
White spirit	0.190	525				

[a] at 25°C

[b] Time Weighted Average Threshold Limit Value set by ACGIH (1991)

[c] Mean overnight value (Wallace *et al*, 1985), or range in buildings (Sterling, 1985)

[d] Level in hairdressing salon (IARC, 1979*d*)

[e] Steady state values, Droz (1986)

[f] Perbellini *et al.* (1985)

[g] Data of Riihimaki (1986) extended

Solvents

Because of their high volatility and wide usage, solvents merit careful consideration as domestic hazards. This has been highlighted by the recognition of the 'Sick' or 'Tight' Building Syndrome by the WHO (Lewtas, 1989). This shows itself as irritation and dryness of the eyes, nose and throat, mental fatigue, hoarseness, wheezing, headache, nausea, dizziness, airway infections and unspecific hypersensitivity reactions (Molhave, 1985), and has shown itself primarily in new office buildings. Although vague and rather non-specific, these signs resemble those produced by respiratory irritants and stress. Since the initial levels of total volatile organic materials in new buildings (0.5 to 19 mg/m^3) can approach the levels which are detectable by sensitive individuals (Molhave, 1985), it is not impossible that this could then be associated by them with previous exposure to higher (irritant) levels and lead to the 'Sick' building syndrome, especially when combined with other work-related stress. Such ideas are, however, speculative, and relate more to air 'quality' than hazards.

Absorption, distribution and metabolism of solvents

Although blood levels of most solvents rapidly reach equilibrium with the inspired air, they are not all equally efficiently absorbed and, once absorbed, they are not all rapidly excreted. Some solvents accumulate with repeated exposure; others do not. Additionally, not all solvents act directly since some require metabolic activation to produce their toxicity. Clearly, some understanding of solvent kinetics and metabolism is required before their toxicology can be understood. Solvent kinetics have been studied both experimentally and by mathematical modelling by a number of workers (Perbellini *et al.*, 1985; Cohr, 1986; Droz, 1986) and a number of critical factors emerge. Lipophilicity is important as it determines the blood:air and blood:tissue solubility ratios (see Table 9.1). Solvents which dissolve readily in blood (blood:air ratio higher than 10 (Cohr, 1986)) are taken up at a rate limited by the volume of air breathed: this increases with exercise. Solvents with a lower ratio are limited by blood flow through the lungs (which also increases with exercise). Skin uptake is negligible for most vapours (Cohr, 1986) except, of course, when solvent is spilt on the skin. Once in the blood, the solvent is distributed to the tissues at a rate that is determined both by their blood:tissue solubility ratios and their tissue blood flows (also increased by exercise). Blood flow is a very important factor as although body fat has the highest tissue:blood ratios (between 104 and 254 for a range of solvents, whereas brain:blood ratios are between 6 and 11 and muscle:blood 2 to 9); (Perbellini *et al.,* 1985) it also has one of the lowest blood flows and only reaches equilibrium very slowly. Thus, although n-hexane reaches 70% of its steady state level in blood after 30 minutes, fat reaches 7% after 8 hours (Perbellini *et al.,* 1985). Body fat also takes a correspondingly long time to lose solvent after exposure. Thus, to successfully model the kinetics of solvents in the whole body, several compartments must be used and such models have proved very successful in predicting human exposure levels. In the case of n-hexane and methyl n-butyl ketone, the active metabolite, 2,5-hexane dione is water soluble and is eliminated largely in the urine, with a half time of 13–14 hours in man (Perbellini *et al.,* 1985) introducing an extra level of complexity.

Metabolic activation or degradation of solvents is a very important factor (Parkki, 1986; Dossing, 1986). Methyl *n*-butyl ketones and *n*-hexane, and the chlorinated hydrocarbons are all toxic via their liver metabolites. Comparison of the metabolic fraction data in Table 9.1 with threshold levels for causing liver damage by the chlorinated hydrocarbons shows a good correlation. Other solvents are largely detoxified by metabolism. The general pattern of hepatic metabolism is oxidation by cytochrome P-450 mono-oxygenase isozymes in smooth endoplasmic reticulum, microsomal oxidation, or hydrolysis. This is followed by conjugation; the overall effect being to transform lipid soluble compounds into polar ones capable of urinary excretion. The liver has generally a very high capacity for solvent metabolism, benzene, styrene, toluene, xylene, n-hexane, and trichloroethylene being almost completely extracted from the blood flowing through the liver at exposure levels up to 2 to 10 times the threshold limit value (TLV) (Riihimaki, 1986). An important aspect of the metabolism of solvents is the capacity of other substances to change solvent toxicity by an action on the liver. Moderate alcohol consumption causes a 1.5 to 2 fold elevation in blood levels of toluene, xylene and trichloroethylene in man by competing for metabolism (Dossing, 1986), but its action is complex since, although it competes initially, it enhances metabolic capacity over the longer term by enzyme induction. This enhanced metabolism predominates in experimental animals but in man the liver enzymes are normally induced; even so in both man and experimental animals ethanol potentiates chlorinated hydrocarbon hepatotoxicity via increased metabolism. The enhanced metabolism and toxicity of n-hexane and methyl n-butyl ketone when combined with methyl n-ethyl ketone is described in the section on specific solvents.

Sedative action of solvents

All lipophilic agents are capable of producing sedation and anaesthesia if given in sufficient quantity provided they do not have toxic actions at lower doses. Indeed, several solvents have found use as anaesthetics, notably chloroform, ether and trichloroethylene, although safer alternatives are now in use. Their anaesthetic actions can be related simply to their oil:water partition coefficients (Meyer, 1937), since in this respect they are acting merely as inert substances to dissolve in, and expand, the nerve membrane, thus depressing excitability. An important consequence of this is that such agents can combine to produce an additive response in molar terms (Hills & Ray, 1977), which is fortunate in that it simplifies estimation of the effect of combined solvent exposure. Lipid solubility has also been found to be a good predictor of irritancy within a series of solvents (Nielsen & Alarie, 1982). Below anaesthetic levels solvents produce sedation, slowed reaction time and decreased vigilance (Gamberale, 1986; Hanninen, 1986; Laine & Riihimaki, 1986; Anger & Johnson, 1985). The exposure level producing this is discussed for each agent in the next section. It is important to note that, as with anaesthesia, such pre-anaesthetic effects should be reversible (see section on long-term effects).

Specific solvents

In this section, most of the solvents to which there is likely to be any significant degree of domestic (as distinct from industrial) exposure (Wallace *et al.*, 1985) are discussed in detail, although at levels of 0.0002 mg/m^3 more than 350 organic compounds can be identified in indoor air (Sterling, 1985). Ethanol and acetone are excluded from the list since, as pointed out by Dossing (1986), consumption of only one glass of wine is equivalent to a 24-hour exposure to the TLV of 1900 mg/m^3 ethanol in terms of dose, and acetone is a metabolite present in normal fasting breath.

The solvents are listed in Table 9.1. Exposure levels are given in mg/m^3 with a conversion to ppm (parts per million by volume). As a guide the volume of air breathed by an adult at rest is about 0.3 m^3/h and can be over

1 m^3/h when working. For each solvent the current TWA/TLV value is given (ACGIH, 1989). This is the Threshold Limit Value set by the American Conference of Governmental Industrial Hygienists for industrial exposure during a standard 40 hour working week. These are approximate levels to be used as general guidelines for maximum safe exposure. Clearly, somewhat lower levels would be appropriate for continuous exposure, but it should also be noted that high work rate, noise and stress all potentiate the adverse effects of solvents, and would be less important in a domestic environment. Domestic air levels are also given in the table. These are taken from values obtained either in a large scale Total Exposure Assessment Methodology study (Wallace *et al.,* 1985) or from a review of data by Sterling (1985), and it can be seen that although they range over at least a factor of 100, all are far below their TLVs. These concentrations represent 2 to 100 times the prevailing outdoor levels (Wallace *et al.,* 1985). Clearly under conditions of actual use, solvents will reach higher levels, but not over long periods, and one would hope that domestic as well as industrial users would take sensible precautions to ensure adequate ventilation whilst using these solvents.

There are many sources of these solvents in the home, most notably: paints, glues and lacquers (toluene, methyl *n*-butyl ketone [being replaced], *n*-hexane [being replaced], methyl *n*-ethyl ketone, xylene); dry cleaned articles (carbon tetrachloride [being replaced], tetrachloroethylene, trichloroethylene); fabric cleaners (trichloroethane, petrol, benzene, toluene); petrol (toluene, benzene); aerosol propellants (dichloromethane); correction fluid (l,1,1-trichloroethane); chlorinated hot-water showers (chloroform); expanded plastics (styrene); paint strippers and solvents (white spirit, dichloromethane, cellosolves) and even cigarette and other vegetable smoke (benzene).

Aromatic hydrocarbons

Benzene

Since benzene is not now commonly used as a solvent, its major source is from petroleum motor fuel which probably accounts for its presence in domestic air in traces. Table 9.1 shows that it has a rather low TLV, which is largely a reflection of its unusual toxic action upon the white blood cell (leucocyte) forming tissues in the bone marrow. Its CNS depressant effects are seen only at much higher levels, 42000 mg/m^3 being required to kill mice (ACGIH, 1971) and 64000 mg/m^3 proving fatal in man within 5–10 minutes (Cornish, 1980). Benzene has little irritant action (Nielsen &Alarie, 1982) and is exhaled with a half time of 2.5 hours in man, with a slower component of 24 hours representing persistence in body fat (Berlin, 1988). The depletion of leucocytes by benzene has been well studied and is reviewed by Kalf *et al.*, (1987). A number of deaths due to bone marrow toxicity resulting in aplastic anaemias occurred as a result of the high level industrial exposures prevalent in the 1930s (300–600 mg/m^3), and experimental exposure of rats indicates a threshold value for this of about 140 mg/m^3 over eight weeks (Cornish, 1980). Benzene is 50–70% metabolised (IARC, 1974), and this oxidative metabolism appears essential for its toxic action. It is possible that interaction of benzene metabolites with tubulin interferes with the structural processes involved in cell division in this very actively dividing tissue. In addition to producing aplastic anaemia, benzene has been implicated as a human carcinogen, largely on the basis of the leukaemia also produced in heavily exposed workers in the 1930s and its interaction with chromosomes, 45 mg/m^3 producing an elevation in bone marrow micronuclei in mice (Andrews & Snyder, 1986). Unlike the aplastic anaemia, the leukaemia has not been produced in experimental animals, nor has it been seen in exposed workers since lower safety limits have been set (Grasso, 1986). It would seem likely that a very pronounced or prolonged degree of damage to the bone marrow is required to produce human leukaemia, although the very large dose of 500 mg/kg benzene by the oral route does induce cancers (not leukaemia) in the mouse (Maltoni *et al.*, 1988*a*). The evidence for a cancer risk from low level inhaled benzene is thus very equivocal (Forni, 1988). In common with other solvents activated by metabolism, simultaneous exposure to ethyl alcohol potentiates benzene's toxicity by induction of the liver cytochrome P-450 mono-oxygenase system whereas competition for metabolism by co-exposure to toluene decreases benzene toxicity (Kalf *et al.*, 1987).

Ethylbenzene

This does not produce the bone marrow effects of benzene, and represents an acute hazard only by way of CNS depression, the threshold concentration for this in man being 8700 mg/m^3 (ACGIH, 1971), but irritation to the eyes and throat being seen at 4300 mg/m^3 (Nielsen & Alarie, 1982). Its toxicity seems to resemble that of toluene except that it is more irritant.

Styrene

This is readily smelt at 250–400 mg/m^3, impairs tests of alertness at 1500 mg/m^3 (Gamberale, 1986; Stewart *et al.,* 1968), and produces mild CNS depression in man at 3400 mg/m^3 (Laine & Riihimaki, 1986), although it is very irritant at 2100 mg/m^3 which limits exposure (Browning, 1965). More subtle behavioural tests show effects of styrene at 300 mg/m^3 in man (Anger & Johnson, 1985). It is extensively metabolised to styrene oxide (IARC, 1979). Styrene is one of the solvents particularly linked with the Scandinavian psycho-organic syndrome which is discussed in the section on long-term toxicity, but considerable exposure of US workers at levels of 85 mg/m^3 has not resulted in any reports other than transient effects (O'Donoghue, 1985*b*), and a report of ten cases of acute styrene poisoning (five unconscious) in the UK gave similar results (Bakinson & Jones, 1985). Some studies of exposed workers have shown paraesthesia of fingers and toes and marginal slowing of nerve conduction velocity (Harkonen, 1979), but these have not been reproduced in animal studies.

Toluene

This can be detected by man at 10 mg/m^3 (Carpenter *et al.,* 1976) and produces CNS depression, headache and fatigue at 200–750 mg/m^3; weakness and confusion at levels of 750–1130 mg/m^3, and more severe effects at 2260 mg/m^3 in experimental human exposures (Cornish, 1980) but little irritation even at 3000 mg/m^3 (Nielsen & Alarie, 1982). Evidence from solvent abuse fatalities suggests that 7500–12300 mg/m^3 can be fatal within 30 minutes (Laine & Riihimaki, 1986). In addition to these acute effects there is evidence that at the high concentration reached during solvent abuse (800–2500 mg/m^3 or more) both a reversible encephalopathy and a cerebellar atrophy leading to irreversible ataxia can be produced. Animal studies also show cochlear damage at levels above 3800 mg/m^3. A survey by King *et al.* (1981) of Glasgow children admitted to hospital showed 4/19 cases of coma and 3/19 cases of acute ataxia. There were also cases of suicide apparently precipitated by hallucinations. When admitted there was frequently slowing of the EEG which reversed on recovery from the acute intoxication, but one case showed a persisting cerebellar ataxia. In all cases blood chemistry and haematology remained normal. In a study of factory workers exposed to 75–750 mg/m^3 toluene no chronic effects were seen (O'Donoghue, 1985*b*), nor after acute industrial poisoning (Bakinson & Jones, 1985), indicating that the cerebellar atrophy is a very high dose effect. Dogs and rats exposed to 1790 mg/m^3 6 hours a day for 13 weeks showed no neuropathological changes (Carpenter *et al.*, 1976). Fatty infiltration of the liver has been seen after solvent abuse or levels of industrial exposure leading to unconsciousness (Klockars, 1986). In one industrial incident three workers developed liver damage, with jaundice and impaired function tests one week after acute exposure to a spill of toluene, carbon disulphide, methanol, acrylonitrile and possibly cyanide. Their longer-term exposure had been to only low levels of these (toluene 57–75 mg/m^3) and the acute exposure did not lead to anything other than drowsiness, so it is possible that toluene may cause liver damage in cases of combined exposure such as this without unconsciousness (Dossing & Ranek, 1984). Toluene is rapidly metabolised in the liver, 80% of the inhaled dose being metabolised in man (Riihimaki, 1986).

Xylene

This exists in the *ortho-*, *meta-*, or *para-* isomeric forms, all with very similar biological actions. It is irritant to the eyes, nose and throat of human subjects at 880 mg/m^3 (ACGIH, 1971), reduces tests of alertness at 390 mg/m^3 (Gamberale, 1986) and is anaesthetic above 22000 mg/m^3 (Laine & Riihimaki, 1986). Early industrial exposures indicated some effect on the bone marrow, but this was probably largely due to the presence of benzene, as in rabbits a protracted exposure to 5060 mg/m^3 was needed to depress leucocyte levels and 3036 mg/m^3 was ineffective (ACGIH, 1971). In two human cases of very severe acute intoxication at an estimated level of 43500 mg/m^3 xylene plus traces of toluene (a further case died), Morley *et al.* (1970) found only transient evidence of liver and kidney dysfunction despite unconsciousness lasting for 15 and 18 hours and a probable degree of hypoxia. No evidence of liver damage was seen in a further 35 cases of acute xylene poisoning reported by Bakinson and Jones (1985), despite ten patients being unconscious.

Aliphatic hydrocarbons

n-hexane

This has a well-documented neuropathic action on the central and peripheral nervous systems, and this, rather than CNS depression or irritation, determines the Threshold Limit Value. Its toxicity has been reviewed by Spencer *et al.* (1980), O'Donoghue (1985*a*), and Couri & Milks (1982). It causes no irritation at 1800 mg/m^3 and CNS depression (dizziness) in experimental human exposures only at 18000 mg/m^3, whilst 100000 mg/m^3 is required to produce narcosis in mice (ACGIH, 1971). Human neuropathies have been shown in workers exposed to levels of 1800–9000 mg/m^3 over several months in Japanese and US factories (ACGIH, 1971) or 850–1670 mg/m^3 over a period of years (O'Donoghue, 1985*a*) and also in solvent abuse cases. This corresponds with the experimental induction of neuropathy in animals at 700–35000 mg/m^3 over periods of weeks. In man, the effect has been categorised as a central peripheral distal polyneuropathy, i.e. it is a degeneration of both sensory and motor nerves that begins distally (i.e. nearest the periphery) and affects both the central and peripheral nervous system. It shows itself as a slowly developing bilateral sensory loss with numbness of the extremities in a 'stocking and glove' distribution and a corresponding motor weakness. This may be associated with headache, nausea, anorexia and calf cramps, also colour vision loss and cranial nerve involvement. It progresses for 4 to 8 weeks beyond the end of acute exposures, but unless the condition has progressed so far as to cause muscle wasting, there can be good recovery over several months to a year (O'Donoghue, 1985*a*). There is a clear progression of histopathological changes with multifocal axonal damage first at the distal end of long axons with accumulation of 10 nm neurofilaments on the proximal side of nodes of Ranvier and swollen axonal terminals in the CNS, particularly in animals. There is then a phase of secondary demyelination which can be quantified *in vivo* by nerve conduction velocity measurement (Jacobs & Le Quesne, 1984).

n-Hexane is readily absorbed, 28% of that inhaled being absorbed in man (Couri & Milks, 1982) and rapidly metabolised. Animal studies suggest 70–90% metabolism (Perbellini *et al.,* 1985) which would not be limited until inhaled concentrations are greater than 3600 mg/m^3. It is rapidly eliminated from the blood after exposure with a half time of 1 hour in rats (Couri & Milks, 1982) or falling to 30% in 30 minutes in man, although elimination from body fat is very slow with a half time of 64 hours (Perbellini *et al.,* 1985). Its toxic actions are however dominated by its metabolites, notably 2,5-hexane dione produced via liver cytochrome P450 dependent hydroxylases. This is 7 to 38 times more toxic than n-hexane (O'Donoghue, 1985*a*) and is much less lipophilic than n-hexane and hence excreted less readily, via the urine, with a half life of 13–14 hours in man (Perbellini *et al.,* 1985). Only hexocarbons metabolised via 2,5-hexane dione are neuropathic, 1-hexanol for example being non-neuropathic in rats (Couri & Milks, 1982). This dependence on hepatic metabolic activation leads to marked interactions with other agents which influence liver metabolism. Toluene is thus mildly protective but the non-neuropathic solvent methyl ethyl ketone (q.v.) potentiates n-hexane toxicity and elevates 2,5-hexane dione levels in rats by an as yet undefined metabolic interaction (Katz, 1985). Good human evidence of this interaction is provided by cases of solvent abuse. Altenkirch *et al.* (1978) noted that, despite extensive solvent abuse involving n-hexane in Berlin, no neuropathies appeared until after the introduction of 10% methyl ethyl ketone, when 19 cases of acute neuropathy presented in 1975. After the methyl ethyl ketone was removed only a further six cases (in which old supplies were used) were seen. In parallel rat experiments 31700 mg/m^3 n-hexane + 2950 mg/m^3 methyl ethyl ketone produced a more intense and more rapidly developing but essentially similar neuropathy than 35000 mg/m^3 n-hexane alone, whereas the methyl ethyl ketone by itself was quite ineffective. Despite the identification of the toxic metabolite as 2,5-hexane dione, the molecular mechanism of n-hexane neuropathy is still not clear, possibly because it is a complex, slowly developing process. There appear to be two major possible actions: a direct action whereby the tertiary structure of the neurofilaments is disrupted by abnormal cross linking or an indirect action by way of depressed neuronal energy metabolism. This secondarily leads to profound slowing of axoplasmic transport mechanism. The direct action mechanism is supported by *in vitro* studies and *in vivo* evidence of 2,5-dimethylpyrrole adduct formation (O'Donoghue, 1985*a*), but not by the occurrence of similar neurofilament accumulations in other conditions such as acrylamide neuropathy. A number of metabolic enzymes, including phosphofructokinase and glyceraldehyde-3-phosphate dehydrogenase are inhibited and, within the CNS, glucose utilisation is decreased prior to development of neuropathy or at lower chronic dose regimes (Planas & Cunningham, 1987).

Methyl n-ethyl ketone (2-butanone)

This is a potentiator of both n-hexane and methyl n-butyl ketone induced neuropathy, but has no neuropathic potential itself, producing no neuropathy in chronically exposed rats up to the lethal dose of 18000 mg/m^3 for 7 weeks (Altenkirch *et al.*, 1978). In man it produces slight nose and throat irritation at 885–1500 mg/m^3, mild sedation at 900–1800 mg/m^3 and narcosis at 2360 mg/m^3 (Laine & Riihimaki, 1986).

Methyl n-butyl ketone (2-hexanone)

This has a very similar pattern of toxicity to n-hexane (q.v.), producing neuropathy via their common metabolite 2,5-hexane dione. At a level of 4100 mg/m^3 it produces transient eye and nose irritation in man (ACGIH, 1971) and is anaesthetic at 16600 mg/m^3 in rats (Couri & Milks, 1982). It produces neuropathy in rats at levels of 410–920 mg/m^3 indicating that it is markedly more toxic than n-hexane and also shows potentiation with co-exposure to methyl n-ethyl ketone (O'Donoghue, 1985*a*). It is less lipophilic than n-hexane and as a result has a longer blood half life (7 hours in rats), and a higher rate of pulmonary absorbtion (75 to 92%), which together with its metabolic conversion of 40% in the rat may explain its higher toxicity (Couri & Milks, 1982).

Chlorinated hydrocarbons

Many of the chlorinated hydrocarbons produce liver toxicity characterised by the accumulation of fatty acids in the hepatocytes due to an impaired release process and a centrilobular necrosis. This is discussed by Andrews & Snyder (1986) in relation to their varying ability to form toxic reactive intermediates. Based on oral dosage, carbon tetrachloride, chloroform, 1,2-dichloroethane, 1,1,1-trichloroethane, 1,1,2,2-tetrachloroethane and 1,1,2-trichloroethylene are capable of producing fatty liver and necrosis; dichloromethane, 1,2-dichloroethylene and tetrachloroethylene are capable of producing fatty liver but not necrosis; whereas others such as methyl chloride have little or no effect. These differences can be related to differential metabolic activation by the hepatic mixed function oxidase system and have formed a major field of study in hepatic biochemistry recently reviewed by Kalf *et al.*, (1987). It is inappropriate to describe this in any detail here other than to note that the probable common mechanism of action of these agents may well lead to additive effects and that their action is potentiated by alcohol (Klockars, 1986).

Dichloromethane (methylene chloride)

This is one of the less hepatotoxic of the chlorinated hydrocarbons, and produces liver damage in rats at a threshold of 4500 mg/m^3 (after 75 days) (ACGIH, 1971). Whilst long-term inhalation of 350 mg/m^3 had no adverse effect on rats (Maltoni *et al.*, 1988) a depression of mouse liver cytochrome P450 enzyme activity but with no pathological change is seen at the same dose level after 100 days exposure (Cornish, 1980), and at 6900 mg/m^3, which killed 50% of mice, increased tumour incidence was seen (IARC, 1986). In a survey of 33 cases of acute solvent intoxication in workers, where exposure was not quantified (13/33 were rendered unconscious and 1 died), narcosis was the major sign, but no abnormal liver function tests were seen (Bakinson & Jones, 1985). This suggests that severe chronic exposure is necessary for liver damage to occur in man. In man about 50% of inhaled dichloromethane is absorbed; and in the rat, elimination is rapid, body fat levels falling to 10% within 2 hours of inhalation (IARC, 1979*d*). It reduces scores in tests of alertness at 700 mg/m^3 (Gamberale, 1986) and is anaesthetic at 35000 mg/m^3 (Browning, 1965).

Carbon tetrachloride (tetrachloromethane)

This is the classic liver toxin and has been extensively studied for many years as an agent for the experimental production of fatty liver degeneration (usually by the oral route). It produces mild liver pathology in animals at 160 mg/m^3 over long term exposures and more severe changes at 315 mg/m^3; metabolic conversion is about 50% in monkeys (IARC, 1979*b*). Human deaths and illness have been reported after inhalation of 350–1400 mg/m^3 (ACGIH, 1971), acute signs of intoxication being headache and lethargy, with liver damage occurring over a longer period. Neuropathological changes have been recorded after chronic exposure, but these (encephalopathy with astrocytic changes and proliferation) are only seen several weeks after the onset of liver damage and are clearly secondary to this, resembling the changes seen in other liver diseases (Politis *et al.*, 1980).

Chloroform

This was formerly widely used as an anaesthetic. It also produces liver damage. This is irreversible in dogs, but reversible in rats, after long term exposure to 120 mg/m^3 and becomes irreversible in rats at 240 mg/m^3 (IARC, 1979*c*). Data from oral administration studies in animals show that hepatic carcinomas are frequently associated with gross liver damage. Acute death from respiratory depression occurs at 40000–60000 mg/m^3 in animals (Cornish, 1980), and levels of 390–1200 mg/m^3 produce lassitude in exposed workers (ACGIH, 1971).

Tetrachloroethylene (perchloroethylene)

This is one of the less hepatotoxic chlorinated hydrocarbons and its TLV is set for eye irritation and CNS depression. Experimental human exposures at 670 mg/m^3 produced headache and mucous membrane irritation, whilst 7370 mg/m^3 caused unconsciousness after 30 minutes (ACGIH, 1971) although industrial exposure to 120–215 mg/m^3 had no significant effect on behaviour (Anger & Johnson, 1985). The anaesthetic level in rats is about 20000 mg/m^3 (Browning, 1965). Liver damage has been reported after some industrial exposures at 1540–2580 mg/m^3 (ACGIH, 1971), although in a report of 44 acute industrial intoxications with tetrachloroethylene in which 3 died and half were rendered unconscious, only one case of an elevation in serum enzymes (indicating marginal liver damage) was seen (McCarthy & Jones, 1983). Long-term exposure of rats to 4000 mg/m^3 resulted in reversible liver damage, but guinea pigs were more susceptible and showed slight degeneration at 1400 mg/m^3 (Browning, 1965). A reasonable explanation of the relatively low toxic potential of tetrachloroethylene is its low metabolic conversion, which is about 1.8% at an exposure of 600 mg/m^3, and saturates out at exposures below 678 mg/m^3 (IARC, 1979*e*). The parent compound is lost on the breath with a half-life of 7 hours in rats.

1,1,1-trichloroethane (methyl chloroform)

This is like tetrachloroethylene in showing only slight hepatotoxic potential. It produced a centrilobular hepatic necrosis with fatty accumulation in mice after 14 weeks exposure to 5500 mg/m^3, but not at 1370 mg/m^3 (McNutt *et al.,* 1975), and rats at 9500 mg/m^3 showed no increase in tumour incidence (IARC, 1979*f*). In man, the odour of 1,1,1-trichloroethane is unpleasant at 5500 mg/m^3 and a loss of coordination is seen at 5000 mg/m^3 (Browning, 1965). At 2700 mg/m^3 for 6.5–7 hours per day over 5 days experimental subjects showed no adverse effects other than transient light headedness and mild sedation (Stewart *et al.,* 1969). Epidemiological studies of workers exposed to 5–1360 mg/m^3 (Klockars, 1986), and also a study of 52 acute solvent intoxications, in which two died (McCarthy & Jones, 1983), showed no evidence of liver damage. As with other less toxic chlorinated hydrocarbons it is poorly metabolised at 2% in rats (IARC, 1979*f*). It is rapidly lost from the blood via the breath with an initial half-time of a few hours, but a slower component with a half-life of 24 hours reflects loss from body fat (Stewart *et al.,* 1969).

Trichloroethylene

Toxicity was complicated by the presence of significant toxic impurities (largely dichloroacetylene) prior to the 1970s. This was the probable cause of the cranial nerve damage seen in cases of earlier industrial exposure since no damage resulted from the extensive use of pure trichloroethylene as an anaesthetic (O'Donoghue, 1985). The pure substance produces relatively little liver toxicity and is hazardous largely as a CNS depressant. The threshold anaesthetic level in man is about 16000 mg/m^3 and experimental exposures to 1070 mg/m^3 caused slight fatigue and dizziness (Stewart *et al.*, 1970) whilst 590 mg/m^3 reduced tests of alertness (Gamberale, 1986). Long term exposure of dogs to 2700-4000 mg/m^3 caused liver dysfunction but also lethargy and anorexia (ACGIH, 1971), and whilst similarly extended exposure at 189 mg/m^3 caused no liver damage to a variety of animals, cats exhibited such damage at only 108 mg/m^3 (IARC, 1979*g*). Metabolic conversion of trichloroethylene is 10–20% in the rat, and respiratory uptake 60% in man (IARC, 1979*g*). In reports of 288 cases of acute trichloroethylene intoxication, of which 12 were fatal, 3 showed evidence of impaired liver function (McCarthy & Jones, 1983). There have also been reports of hepatic injury after solvent abuse with trichloroethylene (Klockars, 1986).

Mixtures

Petrol (gasoline)

This is a variable mixture but its toxicity is considered to be determined largely by its content of aromatic hydrocarbons (ACGIH, 1971). Weaver (1988) suggests that more than 200 compounds, mostly with 4 to 12 carbon atoms, may be present and gives typical values of benzene 1.5%, toluene 4.5%, xylenes 5%, *n*-hexane 4.2%, paraffins 60–70%, olefins 5–15%. The aromatic hydrocarbon content of petrol used at present is higher than that used before the development of catalytic cracking and earlier work should clearly be interpreted with caution (Browning, 1965). Current demands to maintain high engine performance using unleaded petrol without recourse to increased benzene content have led to the use of ethers, such as methyl tertiary-butyl ether, as octane enhancers (Houghton, 1989). The ethers generally have very little toxic potential other than via narcosis. A further factor is the presence of toxic additives such as tetraethyl lead (or organomanganese compounds in unleaded petrol). Controlled human exposure to 2670 mg/m^3 petrol (900 ppm) produced eye irritation and at 8300–20800 mg/m^3 (2800–7000 ppm) dizziness within 14 minutes. Petrol is frequently used in cases of solvent abuse, either by inhalation or by ingestion, and may lead to prolonged unconsciousness (Browning, 1965). A 90-day inhalation of 4600 mg/m^3 of petrol produced no adverse effects in rats (other than some kidney tumours in male rats thought to be related to the interactions of aliphatic hydrocarbons with the unusually high concentrations of protein present in male rat urine) (Weaver, 1988). Under practical conditions of exposure, petrol would not be fully vaporised and so the less volatile (and more toxic) constituents would be under-represented, thus contributing to the absence of increased tumour incidence amongst petrol exposed workers (Weaver, 1988). The maximum content of the additive tetraethyl lead was reduced to 0.15 g/l in 1986 throughout Europe from the previous value of 0.4 g/l used since the 1920s (Houghton, 1989). Tetraethyl lead is a potent neurotoxin acting, after metabolic conversion to triethyl lead, as an excitotoxin and producing selective destruction of neurons in the pyramidal cell band of the hippocampus and in associated structures such as the amygdala and archicortex of experimental animals (Aldridge & Brown, 1988). This is manifested as weakness, headache, hyperactivity and confusion leading to tremor, ataxia, hypotonia and hypothermia which has been seen in many cases of petrol sniffing, and industrial exposure to tetraethyl lead has caused deaths at levels of 1.5 mg/m^3 with prolonged exposure (Grandjean, 1984), although the ready absorption of tetraethyl lead through the skin can make quantification difficult. Excretion of organoleads is slow, resulting in accumulation of dose. Daily exposure of a dog to 42 mg/m^3 caused death after 7 days, and exposure of 2 dogs to 12 mg/m^3 caused death after 24 and 29 days. Rats, however, tolerated 12 mg/m^3 for 7 hours per day for 150 exposures (ACGIH, 1971). Tetraethyl lead is volatile (vapour pressure = 0.3 mm Hg), and the recommended TWA/TLV value is 0.1 mg/m^3, which, if derived entirely from fully vaporised petrol would require air with a very high petrol content of 180000 ppm (18%). Levels of organic lead in the air of parking garages ranged from 0.00045 to 0.0079 mg/m^3 (Nielsen, 1984).

White spirit (Stoddart solvent)

This is a petroleum distillate specified by boiling point range rather than by composition. It consists largely of saturated hydrocarbons such as nonane and isodecane with about 15% of aromatic hydrocarbons which are mainly trimethylbenzenes. Its primary toxic action is as a CNS depressant and eye irritant (ACGIH, 1971). It is detectable at 10 mg/m^3 and transiently irritant at 850 mg/m^3 (Carpenter *et al.*, 1975). Since its saturated hydrocarbon consituents are very lipophilic it has a long half-life of about 180 hours in the body and can readily accumulate with intermittent exposures (Cohr, 1986). Few cases of human intoxication have been described, none with any persisting effects (Browning, 1965). Animal studies showed 10000 mg/m^3 to be acutely lethal to cats and dogs, but 1900 mg/m^3 was well tolerated for 6 hours per day over 13 weeks without giving rise to pathological changes (Carpenter *et al.*, 1975).

Glycol ethers (cellosolves)

These are a group of solvents which are both water and lipid miscible and find wide use in industry, but only limited domestic use in latex paints and cleaners. Ethylene glycol monomethyl ether (2-methoxyethanol) and its acetate (hydrolysed to the parent compound in the body) and ethylene glycol monoethyl ether (2-ethoxyethanol) are commonly used. Vapour pressures range from 1 to 9 mmHg which gives the more volatile cellosolves potential for inhalation toxicity. This has been reviewed by Kalf *et al.*, (1987) who described impaired spermatogensis in animals after exposure to 100–3200 mg/m^3 and embryonic malformation and mortality at 160–1600 mg/m^3 (which is not maternally toxic). Some neurotoxicity has also been reported at these levels. The TWA/TLV of 2-methoxyethanol is 16 mg/m^3 (ACGIH, 1991).

Long-term effects

Whilst the major toxic actions of the relevant solvents have been described in this section, a number of potentially important aspects have been omitted or only briefly indicated. This is because they are all broadly similar and are best considered in one section.

In Denmark, Finland, Norway and Sweden, a condition called the psycho-organic syndrome, or solvent-induced pre-senile dementia, is officially recognised as a pensionable industrial disease. This is largely defined in terms of below average performance in psychometric tests designed to test levels of alertness, cognitive and other intellectual functions (Gamberale, 1986). Such tests are similar to those used to detect stress, threshold brain damage due to disease, and to those used to quantify the reversible pre-narcotic effect of solvents. This long-term solvent-induced condition would appear to be caused by styrene, toluene, tetrachloroethylene, trichloroethylene and white spirit together with mixed exposures, all at, or below,

TWA/TLV levels (Gamberale, 1986; IARC, 1979; Antti-Poika, 1986), but the most frequently implicated agent is styrene. This condition is, however, not universally recognised as a number of methodological criticisms have yet to be fully answered (Gamberale, 1986). Not all studies clearly distinguish between reversible pre-narcotic intellectual impairment (which all solvents produce below the clearly sedative level of exposure) and an irreversible intellectual impairment persisting after the solvent has been eliminated from the body. Solvent elimination may take some time, as discussed in the section on solvent kinetics. This is particularly true of styrene which, although it has a blood half-time of about 4 hours (Stewart *et al.*, 1968), has a half-time of 53–94 hours in body fat and shows increased excretion of urinary metabolites throughout the working week (Droz, 1986). The other problems are common to many studies in which very small impairments are detected, near to their incidence in the 'normal' population. The first point relates to the selection of normal reference groups, which can be difficult as people drawn from one occupation (exposed group) tend to resemble each other and differ from those in other occupations (potential reference groups) in both intellectual ability and psychological profile. A number of apparent differences have disappeared when reference groups of similar intelligence are used. Attempts have been made to overcome this by 'before and after starting employment' studies but these necessarily take a long time to conduct. A second point relates to an element of self-selection for testing from within the solvent exposed population, as in hospital patients who present themselves with solvent induced conditions. Another problem, specific to solvent studies, is that, although mean exposure levels may be low, some workers could be expected to have been exposed briefly to high levels, and may even have had periods of unconsciousness. A final point is that it has generally proved difficult to find evidence of organic damage to the nervous system in these cases, even in computerised tomographic studies (Antti-Poika, 1986).

It must, however, be said that it is difficult to detect small differences within the 'normal' range of brain dysfunction and that most such studies are as yet incomplete. It is also an historical aspect of industrial medicine that adverse effects, even without an obvious organic basis, should not be ignored. In the very first edition of the *British Journal of Industrial Medicine* (1944) Stewart and Witts, in a study dealing with the gastro-intestinal disturbance produced by carbon tetrachloride poisoning (and, incidentally, financed by the Medical Research Council), say 'The fact that

Table 9.2. *Pesticides approved for UK non-professional use*

Pesticide	Vapour pressure[a] (mm Hg)	Max air content[b] (mg/m^3)		LD 50[c] (mg/kg)
Organophosphates:				
Bromophos	1.3×10^{-4}	2.6	(20°C)	3750–8000
Chlorpyrifos	1.9×10^{-5}	0.36	(25°C)	96–163
Diazinon	1.4×10^{-4}	2.3	(20°C)	300–400
Dichlorvos	1.2×10^{-2}	145	(20°C)	56–108
Dimethoate	8.3×10^{-6}	0.10	(25°C)	215–680
Fenitrothion	6×10^{-6}	0.091	(20°C)	800
Iodofenphos	8×10^{-7}	0.018	(20°C)	2100
Malathion	4×10^{-5}	0.70	(30°C)	1375–2800
Pirimiphos methyl	9.8×10^{-5}	1.6	(30°C)	2050
Carbamates:				
Bendiocarb	5×10^{-6}	0.060	(25°C)	40–156
Carbary	4.1×10^{-5}	0.44	(25°C)	300–850
Propoxur	10^{-3}[d]	12[d]	(20°C)	90–128
Pyrethroids:				
Allethrin	1.2×10^{-4}	1.9	(30°C)	685–1100
S-Bioallethrin	1.2×10^{-4}	1.9	(30°C)	200–1545
Cypermethrin	1.4×10^{-9}	3×10^{-5}	(20°C)	250–4000
Deltamethrin	1.5×10^{-8}	5×10^{-4}	(25°C)	52–5000
Permethrin	1.5×10^{-8}	3×10^{-4}	(20°C)	430–4000
Phenothrin	1.2×10^{-6}	0.023	(20°C)	>10000
Pyrethrins (mixture)				584–900
Resmethrin	1.6×10^{-8}	3×10^{-4}	(30°C)	1347–2500
Tetramethrin	2.4×10^{-6}	0.044	(20°C)	>5000
Other:				
gamma HCH (Lindane)	4.2×10^{-5}	0.67	(20°C)	88–270
Penfenate (Plifenate)	1.1×10^{-6}	0.021	(20°C)	>10000
Para-dichlorobenzene	1.2	8010	(20°C)	500
Methoprene	2.3×10^{-5}	0.38	(25°C)	>34600

[a] at indicated temperatures.
[b] 'Max air content' = mass volatalised into 1 m^3 of air saturated with pesticide vapour [calculated from v.p.].
[c] 'LD 50' = Range of acute oral LD 50 values for rats.
[d] approximate value.

they are not due to hepatic or renal disease does not mean that we can afford to overlook their unpleasant nature and significance.' However, even with this rider, the absence of irreversible solvent-induced changes in workers or experimental subjects exposed to much higher levels than those producing the psycho-organic syndrome means that such results should be interpreted very carefully, and that only n-hexane, methyl n-butyl ketone, toluene and those chlorinated hydrocarbons giving rise to hepatic encephalopathy can be considered to have proven irreversible neuropathic potential, even at high exposure levels.

Pesticides

Pesticide exposure comes from those which are used in the home (mainly insecticides) or from the treatment of timber, furnishings, etc with insecticides and fungicides. Table 9.2 shows those pesticides approved for UK 'amateur' (i.e. non-professional) use (MAFF/HSE, 1989). They have been divided into four chemical groups.

The largest group, the organophosphates and carbamates, share a common mode of action, namely the inhibition of the enzyme acetylcholinesterase. This disrupts the function of the insect neuromuscular system leading rapidly to death. Acetylcholinesterase is equally important in man, but the very high metabolic capacity of mammals for these agents means that they are able to survive much higher doses than insects. Acute poisoning, by ingestion of very large doses via concentrates, is possible, and deaths from intentional or, more rarely, accidental drinking occasionally occur. An additional action of these agents is the production of a central/peripheral distal polyneuropathy some weeks after exposure (Davis *et al.,* 1985). This, however, is only produced by certain specific compounds not licenced for domestic use. A convenient early warning of potential toxic risk seen with the organophosphates and carbamates is their ready inhibition of cholinesterases in human plasma at non-toxic levels below those which affect the nervous system.

The pyrethroids (synthetic analogues of the natural plant products, pyrethrins) are another large group also acting on the nervous system, in this case affecting the sodium channel complex. Again mammals are much more resistant than insects, largely due to their very high rate of metabolic detoxification. Pyrethroids can produce acute poisoning in mammals by ingestion (or extensive spillage on the skin) of concentrates, but very few cases have occurred in man and these have all involved the handling of industrial/agricultural concentrates (Ray, 1990). Pyrethroids do not produce any serious longer-term toxic effects but an allergic skin response can develop to the plant impurities present in preparations of the natural pyrethrins; additionally a paraesthesia (a tingling sensation in the skin, especially around the hands and face) lasting 2–12 hours can be caused by direct contact with pyrethroid dusts or concentrates (Ray, 1990).

The other pesticides belong to different classes such as the organochlorine gamma HCH lindane and the insect growth regulator methoprene (Ray, 1990).

It will readily be seen from Table 9.2 that all of the pesticides have very low vapour pressures (that estimated for propoxur is a maximum value), with the exception of dichlorvos and para-dichlorobenzene. This very low volatility means that even under the worst possible conditions of no ventilation and complete saturation of the air with the vapour of the pesticide, only extremely low exposures would be possible. Not surprisingly it has proved impossible to find data for inhalation toxicity for these agents, and oral 50% lethality data has been used instead. These all show extremely large safety margins and could therefore only represent a risk if inhaled in the form of spray particles or droplets. In an experimental study of high strength propoxur/dichlorvos aerosol sprays (Pauluhn & Machemer, 1988) it was found that all the larger particles (greater than 20 μm) fell to the ground within 4 minutes whilst the smallest (mean diameter 5μm) had reduced to 20% of their initial concentration by 20 min. As a result of this, despite being subjected to half second spray bursts at 10 to 90 second intervals for 4 hours (86000 mg/m^3 total can contents) rats showed only mild reversible anticholinergic signs, and effective inhaled insecticide levels were only 2.6 or 5% of the applied values. An additional means of exposure is to fine particulates released from heated vaporisers, usually of gamma HCH. Cases reviewed by Smith (1991) show that exposure levels can approach those of industrial workers (3.2 to 48 mg/m^3). One case of dermatitis was reported, but other cases of poisoning (signs: headache, nausea) are probably attributable either to overheating of the vaporiser producing irritant breakdown products, or to concomitant exposure to DDT. No industrial poisonings have been reported due to inhalation of gamma HCH.

Dichlorvos

This is rapidly hydrolysed by moisture and this breakdown, plus the limitation of slow release by diffusion from polyvinylchloride strips (Vapona) means that the high theoretical air content of Table 9.2 is not reached in practice. Mean levels in US domestic air are 0.0005 to 0.028 mg/m^3 (Sterling, 1985) and an estimated practical use level is 0.06 mg/m^3 (IARC, 1979*a*). Higher levels of 0.5 to 37 mg/m^3 are reached immediately after indoor fumigation spraying, and up to 18 hours of ventilation may be required to reduce these to acceptable levels (WHO, 1989).

Uptake of inhaled dichlorvos was 15–70% over 6h at 0.1 to 2 mg/m^3 in pigs but tissue half-life is short, being only 13.5 minutes in rat kidney (WHO, 1989) and metabolic elimination is fairly rapid, being 27% over 8 hours in man or 42% over 24 hours in rats (IARC, 1979*a*). As in other organophosphates the earliest detectable change preceding any adverse effect is plasma cholinesterase inhibition. In man, exposure to 0.7 mg/m^3 dichlorvos during working hours over 8 months reduced plasma cholinesterase levels by 40 to 60%. The effect had reversed by 1 month after exposure and was not seen after exposures to 0.1 mg/m^3. Acute exposures to 1 mg/m^3 require 6 to 7 hours' inhalation to produce cholinesterase inhibition (WHO, 1989). Similar results have been obtained for animals: monkeys exposed to 0.05 mg/m^3 continuously for 3 months showed moderate plasma cholinesterase inhibition as did rats given 2 years of 23 hours per day exposure to 0.5 mg/m^3 or rabbits given 3 months of 4 hours per day exposure to 1 mg/m^3. The concentrations killing 50% of rats or mice within 4 hours vary from 13 to greater than 218 mg/m^3 (WHO, 1989), but this depends on purity and requires the air supply to be given directly to the rats, otherwise toxicity is greatly reduced by spontaneous breakdown of the dichlorvos in moist air (ACGIH, 1971).

Para-dichlorobenzene

This is commonly used in moth balls, for moth proofing furnishings and also in air fresheners. It has an appreciable smell at levels of 90 mg/m^3 and becomes painfully irritant at 500 mg/m^3 (Hollingsworth *et al.*, 1956). Domestic exposure is limited by the slow rate of vaporisation and would rarely reach these levels except in confined spaces or if the source was heated. Extensive industrial exposure to levels up to 260 mg/m^3 has shown no ill effects except some early reports of cataracts which are attributable instead to simultaneous exposure to high levels of naphthalene (ACGIH, 1971). In animals, exposure to 4800 mg/m^3 for 133 7 hour-exposures produced mild fatty necrosis of the liver, whilst 580 mg/m^3 had no effect (Hollingsworth *et al.*, 1956).

In addition to pesticides, 31 additional agents are approved as wood or masonry preservatives (MAFF/HSE, 1989). Most of these are general biocides, involatile and represent a hazard only during application, mainly via spills or accidental ingestion. One of them, tributyltin oxide (readily converted to the carbonate in the atmosphere), is a potential neurotoxin. Investigation of the alkyl tin derivatives indicates that tri-ethyl, propyl and butyl tins all have the potential to disrupt myelin, causing axonal degeneration and brain oedema (Aldridge & Brown, 1988). Some reports suggest that this is the case in animals dosed orally, but this is difficult to reproduce since the tributyltins are all highly irritant and destructive of the gastric mucosa and absorption is very variable (R.D. Verschoyle, personal communication). The vapour pressure of tributyltin oxide is very low however (approx. 7×10^{-5} mm Hg at 20 °C) and true vapour inhalation exposure would be expected to be negligible. Workers inhaling spray droplets at below 0.05 mg/m^3 experienced nasal irritation, and one worker engaged in tributyltin manufacture suffered a long lasting loss of the sense of smell (WHO, 1980). This would be compatible with the severe skin irritation also seen after spillage on to unprotected skin, but does not suggest any risk from tributyltin once application is complete.

Polychlorinated biphenyls (PCB)

These are a complex range of polychlorinated biphenyls, comprising 209 different structures (Kimbrough, 1987). They are no longer produced for domestic applications but production was very high in the 1970s and they were used for plasticisers, adhesives and (mostly) as large transformer coolants. The major source in the United States is consumed fish (Kimbrough, 1987), but PCBs, probably from electrical fittings, give US indoor air levels of 44–240 ng/m^3, representing a tenfold increase over the outside air. This represents an insignificant source compared to dietary intake, although levels of 0.011 mg/m^3 falling off with a half-lives of 2 weeks were seen after burnout of a fluorescent light power unit (MacLeod, 1981). A key feature of PCBs is their long persistence in the body, two PCBs having half times of 15 and 87 days in the dog (Sipes *et al.*, 1982). As they are extremely lipophilic and slowly metabolised, growth, lactation and parturition are major means of dilution/excretion (Matthews & Dedrick,

1984). Exposures of factory workers at 0.1 to 1.4 mg/m^3 for several months have caused a persistent severe acne-like skin condition (chloracne) and loss of appetite, whilst shorter term exposures at 8 to 16 mg/m^3 have caused transient sensory neuropathies (Kimbrough, 1987).

It would seem that PCBs only represent a potential domestic hazard after electrical fires involving older components, and even then only in unusually poorly ventilated conditions.

Conclusion

The major volatile hazards described in this chapter are the organic solvents. These obviously have acute lethal potential under conditions of addictive abuse or gross over usage, and some produce long-term damage. Their hazard at normal indoor air levels is, however, clearly negligible, although their nuisance and irritant potential should not be neglected. Pesticides only represent a hazard during the rather unlikely circumstances of prolonged direct aerosol inhalation, or by spillage of concentrates during application.

References

ACGIH [American Conference of Governmental Industrial Hygienists] (1971). *Documentation of the Threshold Limit Values for Substances in Workroom Air*. Cincinnati, USA, ACGIH.

ACGIH [American Conference of Governmental Industrial Hygienists] (1991). *Threshold Limit Values and Biological Exposure Indices for 1991 to 1992*. Cincinnati, USA, ACGIH.

ALDRIDGE, W.N. & BROWN, A.W. (1988). The biological properties of methyl and ethyl derivatives of tin and lead. In: *The Biological Alkylation of Heavy Elements*. Craig, P.J. & Glockling, F., eds, pp. 147–63. London: Royal Society of Chemistry.

ALTENKIRCH, H., STOLTENBURG, G. & WAGNER, H.M. (1978). Experimental studies on hydrocarbon neuropathies induced by methyl-ethyl-ketone (MEK). *Journal of Neurology*, **219**, 159–70.

ANDREWS, L.S. & SNYDER, R. (1986). Toxic effects of solvents and vapours. In *Casarett and Doull's Toxicology: The Basic Science of Poisons*. 3rd edition, Klaassen, C.D., Amdur, M.P. and Doull, J., eds. pp. 636–68. New York: Macmillan.

ANGER, W.K. and JOHNSON, B.L. (1985). Chemicals affecting behaviour. In *Neurotoxicity of Industrial and Commercial Chemicals*, vol. I, ed. O'Donoghue, J.L. pp. 51–148. Boca Raton: CRC Press.

ANTTI-POIKA, M. (1986). Symptoms and signs in solvent exposed populations. In *Safety and Health Aspects of Organic Solvents*, Riihimaki, V. & Ulfvarson, U., eds, pp. 255–64. New York: Alan R. Liss Inc.

BAKINSON, M.A. & JONES, R.D. (1985). Gassings due to methylene chloride, xylene, toluene and styrene reported to Her Majesty's Factory Inspectorate 1961–1980. *British Journal of Industrial Medicine*, **42**, 184–90.

BERLIN, M. (1988). Biological monitoring of populations exposed to volatile petroleum products. *Annals of the New York Academy of Sciences*, **534**, 472–80.

BROWNING, E. (1965). *Toxicity and Metabolism of Industrial Solvents*. Amsterdam: Elsevier.

CARPENTER, C.P., GEARY, D.L., MYERS, R.C., NACHREINER, D.J., SULLIVAN, L.J. & KING, J.M. (1976). Petroleum hydrocarbon toxicity series XIII: Animal and human response to vapours of toluene concentration. *Toxicology and Applied Pharmacology*, **36**, 473–90.

CARPENTER, C.P., KINKEAD, E.R., GEARY, D.L., SULLIVAN, L.J. & KING, J.M. (1975). Petroleum hydrocarbon toxicity series III: Animal and human response to vapours of Stoddart solvent. *Toxicology and Applied Pharmacology*, **32**, 282–97.

CLARK, J.M. & LAMBERTSEN, C.J. (1971). Pulmonary oxygen toxicity – a review. *Pharmacological Reviews*, **23**, 37–133.

COHR, K-H. (1986). Uptake and distribution of common industrial solvents. In *Safety and Health Aspects of Organic Solvents*, Riihimaki, V. & Ulfvarson, U., eds, pp. 45–60. New York: Alan R. Liss Inc.

CORNISH, H.H. (1980). Solvents and vapours. In *Casarett and Doull's Toxicology: The Basic Science of Poisons*. 2nd edition, Klaassen, C.D., Amdur, M.O. & Doull, J., eds, pp. 468–96. New York: Macmillan.

COURI, D. & MILKS, M. (1982). Toxicity and metabolism of the neurotoxic hexacarbons n-hexane, 2-hexanone and 2,5-hexanedione. *Annual Review of Pharmacology & Toxicology*, **22**, 145–66.

DAVIS, C.S., JOHNSON, M.K. & RICHARDSON, R.J. (1985). Organophosphorus compounds. In *Neurotoxicity of Industrial and Commercial Chemicals*, vol. II, O'Donoghue, J.L., ed, pp. 1–24. Boca Raton: CRC Press.

DOSSING, M. (1986). Metabolic interactions between organic solvents and other chemicals. *Progress in Clinical Biological Research*, **220**, 97–105.

DOSSING, M. & RANEK, L. (1984). Isolated liver damage in chemical workers. *British Journal of Industrial Medicine*, **41**, 142–4.

DROZ, P.O. (1986). Simulation models for organic solvents. In *Safety and Health Aspects of Organic Solvents*. Riihimaki, V. & Ulfvarson, U., eds, pp. 73–88. New York: Alan R. Liss Inc.

FORNI, A. (1988). Benzene and its biological effects in humans. *Annals of the New York Academy of Sciences*, **534**, 493–95.

GAMBERALE, F. (1986). Application of psychometric techniques in the assessment of solvent toxicity. In *Safety and Health Aspects of Organic Solvents*, Riihimaki, V. & Ulfvarson, U., eds., pp. 203–24, New York: Alan R. Liss Inc.

GRANDJEAN, P. (1984). Organolead exposures and intoxications. In *The Biological Effects of Organolead Compounds*, Grandjean, P., ed, pp. 227–41. Boca Raton: CRC Press.

GRASSO, P. (1986). Cancer hazard from exposure to solvents. In *Safety and Health Aspects of Organic Solvents*, Riihimaki, V. & Ulfvarson, U., eds, pp. 187–202. New York: Alan R. Liss Inc.

HANNINEN, H. (1986). Neurobehavioural assessment of long-term solvent effects on man. *Progress in Clinical and Biological Research*, **220**, 225–36.

HARKONEN, H. (1979). Styrene, its experimental and clinical toxicity. A review. *Scandinavian Journal Work Environment Health*, **4**, 104.

HILLS, B.A. & RAY, D.E. (1977). Inert gas narcosis. *Pharmacology Therapeutics*, **3**, 99–111.

HOLLINGSWORTH, R.L., ROWE, V.K., OYEN, F., HOYLE, H.R. & SPENCER, H.C. (1956). Toxicity of paradichlorobenzene: determinations on experimental animals and human subjects. *Archives Industrial Health*, **14**, 138–47.

HOUGHTON, R. (1989). Unleaded gasoline: a refiner's view of the future. *Chemistry and Industry*, **16**, 523–6.

IARC (1974). Benzene. In *IARC Monographs on the Evaluation of the Carcinogenic Risk of Chemicals to Humans*, **7**, 203–22.

IARC (1974*a*). Ortho-and para-dichlorobenzene. In *IARC Monographs on the Evaluation of the Carcinogenic Risk of Chemicals to Humans*, **7**, 231–44.

IARC (1979). Styrene. In *IARC Monographs on the Evaluation of the Carcinogenic Risk of Chemicals to Humans*, **19**, 231–74.

IARC (1979*a*). Dichlorvos. In *IARC Monographs on the Evaluation of the Carcinogenic Risk of Chemicals to Humans*, **20**, 97–127.

IARC (1979*b*). Carbon tetrachloride. In *IARC Monographs on the Evaluation of the Carcinogenic Risk of Chemicals to Humans*, **20**, 371–91.

IARC (1979*c*). Chloroform. In *IARC Monographs on the Evaluation of the Carcinogenic Risk of Chemicals to Humans*, **20**, 401–27.

IARC (1979*d*). Dichloromethane. In *IARC Monographs on the Evaluation of the Carcinogenic Risk of Chemicals to Humans*, **20**, 449–65.

IARC (1979*e*). Tetrachloroethylene. In *IARC Monographs on the Evaluation of the Carcinogenic Risk of Chemicals to Humans*, **20**, 491–514.

IARC (1979*f*). 1,1,1-Trichloroethane. In *IARC Monographs on the Evaluation of the Carcinogenic Risk of Chemicals to Humans*, **20**, 515–31.

IARC (1979*g*). Trichloroethylene. In *IARC Monographs on the Evaluation of the Carcinogenic Risk of Chemicals to Humans*, **20**, 545–72.

IARC (1986). Dichloromethane. In *IARC Monographs on the Evaluation of the Carcinogenic Risk of Chemicals to Humans*, **41**, 43–85.

JACOBS, J.M. & Le QUESNE, P.M. (1984). Toxic disorders of the nervous system. In *Greenfield's Neuropathology*. 4th edition, Adams, J.H., Corsellis, J.A.N. & Duchen, L.W., eds, pp. 627–98. London: Edward Arnold.

KALF, G.F., POST, G.B. & SNYDER, R. (1987). Solvent toxicology:

recent advances in the toxicology of benzene, the glycol ethers and carbon tetrachloride. *Annual Review of Pharmacology & Toxicology*, **27**, 399–428.

KATZ, G.V. (1985). Chemical and biological interactions affecting neurotoxicity. In *Neurotoxicity of Industrial and Commercial Chemicals*, vol. I, O'Donoghue, J.L., ed., pp. 149–58. Boca Raton, CRC Press.

KIMBROUGH, R.D. (1987). Human health effects of polychlorinated biphenyls (PCBs) and polybrominated biphenyls (PBBs). *Annual Review of Pharmacology and Toxicology*, **27**, 87–111.

KING, M.D., DAY, R.E., OLIVER, J.S., LUSH, M. & WATSON, J.M. (1981). Solvent encephalopathy. *British Medical Journal*, **283**, 663–5.

KLOCKARS, M. (1986). Solvents and the liver. In *Safety and Health Aspects of Organic Solvents*, Riihimaki, V. & Ulfvarson, U., eds., pp. 139–54. New York: Alan R. Liss Inc.

LAINE, A. & RIIHIMAKI, V. (1986). Acute solvent intoxication. In *Safety and Health Aspects of Organic Solvents*, Riihimaki, V. and Ulfvarson, U., eds., pp. 123–32. New York: Alan R. Liss Inc.

LEWTAS, J. (1989). Toxicology of complex mixtures of indoor air pollutants. *Annual Review of Pharmacology and Toxicology*, **29**, 415–39.

McCARTHY, T.B. & JONES, R.D. (1983). Industrial gassing poisonings due to trichloroethylene, perchloroethylene and 1,1,1-trichloroethane, 1961–1980. *British Journal of Industrial Medicine*, **40**, 450–5.

MacLEOD, K.E. (1981). Polychlorinated biphenyls in indoor air. *Environmental Science and Technology*, **15**, 926–8.

McNUTT, N.S., AMSTER, R.L., McCONNELL, E.E. & MORRIS, F.M. (1975). Hepatic lesions in mice after continuous inhalation exposure to 1,1,1-trichloroethane. *Laboratory Investigation*, **32**, 642–54.

MAFF/HSE (1989). *Pesticides 1989: Pesticides Approved Under the Control of Pesticides Regulations 1986*. London, HMSO.

MALTONI, C., COTTI, G. & PERINO, G. (1988). Long-term carcinogenicity bioassays on methylene chloride administered by ingestion to Sprague-Dawley rats and Swiss mice and by inhalation to Sprague-Dawley rats. *Annals of the New York Academy of Sciences*, **534**, 352–66.

MALTONI, C., CONTI, B., PERINO, G. & di MAIO, V. (1988*a*). Further evidence of benzene carcinogenicity: results on Wistar rats and Swiss mice treated by ingestion. *Annals of the New York Academy of Sciences*, **534**, 412–26.

MARSH, D.O. (1985). The neurotoxicity of mercury and lead. In *Neurotoxicity of Industrial and Commercial Chemicals*. O'Donoghue, J.L., ed., vol. I, pp. 159–70. Boca Raton: CRC Press.

MATTHEWS, H.B. & DEDRICK, R.L. (1984). Pharmacokinetics of PCBs. *Annual Review of Pharmacology & Toxicology*, **24**, 85–103.

MEYER, K.H. (1937). Contributions to the theory of narcosis. *Transactions of the Faraday Society*, **33**, 1062–4.

MOLHAVE, L. (1985). Volatile organic compounds as indoor air pollutants. In

Indoor Air and Human Health, Gammage, R.B. & Kaye, S.V., eds., pp. 403–14. Chelsea: M.I., Lewis.

MORLEY, R., ECCLESTON, D.W., DOUGLAS, C.P., GREVILLE, W.E.J., SCOTT, D.J. & ANDERSON, J. (1970). Xylene poisoning: a report on one fatal case and two cases of recovery after prolonged unconsciousness. *British Medical Journal*, **3**, 442–3.

NIELSEN, G.D. & ALARIE, Y. (1982). Sensory irritation, pulmonary irritation and respiratory stimulation by airborne benzene and alkylbenzenes: prediction of safe industrial exposure levels and correlation with their thermodynamic properties. *Toxicology and Applied Pharmacology*, **65**, 459–77.

NIELSON, T. (1984). Atmospheric occurrence of organo-lead compounds. In *Biological Effects of Organo-lead Compounds*, Grandjean, P., ed., pp. 43–62. Boca Raton, USA: CRC Press.

O'DONOGHUE, J.L. (1985). Aliphatic halogenated hydrocarbons, alcohols and acids and thioacids. In *Neurotoxicity of Industrial and Commercial Chemicals*, O'Donoghue, J.L., ed., vol. II, pp. 99–126. Boca Raton: CRC Press.

O'DONOGHUE, J.L. (1985*a*). Alkanes, alcohols, ketones and ethylene oxide. In *Neurotoxicity of Industrial and Commercial Chemicals*, O'Donoghue, J.L., ed., vol. II, pp. 61–98. Boca Raton: CRC Press.

O'DONOGHUE, J.L. (1985*b*). Aromatic hydrocarbons. In *Neurotoxicity of Industrial and Commercial Chemicals*, O'Donoghue, J.L., ed., vol. II, pp, 127–38. Boca Raton: CRC Press.

PARKKI, M.G. (1986). Biotransformation reactions and active metabolites. In *Safety and Health Aspects of Organic Solvents*, Riihimaki, V. & Ulfvarson, U., eds, pp. 89–96, New York: Alan R. Liss Inc.

PAULUHN, J. & MACHEMER, L. (1988). Investigation of the acute inhalation toxicity of spray can contents. In *Environmental and Occupational Toxicology: Proceedings of the Asia-Pacific Symposium on Environmental Occupational Toxicology*, Sumino, K., ed., pp. 407–12. Japan: ICMR, Kobe.

PERBELLINI, L., BRUGNONE, F., CARETTA, D. & MARANELLI, G. (1985). Partition coefficients of some industrial aliphatic hydrocarbons (C5–C7) in blood and human tissues. *British Journal of Industrial Medicine*, **42**, 162–7.

PLANAS, A.M. & CUNNINGHAM, V.J. (1987). Uncoupling of cerebral glucose supply and utilization after hexane-2,5-dione intoxication in the rat. *Journal of Neurochemistry*, **48**, 816–23.

POLITIS, M.J., SCHAUMBURG, H.H. & SPENCER, P.S. (1980). Neurotoxicity of selected chemicals. In *Experimental and Clinical Neurotoxicology*, Spencer, P.S. & Schaumburg, H.H., eds., pp. 613–630. Baltimore: Williams and Wilkins.

RAY, D.E. (1990). Pesticides derived from plants and other organisms. In *Encyclopedia of Pesticide Toxicology*, Hayes, W.J., ed., pp. 585–636. New York: Academic Press.

RIIHIMAKI, V. (1986). Metabolism and excretion of organic solvents. In

Safety and Health Aspects of Organic Solvents, Riihimaki, V. & Ulfvarson, U., eds., pp. 61–72. New York: Alan R. Liss Inc.

SIPES, I.G., SLOCUMB, M.L., PERRY, D.F. & CARTER, D.E. (1982). 2,4,5,2',4',5'-Hexachlorobiphenyl: distribution, metabolism and excretion in the dog and the monkey. *Toxicology and Applied Pharmacology*, **65**, 264–72.

SMITH, A.G. (1991). Chlorinated hydrocarbon insecticides. In *Encyclopedia of Pesticide Toxicology*, Hayes, W.J., ed., New York: Academic Press (in press).

SPENCER, P.S., COURI, D. & SCHAUMBURG, H.H. (1980). *n*-hexane and *n*-butyl ketone. In *Experimental and Clinical Neurotoxicology*, Spencer, P.S. & Schaumburg, H.H., eds., pp. 456–75. Baltimore: Williams and Wilkins.

STERLING, D.A. (1985). Volatile organic compounds in indoor air: an overview of sources, concentrations and health effects. In *Indoor Air and Human Health*, Gammage, R.B. & Kaye, S.V., eds., pp. 387–402, Chelsea: M.I., Lewis.

STEWART, A. & WITTS, L.J. (1944). Chronic carbon tetrachloride intoxication. *British Journal of Industrial Medicine*, **1**, 11–19.

STEWART, R.D., DODD, H.C., BARETTA, E.D. & SCHAFFER, A.W. (1968). Human exposure to styrene vapour. *Archives in Environmental Health*, **16**, 656–69.

STEWART, R.D., DODD, H.C., GAY, H.H. & ERLEY, D.S. (1970). Experimental human exposure to trichloroethylene. *Archives in Environmental Health*, **20**, 64–71.

STEWART, R.D., GAY, H.H., SCHAFFER, A.W., ERLEY, D.S. & ROWE, U.K. (1969). Experimental human exposure to methyl chloroform vapour. *Archives in Environmental Health*, **19**, 467–77.

WALLACE, L.A., PELLIZZARI, E.D. & GORDON, S.M. (1985). Organic chemicals in indoor air: a review of human exposure studies and indoor air quality studies. In *Indoor Air and Human Health*, Gammage, R.B. & Kaye, S.V., eds., pp. 361–78. Chelsea: MI, Lewis.

WEAVER, N.K. (1988). Gasoline toxicology: implications for human health. *Annals of the New York Academy of Science*, **534**, 441–51.

WHO (1980). *Tin and Organotin Compounds: A Preliminary Review*. Geneva: World Health Organisation.

WHO (1989). *'Dichlorvos' IPCS Environmental Health Criteria No. 79*. Geneva: World Health Organisation.

10

VEGETABLE DUSTS AND LUNG DISEASE

P. J. NICHOLLS

Introduction

Organic material of vegetable origin has been closely associated with human activity from earliest times, for plants have been exploited as a source of food and clothing and for a host of domestic and industrial purposes.

At certain times and in certain situations, the inhalation of airborne dust from such materials has given rise to respiratory or other problems. Vegetable dusts are of great concern with respect to the quality of air in the work environment in a range of industrial processes. Thus the clinical problems arising from exposure to such dusts are normally considered as occupational diseases. Experience shows that such a direct inhalation is unlikely in the indoor environment (e.g. offices, homes, leisure areas) mainly because the chance of attaining a sufficiently high dust concentration for long enough or repeatedly often is rare.

Bernardo Ramazzini (1713) in his treatise *De Mortis Artificum* describes many occupational lung diseases including those of hemp workers, bakers, millers, grain measurers and farmers. Today, it is recognised that dusts from textiles, grain (and flour), wood and hay account for a significant number of airway disorders. These include: asthma, hypersensitivity pneumonitis, bronchitis and certain types of systemic febrile reactions. These features may appear singly or in combination as part of a symptom complex such as occurs in workers exposed to vegetable fibre textile dusts.

In general, adverse reactions may arise in the occupational context at any one or more of the stages of harvesting, storage and processing, each of which may be associated with the generation of a particulate cloud of the appropriate economically important crop.

In contrast to most chemical or inorganic pollutants, dusts of vegetable origin are not simple homogeneous systems but a heterogeneous mixture of vegetable particles and contaminants. There may be botanical variations in individual plants from which the product is harvested and variation in environmental factors, e.g. climate and soil may enhance biochemical differences between one crop and another even when from the same location. Depending on the plant and the crop, there is also a wide range of materials that may be acquired during the growing, harvesting, storage and processing phases. These include fertilisers, insecticides, soil, bacteria, fungi and processing chemicals.

In view of this, it is hardly surprising that studies into the mechanisms underlying such occupational diseases have been difficult to undertake. Even today much uncertainty exists regarding the nature and identity of the active agents in vegetable dusts and of the precise pathogenic mechanisms at work. Despite this, in industrialised countries occupational exposure has been recognised as an important factor in assessing respiratory disorders arising during work practice, and this has been greatly aided by the development of sensitive lung function tests and epidemiological instruments. It is of interest to note that the dusty trades, notably the textile industry, were among the earliest to respond to the demand for a cleaner work environment.

Among the many factors that must be considered in studying these diseases are:

(a) The concentration of airborne dust in terms of both total and respirable fractions. In this regard, it is now recognised that measurement of dust concentrations in the personal zone of the worker is required to accurately determine dose–response relationships. This approach has greatly refined epidemiological studies amongst cotton textile workers.

(b) The particle size distribution of the inhaled dust. This can determine the level in the respiratory tract at which the patho-physiological response is initiated.

(c) The pattern of exposure. In the vegetable textile worker, exposure is normally continuous throughout the working life while in some occupations the offending material may be present only intermittently, as occurs with grain and mouldy hay dusts for farm workers.

(d) Factors which relate to the individual worker. Of this category the more important are age, length of employment, type of employment in that particular industry, atopy and smoking history.

Types of disease

The respiratory tract is able to respond in a variety of ways to inhaled vegetable dusts and this leads to a number of different types of disease. These can be classified broadly as immunologically specific and non-specific and as indicated earlier their symptoms may occur singly or in combination.

Mechanisms

A currently acceptable proposal for a general mechanism of the induction of inflammatory diseases caused by inhalation of organic dusts gives a central role to cell activation. The hypothesis is that such dusts contain activators of several types of cell which are stimulated to release a plethora of inflammatory mediators and/or are recruited to (i.e. accumulate in) the airway. The specificity of the disease is determined largely by the nature of the activator and the types of cell affected by it.

For some diseases of this type, the activation appears to be antigen-independent and directed at lung lymphocytes. Under such an influence these cells proliferate and release a high level of lymphokines which increase permeability of the alveolar epithelium and the capillary endothelium. As a result, inflammatory cells enter the airway where, under the continuing influence of the activator, inflammatory mediators are liberated. The increased capillary permeability also allows escape of plasma from the circulation leading to the oedematous component of the inflammatory response.

Acute and chronic inflammation

Irritation of the upper airways may induce acute reactions in the epithelium. These appear within a short time after exposure and are evident as rhinitis, cough and increased mucus secretion. Associated with these symptoms there may be a gradual decline in ventilatory function as has been observed in grain handlers. Chronic bronchitis is a consequence of long-term exposure and is characterised by persistent cough and sticky phlegm, both of which are readily aggravated by infection of the upper airways. In the chronic condition, airflow is permanently reduced and the airways are insensitive to bronchodilators.

Non-specific irritation of the airway is believed to occur following the inhalation of a number of vegetable fibre dusts such as hard hemps and sunn (Indian hemp) in rope- and twine-makers, kapok in bedding and upholstery workers and coir (coconut husk fibre) in brush and rope-makers. However, the risk is very low for exposure to coir. In both acute and chronic inflammatory conditions, upper airway narrowing (bronchial hyper-reactivity) may arise in response to a wide range of agents mostly of a particulate nature. This response to such non-specific stimuli thus renders individuals more susceptible to respiratory difficulties from a variety of environmental contaminants such as house dust, irritant volatile organic compounds and tobacco smoke. Therefore when considering complaints of respiratory difficulty in indoor environments, it is important to consider possible outdoor occupational exposure to the agents discussed in this chapter.

Hypersensitivity pneumonitis

In many cases of disease caused by exposure to vegetable dusts, an immunological mechanism has been implicated and this has allowed the disease classification to be either hypersensitivity pneumonitis or occupational asthma. The former category, also known as extrinsic allergic alveolitis, represents an inflammatory disease of the peripheral airways and alveoli. It may be caused by inhalation of particulate organic matter in the particle size range of 1–7 μm aerodynamic diameter (this allows penetration of the offending agent to the terminal respiratory passages where lesions occur after repeated inhalation of the agent). The acute phase occurs 4–8 hours after (typically) a heavy level of exposure. It is characterised by symptoms such as rhinitis, cough, headache, sweating, fever and joint and muscle pain. Chest tightness is commonly experienced and there is breathlessness on effort which may be severe. A reduction in airflow reaches its peak in about 8 hours. The disease presents in a number of different clinical forms depending on the immunological responsiveness of the subject and the intensity and frequency of exposure. In most cases symptoms gradually disappear (ranging from overnight up to several weeks) when exposure is terminated. At the cellular level there is evidence of lymphocyte infiltration of the lung and sarcoid-like granulomas in the alveolar and small airway walls. In the chronic phase, with repeated exposure the acute symptoms become permanent and dyspnoea worsens in the absence of the inhaled material. A progressive diffuse intrapulmonary fibrosis often occurs. This is believed to be the result of activation of alveolar macrophages, the latter activating fibroblasts to secrete increased levels of collagen. In advanced cases of the chronic condition, there may also be signs of pulmonary heart disease and chronic heart failure.

In many cases of hypersensitivity pneumonitis, precipitating antibodies have been detected in the blood. However, the presence of precipitins is not necessarily an indication of their involvement in the pathogenesis of the disease and evidence for an immunological mechanism is often either incomplete or lacking. The observation of increased numbers of suppressor T-lymphocytes in the bronchio-alveolar lavage fluid of patients with hypersensitivity pneumonitis may be indicative of a cell-mediated immune reaction.

Although a specific causal agent has been identified only for a small number of the disease types in this category of occupational conditions, it is clear that high airborne concentrations (up to 10^{10} organisms/m^3) of moulds of the thermophilic actinomycete species e.g. *Thermoactinomycetes vulgaris* and *sacharii* and *Micropolyspora faeni* are important. However, it is believed that other agents need to be present as well to initiate the sensitisation mechanism. Most of the inhaled materials giving rise to a hypersensitivity pneumonitis also contain varying amounts of bacterial endotoxins. While these are not antigenic *per se*, they have adjuvant properties and this may be a further factor in the disease. In relation to these observations it is interesting to note that the majority of affected individuals is non-atopic. In addition, the disease is more common among non-smokers. It has been proposed that the apparent protective effect is related to the airway narrowing caused by smoking which makes it more difficult for the dust to penetrate to the distal parts of the respiratory tract.

Table 10.1 lists some of the main types of hypersensitivity pneumonitis. The alveolitis caused by inhalation of a pyrethrum aerosol is one of the few examples of a hypersensitivity pneumonitis where there was no apparent microbial link. There is also the single case of an orchid grower occupationally exposed to fir bark dust who experienced pulmonary irritation. A mild haemorrhaging pneumonitis was discovered but a bacterial cause was considered not to be responsible for the condition.

Table 10.1. *Examples of hypersensitivity pneumonitis caused by exposure to vegetable dusts*

Disease	Material responsible (dust of:)	
Bagassosis	Mouldy sugar cane	(Sugar cane workers paper and board makers)
Farmers' lung	Mouldy hay and straw	
Malt workers' lung	Mouldy malt or barley	(*Aspergillus clavatus* spores)
Maple bark disease	Mouldy maple bark	(*Cryptostroma corticale* spores; bark strippers)
Mushroom pickers' lung	Mushroom compost	
Paprika splitters' lung	Mouldy paprika	(*Mucor* sp.)
Pyrethrum alveolitis	Pyrethrum insecticide	(Heavy use of the aerosol)
Sequoiosis	Mouldy redwood dust	(*Graphium* sp.; woodworkers)
Suberosis	Mouldy cork bark	

Occupational asthma

Approximately 2% of all cases of asthma are occupational which is an indication that only a minority of workers exposed to industrial dusty atmospheres develop an extrinsic asthma. However, the disease is well recognised as occurring in response to the inhalation of a wide range of vegetable dusts (Table 10.2), including several pollens.

Occupational asthma is characterised by a bronchoconstriction which may vary in severity and is reversible either spontaneously or with therapy. This major symptom may occur either rapidly (immediate reaction) or after a delay (late or non-immediate reaction) following exposure. The immediate reaction develops within minutes of inhaling the offending dust, becoming maximal after approximately 20 minutes. Generally, the bronchoconstriction is accompanied by wheeze and chest tightness. Recovery is usually within 2 hours. The non-immediate reaction normally appears several hours after exposure, proceeding to a maximum within 4–8 hours. However, in some cases it may commence late in the night following exposure. With this type of reaction, wheeze may be slight or absent and, in some instances, the only symptoms are mild breathlessness, cough and a little sputum.

Table 10.2. *Examples of occupational asthma*

	Material (dust of:)	Worker affected
Wood:	Abirnana, African Zebra, Californian redwood, Cedar of Lebanon, Cinnamon, Iroko, Mahogany, Oak, Ramin, S.African boxwood, Western red cedar	Lumber workers, wood workers, carpenters
Grain:	wheat, buckwheat, barley, corn, flour, rice	Grain handlers, grain elevator workers, dockers, farmers, millers, bakers, pastry cooks
Other seeds and beans:	coffee (green and roasted), castor bean, tamarind seed	
Leaves:	tea (common and herbal), tobacco	Processors, cutters, blenders, packers
Plant products:	gums (acacia, tragacanth)	

Occasionally, affected subjects may experience muscle and joint aches and pains together with fever.

While acute massive exposure to an offending vegetable dust may give rise to asthma on a first occasion, there is normally a latent period ranging between several weeks to many years during which sensitisation occurs and in which further exposure normally continues at a low level. However, intensity and duration of exposure are not equivalent and it is believed that short periods of high exposure are more important than the equivalent dose accumulated at a lower exposure over a longer time.

During the initial acute phase of occupational asthma there is evidence of perturbation of the airway epithelium surface and an increase in lumenal eosinophils. With repeated attacks, smooth muscle in the small bronchi enlarges and the basal layer becomes thickened. Although not of a high prevalence, occupational asthma is a disabling condition. There is a risk of death from a severe acute attack and there is also the risk of gradually worsening symptoms leading to chronic impairment.

Depending on the cumulative effect of repeated exposure and the duration of recovery period, it has been possible to define four patterns amongst the mixed group of reactions that constitutes occupational asthma. These are: (a) deterioration progressing during the working week with the severest symptoms and ventilatory reductions occurring at the end of the week and subsequent recovery taking place during the weekend break; (b) similar symptoms and their intensity each working day with essentially complete recovery by the following day; (c) long-term progressive deterioration, where recovery is so slow that even after a break from work the worker commences each week with a significant reduction in lung function; (d) symptoms occurring maximally on the first day of the working week with recovery during the rest of the week.

For some vegetable dusts, e.g. grain dusts, there is an association between increased risk of occupational asthma and atopy of the workers. However, in other cases, the relationship is either poor, absent (e.g. as with red cedar wood) or remains to be adequately assessed. Smoking also appears to be a variable influence. Thus it has been found to be a factor in grain handlers but, amongst farm workers, smokers and non-smokers are equally affected. A possible mechanism that may operate when smoking is a factor is that the smoke may increase lung permeability which would promote the penetration of the active agent(s) to the submucosal cells where the sensitisation reaction may occur.

There are several mechanisms that have been variously considered to play a role in the manifestation of occupational asthma. For some types of the disease, Type I and Type III hypersensitivity reactions have been implicated. It is believed that grain dust and castor oil bean may be allergens acting via Type I (IgE) sensitisation reactions. The evidence for Type III reactions, however, is not clear (although it is suggested by the non-immediate type of asthma) and there has been a singular lack of association between specific precipitating antibodies and Type III skin reactions to the suspected agent. One of the problems is undoubtedly the difficulty of identifying the allergen in a complex mixture. Even where this has been achieved, the allergen may be difficult to quantify at its low active concentrations in the dust. This is probably the reason why there is little information on exposure–response relationships in this area.

Activation of complement by either the classical or alternative pathways has also been suggested as a possible mechanism. However, such evidence derives from *in vitro* evaluation of the dust and this is difficult to extrapolate with confidence to the *in vivo* situation.

A non-specific irritation by dusts has also been proposed but this is likely to be of significance only when exposure occurs at very high dust concentrations. The mechanisms involving cell activation and recruitment to the airway are certainly important but further evaluation of their role is required. This also applies to the various mediators that can be derived from membrane phospholipid, e.g. prostaglandins, leukotrienes and platelet-activating factor. The latter may play a central role in the genesis of the bronchial hyperreactivity that is known to occur in cases of occupational asthma caused by inhalation of vegetable dusts. An antidilator factor derived from normal bronchial epithelium may be absent where this cell layer has been damaged by a sensitisation reaction and this may also contribute to the hyperreactivity of the airway to bronchoconstrictors.

Wood dusts

Occupational asthma can be caused by exposure to dusts from a large variety of hardwoods. Although immediate reactions do arise, the non-immediate type is more common, particularly with a nocturnal onset and a slow recovery over several days. In many cases a long latent period is required but for Western red cedar sensitisation occurs within a few months of exposure. This Canadian wood is used for both indoor and outdoor construction and the asthma and rhinitis arising from exposure to its dust have an allergic basis. A non-volatile component of the wood, plicatic acid, gives positive results on inhalation in susceptible workers and this substance may be the allergen. The dust also releases histamine and possibly other mediators from lung tissue by a non-immune mechanism. However, the role of this action in the symptomatology has not been assessed.

The African woods, ukola and abura and the tropical wood, okumen, used to make furniture, cause symptoms of rhinitis, wheeze and breathlessness in carpenters exposed to their dusts. All three types release mast cell histamine non-immunologically. For ukola, this is believed to be caused by saponins present in the wood. Although skin tests with extracts of ukola are negative, it is not possible to exclude a hypersensitivity response as contributing to the overall effects on the respiratory tract.

Ramin (*Gonystylus bancanus*) is a hardwood from Borneo that is employed as a substitute for oak. About 20 cases of hypersensitivity to the material have been described but of these only five presented with respiratory symptoms. One of these was typical of a Type III allergy.

In addition to occupational asthma, another hazard of exposure to wood dust is cancer. There is some evidence of a small but definite increased risk of nasal adenocarcinoma from exposure to hardwood dust and of nasal and sinus paranasal cancer from softwoods, e.g. pine and spruce. However, it is not proven that amongst furniture workers dustiness is proportional to the incidence of bronchial cancer. Neoplasia may be associated with the presence, in woods, of pesticides such as chlorophenols, aldrin and chromates. In sawmills, vapours may be generated that contain monoterpenes. These could possibly react with ozone to give rise to allergens and hydroperoxides. In addition, the resin acids of woods may have a role, although the cancer risk they pose is unknown.

In pulp and paper mills, there is an association between employment and excess mortality due to ischaemic heart disease and to cancers of the digestive system, lymphopoietic tissues and the respiratory tract. The industry uses a very great number of chemicals and identification of a single toxic agent has not been possible. Presently, it is thought that sulphur compounds such as sulphides and sulphites may be common aetiological agents rather than wood-derived compounds.

Grain and flour dusts

A wide range of workers may be affected by the inhalation of grain and/or flour. Thus farm workers are exposed to grain dust both during harvesting and in the winter when handling stored grain. Also at risk are grain elevator workers, dockers and millers. The latter group plus bakers and pastry cooks are exposed to flour. It is interesting to note that, even within one group of workers, disease prevalence may vary depending on the occupational details. Thus the incidence of respiratory attacks is greater in oven handlers than in dough handlers and general bakers.

Inhalation of these cereal dusts is capable of causing rhinitis and asthma (both immediate and non-immediate reactions occurring) in up to approximately one-third of exposed workers. Within the first three months of employment grain handlers develop a substantial increase in cough, sputum and eye irritation together with small changes in pulmonary function. Amongst grain elevator workers, the symptoms may be more pronounced and falls of peak expiratory flow rate of greater than 20% may be recorded in a high proportion of subjects. A recent survey has shown that even in a modern bakery, wheeze, chest tightness and difficulty in breathing are common (13% of the workforce). Atopy is considered to be a predisposing factor and smoking habit may interact synergistically in the pathogenesis of the chronic respiratory condition. Bronchial hyperreactivity to inhaled spasmogens occurs and there is an increased prevalence of chronic bronchitis in these affected workers.

Mechanisms involving IgE and the mast cell have been implicated. However, precipitins to flour components have also been found and aqueous extracts of grain and flour can activate the classical and alternative pathways of the complement cascade. Of the various dusts of this group, wheat is the most frequently recognised allergenic material. In flour, the relative importance of potential allergens may vary according to the source of material, conditions of storage and intensity of exposure. Thus in Australia where grain has a low moisture content, grain components are important allergens. In other climates, where there is a higher moisture content or storage is for longer periods, microorganisms, mites and insects (grain weevil) may be important. It is also relevant to consider whether material added to flour before baking, e.g. amylase and yeast, may possess allergenic properties.

There is an excess of lung cancer in bakers, pastry cooks and cooks, with smoking habit probably being a confounding factor. Suspect carcinogens in bakeries are polycyclic hydrocarbons arising from pyrolysis of the various products in baking ovens. However, this is likely to be very low in modern ovens although some restaurants and hotels may possess less 'clean' ovens. It is thus of importance to note that the excess cancer rate may be slightly higher in skilled bakers working in such establishments compared with employees of modern large bakeries. Also suggested as possible carcinogens in this environment are free radicals (from fat), nitrosodimethylamine (from microorganisms in the raw materials) and aflatoxin and sterigmastocystin (from various bakery moulds). However, there is currently a dearth of adequate information to properly address this problem.

Rice dust

In many Asian countries rice is a major agricultural crop. After harvesting, the rice is dried and milled and this generates large amounts of dust. Exposure to this is known to cause an occupational asthma and keratitis nummularis. The risk factor appears to be the husk of the rice grain. The syndrome includes acute and chronic irritant effects (to the eye, skin and upper respiratory tract) and allergic reactions (nasal catarrh, chest tightness, asthma and eosinophilia). Radiological opacities (fine nodulation and increased reticulation) are suggestive of either early silicosis or hypersensitivity pneumonitis. However, fever and limb pains, characteristic of hypersensitivity pneumonitis, are absent and the husk is known to have a high silica content. Overall, it is considered that both non-specific and allergic mechanisms operate. In respect of the former, it may be significant that the husk bears microscopic needle-like spines and that, after long-term exposure to rice dust, corneal scarring is readily detectable.

Tea dusts

Acute and chronic respiratory symptoms may occur in cultivators, handlers, blenders, grinders, millers, sifters and packers of common tea (*Camellia sinensis*), an incidence of 5–6% being reported for the workforce in Sri Lanka. In addition, a tea workers' asthma has been described in workers exposed to herbal tea dusts, e.g. sage, camomile, dog rose, mint, chaparral and red clover. It is considered that the risk from herbal products may be greater than from common tea because these tend to be highly blended products and are thus likely to contain a greater variety of potential allergens. However, it is not clear whether the immunogenic stimulus is of fungal or plant origin.

There is a marked incidence of chronic cough and chronic bronchitis in workers handling sage, dog rose and gruzyan tea. However, while sage has a high allergenicity for skin reactions, it has a relatively low potential for inducing spirometric changes. Despite tea workers reporting a higher prevalence of respiratory symptoms on the first working day than later days in the week, the decrements in maximum expiratory flow volume and FEV are very similar each day. The attenuation of the airway response to inhaled tea dust in susceptible subjects by prior administration of cromoglycate suggests that such reactions may be IgE-based. However, cromoglycate does not affect the bronchoconstriction induced by inhalation of Indian tea.

Coffee bean dusts

Workers exposed to both green and roasted coffee beans may experience symptoms of asthma, rhinitis, conjunctivitis and chronic bronchitis. An acute reduction in flow rate occurs, which is greatest on the first day of the working week. This can be diminished by cromoglycate, suggesting that the respiratory symptoms are IgE-mediated hypersensitivity reactions. In addition, the green beans (which are the greater hazard) contain a weak direct-acting bronchoconstrictor activity. As for tea dusts, the respiratory effect of inhaled coffee bean dust is thought to be mainly on the small airways.

Tobacco dust

There is evidence that respiratory symptoms such as dyspnoea, emphysema, chronic bronchitis and asthma occur in tobacco workers. In a recent survey in a cigar factory, there was a significant increase over control subjects in complaints of asthmatic symptoms in exposed workers and a decrease in lung function capacity was found. There was also an association between chronic obstructive lung disorders and long term exposure to the tobacco dust. Dust concentrations (about 0.5 mg/m^3) and bacterial counts in the air of the work areas were low but it was considered that the dust contained a high proportion of macromolecules with potential immunological activity. Hypersensitivity pneumonitis in tobacco workers has also been described where the airborne tobacco dust has contained high concentrations of actinomycete moulds.

Cinnamon dust

Cinnamon is the bark of the *Cinnamonum zeylanicum* tree. Experience of occupational disorders arising from exposure to the dust of the bark is limited. However, one study conducted in a cinnamon export store in Sri Lanka revealed that 88% of the work force had one or more adverse effects. Among these were asthma (in 22.5%), skin irritation (in 50%), eye irritation (in 22.5%), hair loss (in 37.5%) and weight loss (in 65%). The mechanisms for these various effects of very high incidence have not yet been determined, but may be associated with the presence of the irritant cinnamic aldehyde in the cinnamon bark.

Pollens

In this section it is also appropriate to consider these plant materials as they are amongst the many substances that can act as allergens when inhaled. Pollens are the microspores produced by the anthers of flowering plants. The most allergenic are from wind-pollinated plants such as trees, grasses and weeds. However, only a limited number of them is of clinical importance, e.g. elm, oak and maple (trees), timothy, orchard, fescue and sweet vernal (grasses), and nettle. Although weather conditions can affect pollen density and distribution, their release is seasonal, trees pollinating in early spring, grasses from spring to midsummer and weeds from midsummer to autumn. Thus the allergic manifestations (rhinitis and asthma) that they precipitate also occur seasonally. The treatment of these conditions depends on the identification of the offending allergen and then of attempting to avoid exposure to it by strategies such as staying indoors at times of high airborne concentration of the relevant pollen, ensuring windows are closed and by the use of air-conditioning. Personal air filtration is also a possibility which allows a greater freedom of movement or, in extreme cases, holidaying away from the exposure area at the height of the pollen season. Hyposensitisation is another approach where the antigen is injected into the patient over a period of months in a regime of gradually increasing doses. However, this treatment only renders the patient less sensitive and not insensitive. There are also several drug therapies available for the various manifestations of the allergic reaction, e.g. H_1-receptor antagonists, peripheral vasoconstrictors, $ß_2$-adrenoceptor agonists and prophylactic agents such as steroids, cromoglycate and nedocromil.

Fungal spores are significant airborne contaminants throughout the year although some seasonal peaking occurs. Some of these are highly allergenic and often they outnumber pollens. *Cladiosporium* and *Alternaria* are probably the most important.

Organic dust toxic syndrome

An acute fever accompanied by chills, joint pains and other influenza-like symptoms may develop in subjects exposed to vegetable and other organic dusts. The symptoms develop in the afternoon or evening on the first exposure to the material and rarely persist for more than 24 hours. Tolerance rapidly develops on repeated exposure. However, it may recur after a prolonged absence from the work environment or after a heavy exposure. The symptoms have been associated with the presence of endotoxins in the inhaled dust. However, convincing evidence is lacking in every case and other agents may be involved. Thus in farm environments, some instances of febrile symptoms have been associated with dusts of low endotoxin content but high mould spore levels. Whatever the causative agent, it is likely that the fever is mediated by the release of interleukin-l from activated alveolar macrophages.

Byssinosis

The pulmonary disease arising from exposure to vegetable fibre textile dusts is byssinosis. In 1941 the disease became compensatable in the UK. It is, however, a worldwide problem recognised as occurring in at least 16 countries where such fibres are processed.

The manifestations of byssinosis are essentially those of a bronchial disease. Its characteristic symptomatology is chest tightness, dyspnoea, cough and wheeze. The chest tightness usually starts some 4 hours after commencement of the working shift and increases in severity over the shift. In addition, some affected individuals experience fever, chills and nausea on first exposure to the dusts or on returning to the mill after a long absence. Traditionally, this latter syndrome has been described as mill fever and normally disappears after a few days or even hours. Typically, the byssinotic symptoms are rarely seen in workers with less than 5 years' exposure to the dust and they are most prevalent on the first day back to work after a weekend break (the 'Monday syndrome' of the West). These effects either abate overnight or are milder next day. Later in the week, the workers are usually symptom-free. However, the whole cycle of events is repeated on return to work after the next period of absence (i.e. the following weekend). As the disease progresses, the symptoms begin to occur on other days of the week in addition to Mondays and in the final stages there is evidence of permanent impairment of ventilatory capacity. In this latter phase, the disease is indistinguishable from the chronic bronchitis arising from non-occupational causes. Indeed, at this stage it is only the previous occupational history of the individual and evidence that other workers in the same mill will also have the disease that enable the identification of byssinosis.

One of the difficulties in assessing the chronic stages of the disease has been the interaction with other factors such as smoking. Many of the cotton mill workers suffering from byssinosis have been cigarette smokers. In addition, more pronounced reductions in air flow over a workshift have been found in smokers compared with non-smokers exposed to cotton dust and there is an increased prevalence of chronic bronchitis in cotton mill workers who smoke over that observed in non-smoking cotton mill operatives. Also from the autopsy findings, it appears that centrilobular emphysema is not occupationally associated with the textile industry, while bronchitis and bronchiolitis are.

Pulmonary function tests have provided objective evidence for the concept of byssinosis as a separate disease entity distinct from chronic bronchitis. There is a linear fall in indirect maximum breathing capacity and an increase in airway resistance during the working day in cotton workers with clinical byssinosis FEV_1 also decreases significantly during the shift on 'Mondays'. These functional changes can be reversed by the administration of bronchodilator drugs. Such bronchoconstrictor changes do not occur in workers in non-dusty jobs in the same mill or in workers exposed to other dusts, e.g. coal and rayon, suggesting that mechanical irritation of the airway is unlikely to be involved in byssinosis.

Although the acute FEV changes and chest tightness often occur together this is not invariably so and they may appear independently of each other. Chest tightness accompanied by bronchoconstriction evident within half an hour of commencing work and increasing in severity over the rest of the week may occur in a small proportion (<1%) of textile workers. This is considered to represent a type of occupational asthma and most of this affected group are forced to leave the textile industry at an early stage.

While the decrease in FEV_1 in volunteers exposed to cotton dust is independent of atopic status, the effects tend to be greater in atopic subjects and in those with a greater degree of atopy. It is possible that atopy acts as a marker for non-specific bronchial hyperresponsiveness and the latter is responsible for the larger effect observed in atopics. The significance of atopy for the long-term risk of exposure to cotton dust is unknown. Acute exposure to cotton dust over a work shift induces a hyperreactivity to inhaled methacholine. Such an increased reactivity is also seen in workers repeatedly exposed to the dust and the phenomenon may be an important factor for the development of chronic disability. Persistence of airway hyperreactivity after cessation of chronic exposure to cotton dust would most likely prime an ex-mill worker to everyday bronchoconstrictor cues such as smoke, dust and cold air. This would have the effect of maintaining a state of airflow limit in such an individual for several years into retirement.

Several mechanisms have been suggested as an explanation of the symptoms of byssinosis. Among these, release of pharmacological agents is considered to be important. Mediator release is currently viewed in the context of cell infiltration into the lungs after exposure to cotton dust.

In mill workers, the acute airway changes are concomitant with an influx of neutrophils into the nasal mucosa, an increase in polymorphonuclear leucocytes (PMN) in blood and a decrease in circulating platelets. Nebulised extracts of cotton bracts (the small modified leaves at the base of the boll which make a major contribution to the composition of cotton dust) inhaled by volunteers have been found to cause a decrease in the alveolar macrophage population and an accumulation of PMNs in the bronchio-alveolar lavage fluid. Acute exposure of rabbits to an aerosol of cotton dust extract produces an inflammatory PMN alveolitis within a few hours. The influx of cells is paralleled by the appearance of several bronchoconstrictor agents, prostaglandin (PG)$F_{2\alpha}$, thromboxane (A_2), B_2 and 5-hydroxytryptamine (5HT). Although $PGF_{2\alpha}$ is a potent bronchoconstrictor, cyclo-oxygenase inhibitors do not attenuate the acute airway response in guinea pigs and human volunteers exposed to cotton dust extracts. However, interference with the formation and receptor interactions of leukotrienes (LT) does, suggesting that bronchoconstrictor LTs may have a significant role in the acute respiratory effects of inhaled cotton dust.

Tannins (condensed polyphenols) derived from cotton bracts are a major component of cotton mill dust (5% of the water-soluble fraction). They release chemotatic factors from alveolar macrophages and are potent recruiters of neutrophils in the lungs of hamsters. They also release $PGF_{2\alpha}$ into the airway, cause platelet aggregation and release 5HT from circulating platelets and *in vitro* inhibit airway epithelial chloride ion secretion. In a monocyte dependent process they act like a polyclonal cell activator. Activation of T-lymphocytes is coupled with release of lymphokines. The latter have the potential to induce changes in the airway such as goblet cell hyperplasia and mucus gland hypertrophy which are associated with chronic byssinosis. However, the role of tannins in this disease has yet to be established.

In a survey of a large number of cotton mill operatives, while no relationship could be found between airborne dust concentration and acute fall in FEV_1, there was a clear exposure–response relationship between airborne endotoxin concentration and the airway effect. From this study it was estimated that the threshold above which FEV_1 changes would occur was 9 ng/m^3. However, endotoxin levels in the range 1–20 ng/m^3 have been found associated with chronic bronchitis and an increased prevalence of byssinosis amongst cotton workers. In addition, different airway responses from cottons from different growing areas match their endotoxin content. It is therefore of interest that weavers in a carpet factory, almost exclusively using wool, were found to have experienced byssinotic (Monday) symptoms. In this factory, the endotoxin levels were equivalent to those arising from a heavily contaminated cotton.

In animal models, inhalation of endotoxin causes activation of alveolar macrophages and chemotaxis of neutrophils to the airway. Endotoxin can also cause activation of complement, platelet activating factor (PAF) release from macrophags and histamine release from mast cells.

Currently, the balance of opinion is in favour of assigning a major role to endotoxins in the pathogenesis of byssinosis. As airborne endotoxin occurs in many of the environments where organic material gives rise to adverse respiratory effects e.g. with grain, swine and poultry confinement, composting, sewage processing and silo unloading, an endotoxin-mediated process is an attractive unifying mechanism.

The vegetable fibre dusts which produce the disease are:

Cotton

Cotton fibres (linters) are the epidermal trichomes of the seeds of *Gossypium herbaceum*, a shrub which, after flowering, produces capsules (bolls) containing many seeds. When the bolls open, cotton is picked from the plants and at this stage the raw fibre acquires various contaminants such as fragments of seed, bract, leaf and stem. Most of these are removed during the processing of the linters to form yarn. The main step in this manufacture is the process of carding in which the linters are combed into parallel alignment and the 'trash' is removed. This and other early stages, e.g. opening and blowing, are very dusty operations and workers in these areas are at a greater risk than in less dusty zones, e.g. spinning and winding. Typical figures for the prevalence of byssinosis from surveys conducted in the 1970s in cotton mills are 20–50% amongst cardroom workers and 5–10% in spinners.

Byssinosis is not confined to mills processing raw cotton and it has been found in workers involved in cotton weaving, waste cotton processing, quilt preparation, cotton garnetting and the lint-removing and seed crushing industries. A notable exception to this catalogue is the preparation of medical grade cotton wool where byssinosis is absent despite very high airborne dust concentrations. In this case, the dust is virtually free of plant debris because the cotton is washed in aqueous alkali at an early stage in the manufacturing process. The quality of cotton is a significant factor in influencing the prevalence of byssinosis, being higher in mills spinning coarser yarns from lower grades of cotton than in mills using finer and better quality yarns.

Cotton linters are a major component of house dust. This complex grey mixture consists largely of deteriorating cellulose (perhaps to glucopeptides) from cotton and kapok that arises from furniture and furnishings in the home. As is well known to anyone, it is present all year round and in virtually all homes. Allergy may develop to the dust causing a variety of symptoms such as rhinitis, wheeze, eczema and contact dermatitis. While allergy to the house dust mite is considered to be very important in this context, some individuals highly sensitive to house dust are not sensitive to mite allergens.

The other vegetable fibres of industrial importance are produced by retting. Traditionally in this process the plant material has been steeped in water, and fungal and bacterial growth and action have caused digestion of plant cells freeing the fibres. This process can now be achieved chemically. The dried material from natural or chemical retting is then subjected to various processes such as breaking and scutching (roughly equivalent to carding) during which much airborne dust is generated.

Flax

Flax fibres are the pericyclic fibres of the stem of the plant *Linum usitatissimum*. Flax is an ancient crop and respiratory symptoms associated with its preparation have been recognised for centuries. The studies of the past 30 years have confirmed that byssinosis occurs in such workers. In one study, as many as 70% of the workforce in a Yugoslavian flax mill were found to have the disease. However, no byssinosis has been observed in workers handling flax prior to retting and the incidence is low or zero when chemically retted flax is processed.

Hemp

There are two types of hemp fibre of industrial importance, namely hard fibres from the leaf and soft fibres from the stem of the European plant *Cannabis sativa*. Although fewer studies have been made of the occurrence of respiratory symptoms in workers in this industry, it is clear that it is only the soft fibres that are associated with the appearance of byssinosis (originally termed cannabosis in this industry). Chemically retted hemp or hemp retted in running water appears to be better tolerated than hemp retted in stagnant water.

Sisal

Sisal fibres are derived from the thick fleshy leaves of *Agave sisalana*. The leaves are decorticated by a wet retting process and then brushed clean. The resultant long fibres are baled and sent to the factory for processing in a manner similar to that for flax and hemp. While respiratory symptoms have been known to occur in the highly dusty areas of brushing and breaking, it has only been in the past 10 years that byssinosis has been identified in such mills using this hard fibre. However, it appears to be much less common than in workers handling fibres of hemp and flax.

Jute

Jute is obtained from the stem of two related plants, *Corchonus capsularis* and *Corchonus olitarius* and less is known about the respiratory effects of inhaled jute dust than the other fibre dusts. While some such effects have been recorded, e.g. a fall of maximum breathing capacity over the working shift, the pattern is not typical of byssinosis and the consensus view is that the disease does not occur in jute workers.

Prevention

Given that exposure to certain vegetable dusts carries a risk of lung disease, avoidance is the best preventive measure. However, as the activities creating such dusts are economically important, a pragmatic approach is to lower the concentrations of inhalable airborne dust. This may be achieved by improving general ventilation in the workplace, by developing machinery with improved exhaust systems or by altering the nature of the dust generating steps.

Other than for the textile industry, there are presently no guidelines for exposure limits to vegetable dusts although a discussion document on cereal dusts is currently being considered by the Health and Safety Executive in the UK. The disadvantage of using dust concentration is that it does not necessarily reflect the level of the causative agents. Where these can be identified, they should form the basis for assessing the risk of exposure. As yet, this has not (or cannot) be done. In addition, in some work environments such as farms, measuring exposure levels and maintaining this, at or below, a limit value is presently impracticable. However, in cases of hypersensitivity pneumonitis where a mould may have been demonstrated to be important, measures to prevent the development of such a fungal organism should be effective.

Education of worker and employer is essential so that both are fully aware of the risks and of the general local health regulations (e.g. the Control of Substances Hazardous to Health regulations in the UK) regarding their particular work environment. It would then be possible for appropriate action, such as respiratory protection of the individual worker, to be taken. Where atopy is a proven risk factor, it may be possible to utilise a

Table 10.3. *WHO[a] recommended permissible exposure limits for inhalable[b] vegetable fibre textile dusts*

Fibre		Limit (mg/m³)
Cotton	– ginning	0.5
	– yarn processing	0.2
	– carding	0.2
	– weaving	0.75
Cotton waste processing		0.5
Flax	– dew and water retted	2.0
	– green and chemically retted	5.0
Soft hemp		2.0

[a] WHO (1983), Tech. Rep. Ser. 684.
[b] As measured by vertical elutriator with a 15 μm aerodynamic diameter cut-off point.

pre-employment screen for this. However, in the case of occupational asthma, it has been warned that a screening programme may direct attention away from improving the environment that is the cause of the problem.

There has been a marked reduction in the prevalence of byssinosis over the last 30 years and incidences of only 2–10% have been reported recently for the UK. This improvement has been related largely to the progressive reduction in cotton dust levels in this period and to the greater use of cleaner cotton. The report of a WHO study group (1983) acknowledged that there was no clear evidence of a threshold cotton dust concentration below which symptoms or functional changes do not occur. Even at concentrations of 0.1 mg respirable dust/m³, a small proportion of cotton workers may complain of mild symptoms. The recommendations (Table 10.3) of the study group for permissible occupational exposure limits for inhalable dust, taking into account the probable differences in 'byssinogenic potency' of the dust at various stages of production, were made with the proviso that regular medical or health surveillance of the workforce would continue to be necessary. Thus the disease may be detected at an early stage in highly susceptible individuals and this could lead to appropriate action, such as movement of the affected worker to other (non-dusty) jobs in the mill or retirement from the workforce. It would appear sensible to adopt a similar strategy of health monitoring and surveillance for all workers exposed to the various vegetable dusts described in this chapter.

References

HOWARTH, R.F.B. (1989). Grain dust; some of its effects on health. *HSE Research Paper* 28. London: Health & Safety Executive.

PARKES, W.R. (1983). *Occupational Lung Disorders*. London: Butterworths.

RAMAZZINI, B. (1713). Diseases of workers. Translated from the Latin text *De Morbis Artificum* of 1713. New York and London, Hafner, 1964.

RYLANDER, R. (1986). Lung diseases caused by organic dusts in the farm environment. *American Journal of Industrial Medicine*, **10**, 221–7.

RYLANDER, R. *et al.* (1987). Effects after acute and chronic exposure to cotton dust: the Manchester criteria.

SPEER, F. (1982). *Handbook of Clinical Allergy*. John Wright, Littleton.

WILLOUGHBY, W.F., WILLOUGHBY, J.B. & GERBERICK, G.F. (1985). Polyclonae activators in pulmonary immune disease. *Clinical Reviews in Allergy*, **3**, 197–216.

WORLD HEALTH ORGANIZATION (1983). Recommended health-based occupational exposure limits for selected vegetable dusts. *Technical Report* Series 684. Geneva: WHO.

11

DANDERS, etc FROM DOMESTIC AND LABORATORY ANIMALS

BY G. CLOUGH

Exposure to airborne particles is a fact of everyday life and our bodies are generally well equipped to cope with them. Many diseases are transmissible by airborne means and some infections that can pass between animals and man – known as zoonoses – can be transmitted in this way. Probably the most well known of human diseases for which animals can act as a reservoir of infection include things such as rabies, ring-worm, Salmonellosis, Psittacosis and tuberculosis. Other potential hazards from airborne sources are indicated by a recent survey carried out in Holland (Holst, Kromhout & Brand, 1988). In a case-control study they made in four main hospitals in The Hague these authors found that bird-keeping was significantly and independently related to the incidence of lung cancer. Analysis of the data derived from the 49 patients in the study (all under the age of 65) suggested that people who kept birds had a risk of lung cancer 6.7 times greater than those who did not keep birds. That diseases arising from animal sources are not uncommon is clear from the results of another survey, this time of laboratory associated infections, which showed that animals were responsible for 10–17% of identified infections in laboratory staff (Pike, 1979). In overall terms, however, probably the greatest medical problem arising from airborne particles is that of allergy and this is now recognised as one of the major diseases in the western hemisphere. About 20% of the entire population are thought to suffer from this problem in one form or another, the symptoms usually being brought on by the inhalation of particles which are antigenic. Whereas in a normal person these substances would be destroyed by antibodies, in sensitised individuals they produce an abnormal immune response stimulating the release of histamine and other substances leading to symptoms characteristic of allergy such as asthma or a skin rash.

Probably the most well-known form of allergy is 'hay fever'. This generally results in a blocked nose and sneezing and is often associated with irritation and watering of the eyes. Skin rashes are also common, varying in severity from a few itchy red patches (mild erythema) to a more extensive and irritating nettle rash (urticaria) which may become blistered and highly sensitive (papular or vesicular erythema). Wheals may develop in the skin around bites or scratches, especially if the area is contaminated with saliva, urine or other allergenic material.

About 5% of the general population develop a more serious and debilitating condition known as allergic asthma. This condition is characterised by recurrent attacks of increased resistance to airflow in the lungs resulting in dyspnoea, coughing, wheezing and tightness of the chest.

Diseases such as 'bird-fanciers lung' and 'humidifier fever' are merely descriptive terms of a specific allergy and it is now well known that such allergic responses can be caused by a variety of materials including pollens, fungal spores, proteins from grain and house dust mites, proteolytic enzyme preparations, food stuffs, insects, plant parts and animal dander (small particles of dead skin often found amongst animal body hairs). Indeed, it was reported over 120 years ago that '...epidermal scales or dander of animal hair can serve as inhalant allergens' (Salter, 1864; Blackley, 1873) and numerous studies since then have shown that, as might be expected, allergic disease of the airways is common in pet owners. One such report (Wuthrich, Guerin & Hewitt, 1985) states that sensitivity to cats and dogs occurs in 20–40% of exposed atopic individuals (that is, people who show a positive response to a skin prick test involving common allergens such as extracts of house dust mites or cat, dog or horse hair).

Poulsen and Hau (1989) suggested that allergy may be classified as follows:

1. *Contact allergy* This is the most common form of allergic reaction against small molecules such as nickel and food additives, including artificial colours and smells. The symptoms of contact allergy are allergic eczema in the area of contact and, in the case of allergy to food additives, gastrointestinal reactions.
2. *Inhalation allergy* The most common allergens here are pollen and house dust mites and symptoms include irritation of the mucous membranes of the eyes, nose and oral cavity and allergic asthma. In severe cases of the last, respiration is obstructed and the reaction may be fatal unless medical treatment is given immediately.

3. *Food allergy* The commonest allergens here include milk proteins, egg albumin, fish, flour containing gluten and various plant proteins. Soy proteins commonly cause such problems in early childhood and between 2 and 5% of all newborn children develop an allergy to cow's milk within the first year of life. The symptoms of allergy to food proteins are often immediate gastrointestinal reactions which may be associated with immediate atopic eczema as well as delayed whole body reactions of allergic eczema.

Clinically, the allergic reactions are often divided into two groups brought about by different immunological mechanisms. The first of these are the so-called immediate reactions which occur within a few minutes to about one hour after allergen challenge; they are also known as atopic reactions, anaphylactic reactions or IgE (immunoglobulin class E) mediated hypersensitivity. The second are the delayed reactions, the symptoms of which normally develop 24–48 hours after allergen exposure, also known as cell-mediated hypersensitivity resulting most commonly in eczema and other inflammatory reactions.

There are many causes of asthma and both genetic and environmental factors contribute to its development. The importance of the disease was emphasised some years ago when it was recognised that it could be induced by exposure to a variety of substances which people come across only in their place of work. Now known as 'Occupational asthma' this is defined as:

> '...asthma which develops after a variable period of symptomless exposure to a sensitising agent encountered at work.'

Allergies induced by animal danders are prevalent throughout the community due mainly to keeping domestic pets and are also found in farmers, poultry breeders, etc. However, most quantative data have been generated from investigations on staff working in laboratory animal houses. For this reason the bulk of this chapter is focused on this situation.

Over recent years, allergies in people who are exposed to animals or animal products in their work seem to have become increasingly common. The importance of this problem was emphasised and interest in it stimulated when occupational asthma resulting from exposure to (among other causative agents) 'animals and insects' in laboratories became a prescribed industrial disease in the United Kingdom in 1982 under the Social Security Industrial Injuries Prescribed Diseases Amendment Regulation.

The incidence of animal-induced-allergy in susceptible animal house staff has been reported to range from 11 to 32%. As one such survey reports, however, these figures are derived from 'survivor' populations and in an entire working population it has been estimated that the prevalence might be as high as 50%. In one case, when, with a certain procedure, workers were exposed to high levels of allergenic particles at face level, it is known that 100% of them became sensitised.

The size of the problem can be appreciated when it is realised that some 32000 people in the UK are known to work with animals in laboratories and several million people are exposed to small domestic pets. In the year ending June 5th 1978 the Department of Health and Social Security records show 51050 spells and 2495200 days of certified incapacity due to asthma; it is not known how much of this could be attributable to occupational causes. In America, tens of millions of mammals are used in research each year and reports available from America, Denmark, Finland, France, Japan, Poland, The Netherlands, Sweden and Switzerland show that similar problems are encountered in those countries also. In the UK, it has been estimated that 48% of those occupationally exposed to laboratory animals who have symptoms (9% of the total exposed population) may have to cease working, at least temporarily, with animals. The incidence of symptoms due to keeping domestic pets is not known, but in view of the large number of pet keepers, the total number of cases is significant.

Among those with this sort of allergy there is commonly a period of symptomless exposure before overt signs of sensitisation appear. Documented cases range from a few days to as long as 22 years though the majority of workers who become sensitised do so between 1 month and 3 years after their first exposure. Once this happens, symptoms may occur singly or in various combinations. Most common is rhinitis, frequently associated with conjunctivitis. Skin rashes and the development of wheals around bites and scratches are also common. In those who develop the clinically more serious condition of allergic asthma, this can be brought on merely by entering an animal room and some individuals experience asthmatic attacks even when not in direct contact with the animals. Such sensitisation once developed may be permanent, though when the person is removed from the source of the sensitising agent, the attacks usually abate. Further attacks frequently occur, however, when such people are re-exposed to the sensitising agent and some of them may even suffer attacks in response to exposure to general irritants or extremes of temperature. Evidence also

Table 11.1. *Incidence of hypersensitivity to laboratory animals by species*

Species	% of reported hypersensitivity
Rat	56.5
Rabbit	37.0
Mouse	37.0
Guinea-pig	24.0
Cat	12.9
Dog	9.6
Hamster	1.6

exists to show that in some cases a degree of disability may persist for as long as two or three years after removal from exposure to the sensitising agent (Association of the British Pharmaceutical Industry, 1987).

There is not much information indicating the frequency of sensitivity to different species. An American survey (Lutsky & Neuman, 1975), however, showed that among 191 known sufferers from 39 different laboratory animal facilities, 55% were allergic to two or more species and several were sensitive to four or more species. Their frequency of hypersensitivity to seven common species is shown in Table 11.1.

In contrast, a Scandinavian report (Agrup *et al.*, 1986) showed that among 76 exposed persons, the highest relative frequency of allergy was to cats (31%) and the lowest to rabbits (11%).

It is clear that allergens are present in the dust of animal rooms and it is known that fur and dander from mice, rats, guinea-pigs, rabbits, cats, dogs and cattle can be allergenic. Also from the laboratory animal field, locusts – and their faeces – and house-fly and fruit fly rearing rooms are known sources of allergens. Research has shown that in rats and mice the major source of allergens is the urine and it is known that the proteins involved are of low molecular weight: an alpha-2 globulin in the rat and a prealbumin in the mouse.

The majority of animal rooms are provided with turbulent flow air conditioning systems because these provide quite good temperature and humidity distribution patterns. With such systems, however, airborne dust (including any allergens it may contain) is necessarily dispersed throughout the room and it is thus impossible for personnel to avoid contact with it. The uptake of allergens by personnel depends on closeness of contact as well as particle size and the numbers of particles present in the air they are breathing (Yamauchi *et al.*, 1989).

Particles suspended in air tend to fall due to gravity, the speed at which they fall being determined partly by their size, density and shape and partly by the opposing aerodynamic drag of the atmosphere. Thus the particle falls with a steady velocity known as its 'terminal settling velocity'. In order to simplify the very complex problems related to particle shape, for comparative purposes it is usual to assign them an 'aerodynamic equivalent diameter'; this is defined as the diameter of a hypothetical sphere of unit density (1 g/cm^3) having the same terminal settling velocity in air as the particle in question. In order to penetrate into the alveoli of the lung, particles must have a falling speed of less than about 9 mm/s which is equivalent to that of a 7 micron diameter sphere of unit density. Particles with an aerodynamically equivalent diameter above 10 microns do not usually penetrate beyond the nasopharynx and for particles above 50 microns the suction at the mouth and nose originated by the action of breathing would not normally be enough for inhalation (WHO Report, 1984).

It is known that many 'aeroallergens' including animal dander and dust mites and their faeces are normally carried on particles and constitute part of the total airborne dust load. It has also been shown that most guinea-pig allergen activity was associated with particles of a diameter either greater than 5 microns (89%) or less than 0.8 microns (10%) (Swanson *et al.*, 1984). Another study showed that, in contrast, during disturbance of rat litter, 46% of rat urinary allergen was associated with particles of mean size 7 microns; 15–30 minutes after disturbance 16% of these were still airborne (Platts-Mills *et al.*, 1986).

That such particles are capable of wide dispersal within a building is shown by the results from the guinea-pig work referred to above. In this same study, air was sampled from the guinea-pig room (in which an automatic flushing system cleaned away the faeces every 24 minutes) as well as from a laboratory and a library on the same floor 6 storeys below. Guinea-pig urine allergenic activity was detected in all three samples, though not in the outside air which was used as a control (Swanson *et al.*, 1984).

It has been shown that, as might be expected, the dust load in the air of animal rooms increases with animal density. Hence in a room with 475 cages the dust load was 375 μg/m^3 of air, whereas with 800 cages the level rose to 51 μg/m^3. Measurements of crude allergen content show a similar relationship, the 24 h sampling level in a mouse room with 2000 animals being around 825 ng/m^3 of air, that in an immunology laboratory with between 5 and 100 mice being only 1.8 ng/m^3 (Lincoln, Bolton & Garrett, 1974).

There are also known to be within-room variations related to the direction of air flow; hence the distribution of crude allergen in the air of a mouse room was shown to vary from 54 to 315 ng/m^3; in a rat room from 274 to more than 44000 ng/m^3; in a rabbit room from 9900 to 35200 ng and in a Guinea-pig room 22300 to 50600 ng/m^3. In the same study it was shown that increasing the ventilation rate from 7 to 14 air changes per hour reduced the allergen level from 1400 to 10800 to 116 to 3000 ng/m^3, whilst increasing the relative humidity in the same room from 55% to about 77% reduced the level still further to 40 to 430 ng/m^3 (Edwards *et al.*, 1983).

The methods used and the degree of control required to reduce the incidence of such allergies are related to the sensitivity of the personnel and their particular allergic response. For minor cases, control may be effected by the use of simple devices such as protective gloves (with or without a suitable barrier cream) to protect the skin from contamination with saliva, urine, blood and other body fluids and paper or gauze masks over mouth and nose. When these do not prove adequate, better fitting and more sophisticated respirators of various sorts can be useful (e.g. rubber mask with replaceable filters covering mouth, nose and eyes). Alternatively, proprietary safety helmets incorporating a small fan, filter and vizor to provide a stream of clean air over the front of the face can give equivalent protection with greater comfort. It has been shown, however, that, although this type of helmet helps in suppressing severe rhinitis and conjunctivitis, the symptoms are not completely suppressed nor the progression towards asthma prevented.

Antihistamines can be, and are, widely used but these only relieve the symptoms. Attempts at desensitisation by immunotherapy have met with mixed success although in some cases this can provide long-lasting relief.

In more extreme cases it may be necessary to protect personnel further by the use of local barrier devices such as filter racks, safety cabinets, glove boxes or flexible film isolator systems (Clough, 1987). Specially developed air conditioning systems have achieved considerable success in keeping particle counts to a very low level (Federal Standard 209C, Class 100) in the personnel working area of animal rooms, though the counts do, of course, rise dramatically during cage changing (Association of the British Pharmaceutical Society 1987).

As would be expected, there is considerable interest among employers to try and identify susceptible applicants in order to exclude them from occupations where they are at increased risk. It has been suggested that atopic people should not be allowed to work with animals because they are at increased risk of developing allergic symptoms. Although there is considerable evidence that an association does exist between the atopic state and the likelihood of an individual developing allergy to laboratory animals (Beeson *et al.*, 1983; Gross, 1980; Slovak & Hill, 1987; Botham *et al.*, 1987), not all studies support this (Cockcroft *et al.*, 1981; Davies & McArdle, 1981) and the ABPI (Advisory Note on Allergy to Laboratory Animals (1987)) concluded that the exclusion of such people from work animals merely on that basis 'would impose a considerable and unjustifiable penalty upon this group'. The rationale behind this conclusion is that, although some 33% of individuals will be deemed to have an atopic predisposition judged by family history and skin testing, current knowledge shows that only about 30% of those are likely to develop allergy to laboratory animals (3–12% being likely to experience asthmatic symptoms) whilst the other 70% will not.

Thus, because (l) excluding atopic individuals will only reduce but not eliminate the problem, (2) the average exposure period before symptoms develop is long, and (3) skin reactivity to animal extracts is rarely present without symptoms, future efforts to prevent or reduce allergy among animal workers will have to be concerned mainly with finding ways of reducing exposure (Cockcroft *et al.*, 1981). An alternative approach may be to develop improved methods for identifying sensitive individuals.

References

AGRUP, G., BELIN, L., SJÖSTEDT, L. & SKERFVING, S. (1986). Allergy to laboratory animals in laboratory technicians and animal keepers. *British Journal of Industrial Medicine*, **43**, 192–8.

ASSOCIATION OF THE BRITISH PHARMACEUTICAL INDUSTRY (1987). *Advisory Note on Allergy to Laboratory Animals*. ABPI, 12 Whitehall, London, SW1A 2DY, England.

BEESON, M.F., DEWDNEY, J.M., EDWARDS, R.G., LEE, D. & ORR, R.G. (1983). Prevalence and diagnosis of laboratory animal allergy. *Clinical Allergy*, **13**, 433–42.

BLACKLEY, C.H. (1873). *Experimental Researches on the Causes and Nature of* Catarrhus aestivus *(Hay-fever or Hay-asthma)*. London: Baillière Tindall & Cox.

BOTHAM, P.A., DAVIES, G.E. & TEASDALE, E.L. (1987). Allergy to laboratory animals: a prospective study of its incidence and of the influence of atopy on its development. *British Journal of Industrial Medicine*, **44**, 627–32.

CLOUGH, G. (1987). The animal house: design, equipment and environmental control. In *The UFAW Handbook on the Care and Management of Laboratory Animals*, Poole T.B., ed., pp. 108–43. Harlow: Longman Scientific & Technical.

COCKCROFT, A., McCARTHY, P., EDWARDS, J. & ANDERSSON, N. (1981). Allergy in laboratory animal workers. *Lancet*, **i**, 827–30.

DAVIES, G.E. & McARDLE, L.A. (1981). Allergy to laboratory animals: a survey by questionnaire. *International Archives of Allergy and Applied Immunology*, **64**, 302–7.

EDWARDS, R.G., BEESON, M.F. & DEWDNEY, J.M. (1983). Laboratory animal allergy: the measurement of airborne urinary allergens and the effects of different environmental conditions. *Laboratory Animals*, **17**, 235–9.

GROSS, N.J. (1980). Allergy to laboratory animals: epidemiologic, clinical and physiologic aspects, and a trial of cromolyn in its management. *Journal of Allergy and Clinical Immunology*, **66**, 158–65.

HOLST, P.A., KROMHOUT, D. & BRAND, R. (1988). For debate: Pet birds as an independent risk factor for lung cancer. *British Medical Journal*, **297**, 1319–21.

LINCOLN, T.A., BOLTON, N.E. & GARRETT, Jr. A.S. (1974). Occupational allergy to animal dander and sera. *Journal of Occupational Medicine*, **16**, 465–9.

LUTSKY, I.I. & NEUMAN, I. (1975). Laboratory animal dander allergy: I. An occupational disease. *Annals of Allergy*, **35**, 201–5.

PIKE, R.M. (1979). Laboratory associated infections. Incidence, fatalities, causes and preventions. *Annual Review of Microbiology*, **33**, 41–66.

PLATTS-MILLS, T.A., HEYMANN, P.W., LONGBOTTOM, J.L. & WILKINS, S.R. (1986). Airborne allergens associated with asthma: Particle sizes carrying dust mite and rat allergens measured with a cascade impactor. *Journal of Allergy and Clinical Immunology*, **77**, 850–7.

POULSEN, O.M. & HAU, J. (1989). Animal models for the study of allergy. Proceedings from the 19th Annual Symposium of the Scandinavian Federation for Laboratory Animal Science, Hau, J., ed., *Scandinavian Journal for Laboratory Animal Science* (Suppl. 1), 96–101.

SALTER, H.H. (1864). *On Asthma: Its Pathology and Treatment*. Blanchard & Lea, Philadelphia.

SLOVAK, A.J.M. & HILL, R.N. (1987). Does atopy have any predictive value in laboratory animal allergy? A comparison of different concepts of atopy. *British Journal of Industrial Medicine*, **44**, 129–32.

SWANSON, M.C., AGARWAL, M.K., YUNGINGER, J.W. & REED, C.E. (1984). Guinea-pig-derived allergens: clinicoimmunologic studies, characterization, airborne quantitation and size distribution. *American Review of Respiratory Diseases*, **129**, 844–49.

WORLD HEALTH ORGANIZATION (1984). *Evaluation of Exposure to Airborne Particles in the Work Environment*. Offset Publication No. 80. Geneva: World Health Organization.

WUTHRICH, B., GUERIN, B. & HEWITT, B.E. (1985). Cross-allergenicity between extracts of hair from different dog breeds and cat fur. *Clinical Allergy*, **15**, 87–93.

YAMAUCHI, C., OBARA, T., FUKUYAMA, N. & UEDA, T. (1989). Evaluation of a one-way airflow system in an animal room based on counts of airborne dust particles and bacteria and measurements of ammonia levels. *Laboratory Animals* **23**, 7–15.

12

ENVIRONMENTAL TOBACCO SMOKE

D. F. WEETMAN

The possible health effects of exposure to environmental tobacco smoke (ETS) have been subjected to considerable investigation. When people smoke, ETS can accumulate in the air around them. This is not noticeable outdoors because of rapid dilution, but when the smoking occurs indoors, the concentration of ETS in the air may rise. Some may find ETS to be unpleasant, but does it damage health? In addressing this question we must first consider the nature of ETS, then the possible extent of exposure, and finally the uptake of components in man. With this information, it is possible to evaluate any health effects of ETS on man.

Chemistry of environmental tobacco smoke: what is ETS?

ETS has three specific sources: first, a cigarette burns to produce a plume of side-stream smoke (SS) between puffs. Second, there is the smoke exhaled by a smoker, which can be envisaged as having two components: first, simple exhaled main stream smoke (MS), and second, waste smoke (WS) (i.e. mainstream smoke that is taken into the mouth, but escapes without being taken into the lungs). The third main contribution to ETS consists of gases and other materials that may diffuse through the cigarette paper or filter. SS is the major component of ETS.

The chemistry of smoke generated by the combustion of tobacco is extremely complicated. Combustion oxidises the organic matter of the tobacco, but in general, this oxidation is incomplete due to an insufficiency of oxygen, and as a result, a large number of complex molecules are produced[1]. Combustion and other processes, such as pyrolysis (in which a parent compound is transformed into one or more other substances by heat alone, i.e. without oxidation), distillation, condensation, and filtration, cause the more than 2000 individual compounds that have been identified in tobacco leaf to yield at least twice as many compounds in tobacco smoke.

Commercial tobacco used in cigarettes, pipes, and cigars is not uniform, but rather consists of various blends of leaves from different varieties of tobacco plants which may have been cured by distinct processes. Once the tobacco leaf is harvested, it is dried. Although there is much variation from brand to brand, virtually all cigarettes utilise a high proportion of tobacco prepared by a process of flue-curing, where warm, dry air is passed over the leaves. With pipe and cigar tobacco, the leaves are allowed to dry more slowly by a process of natural ventilation. During the process of drying, fermentation occurs, so that the sugar content of the product depends in part on how much starch is converted to sugar. With flue-cured tobacco, the initial contact with heat inactivates some enzymes that normally metabolize sugars, so this product is rich in sugar (up to 20% of the dry weight), whereas the air cured product may contain less than 1% sugar. Thus, while the components of ETS produced from all tobacco are functionally similar, the specific material burnt, and thus the smoke produced, is not uniform, but rather is a complex and varying mixture. In order to overcome the complexity resulting from this heterogeneity, a standardised 'research' cigarette containing a specific blend of tobacco has been developed. It is normal in research on cigarette tobacco to control the rate of burning, and therefore the temperature of combustion, and the rate of dilution of the smoke with air. Most of the chemistry reported in the literature relates to this non-commercial cigarette; much less information is available about cigar and pipe tobacco.

Analysis of the smoke generated from a standard cigarette by a smoking machine shows it is made up from three main physical classes of chemicals: a gas phase containing nitrogen, nitrogen oxides, carbon monoxide and dioxide, and unused oxygen; a vapour phase containing a large number of volatile organic molecules produced by the burning; and an aerosol of particles (containing all the tar) described as the particulate phase. For most research purposes, the vapour and gas phases are considered together. The particulate phase makes tobacco smoke visible.

Although the individual chemicals in SS and MS are for the most part the same, the concentration of the components differs. Table 12.1 contains concentrations for some of the main ingredients of MS and SS, obtained from analysis of undiluted fresh smoke. ETS represents yet a third mix of chemicals, because two further processes occur in its formation. First, the SS is diluted into the volume of room air, so that a non-smoker receives a much

Table 12.1. *Main components of mainstream and sidestream smoke*

Component/characteristic (mg/cigarettes, unless unit stated)		Type of smoke MS (Mainstream)	SS (Sidestream)	Ratio SS/MS
pH		6.1	7.5	–
Tobacco burnt		347	411	1.2
Burning time (seconds)		20	550	27
Max temperature (°C)		900	600	–
Tar	(no filter)	20.8	44.1	2.1
	(filter)	10.2	34.5	3.4
Particles	Rates of generation	5×10^{11}/s	6×10^{9}/s	0.012
	Median size (μm)	0.26	0.1	–
	Standard deviation (μm)	1.2	1.7	–
	Total amount (mg/cig)	25	40	1.6
Nicotine	(no filter)	0.92	1.69	1.8
	(filter)	0.46	1.27	2.8
Carbon monoxide		19	88	4.7
Carbon dioxide		50	490	9.8
Ammonia		0.16	7.4	46
Nitrogen oxides		0.014	0.051	3.6
Phenols		0.228	0.603	2.6
Catechol (μg/cigarette)				
	(no filter)	42	58	1.4
	(filter)	30	70	2.3
Benzo[*a*]pyrene (ng/cigarette)				
	(no filter)	26	67	2.6
	(filter)	15	51	3.4

The values in this table have been taken from several sources. In recent years, there has been a reduction in the tar and nicotine content of cigarettes, so the values are representative rather than absolute. It should be noted that adding a filter to cigarettes reduces the consumption of potentially toxic material by the smoker, but has much less effect on SS. The composition of MS depends upon the style of smoking, but SS is less affected. The concentration of virtually all material is higher in SS than in MS. SS is diluted and aged before it becomes ETS. The smoker receives the highest dose of ETS (or SS) from his own cigarette.

lower dose of ETS than does the smoker situated near his cigarette. Second, the composition of ETS varies with the time since its generation by a process called ageing[2]. Many of the gas and vapour products of combustion of tobacco undergo further chemical reaction, and, because the half-life of each differs, the composition of ETS is continually changing.

The individual particles in tobacco smoke generated by combustion are small globules of matter of different sizes, and are much like those arising from any organic material burnt at a similar temperature. But, whereas the particles in MS are taken directly into the lung, particles present in ETS can only be inhaled for as long as they remain suspended in the air. Those that are inhaled are referred to as respirable particles. Like gases and vapours, the particles in ETS behave in a time-dependent manner. During the process of dilution with room air, the particles shrink, and some volatile substances adsorbed on their surface dissociate. The adsorption of chemicals on the surface of particles may be important, for when particles are inhaled some are retained in the respiratory tract and substances adsorbed on them may be delivered to the lung. Small particles will penetrate further into the respiratory tract than will large ones.

How are ETS exposure levels determined?

The section above suggests some of the problems in measuring ETS. Although individual components of ETS can be measured, the continually changing mixture of ingredients presents the problem of what substance or substances, usually called surrogates, should be measured to monitor ETS levels. The National Research Council's Committee on Passive Smoking (National Academy: Hulka, 1986) has proposed four criteria for such a surrogate:

1. The surrogate for ETS should be uniquely attributable to tobacco smoke.
2. The surrogate should be present in sufficient concentration in ETS to be detectable with low rates of smoking.
3. The surrogate should be generated at similar rates with a variety of smoking materials.
4. The concentration of the surrogate should bear a constant ratio to that of each compound of interest under a variety of environmental conditions.

No substance satisfies all the criteria. Indeed, because the half-lives of different components of ETS vary, criterion 4 is impossible to achieve. Table 12.2 lists substances that have been used as a surrogate for ETS for the purpose of measuring its concentration in buildings and the like.

Table 12.2. *Surrogates for ETS used in monitoring*

Surrogate	Phase	Method of assay
Carbon Monoxide	G	Electrochemical
Nicotine	G+P	GC, Immunoassay
Particles	P	Gravimetric
		Piezobalance, optical
Nitrogen oxides	G	Chemiluminescence
Ammonia	G	Ion-selective electrode
Hydrogen cyanide	G	Ion-selective electrode
Formaldehyde	G	Derivatisation + GC
Acrolein	G	GC
Solanesol	P	GC

Note that only solanesol and particles can be used to monitor the particulate phase of ETS. Only nicotine and solanesol are likely to be specific to tobacco smoke. P is the particulate phase, G the gas phase, and GC gas chromatography.

There are various ways of measuring ETS components using surrogates. One method is by means of fixed detectors, i.e. instruments placed in a particular position in a building to monitor the component under different conditions. After collection, analysis and quantification of the surrogate varies depending upon the sensitivity of methods of detection available to the chemist doing the work. Because of the possibility that a surrogate has migrated from one phase of ETS to the other, it is necessary that basic experimental work be conducted under strictly controlled conditions so as to have simultaneous determination of the surrogate in all phases. Such calibration of the distribution of the surrogate between the phases would be a prerequisite for mathematical modelling of ETS in indoor air.

While at this stage it is possible to use fixed detectors to obtain the information needed to verify the appropriateness of a mathematical model of ETS in a building, a general model suitable for all buildings has not yet been developed. Any such mathematical model would need to be calibrated against a series of surrogates for ETS. At present, such modelling gives good agreement for single substances (e.g. carbon monoxide, nicotine or particles), but is not yet developed to an extent that predicts accurately the behaviour of all the surrogates for ETS.

A second approach to monitoring ETS exposure within a building focuses on the exposure of individuals who move about freely within the building. Such monitoring is best achieved by means of a personal monitor – a device worn by the subject that samples the ambient air either continuously or intermittently, for collection either for subsequent analysis in the laboratory or direct measurement within the monitor on a minute-by-minute basis. The latter course of action is preferable to avoid ageing of the surrogate in the monitoring equipment. Furthermore, peak levels of exposure to ETS can only be measured reliably by means of a personal monitor.

Fixed monitors and personal monitors have tended to use the same constituents of ETS as surrogates. In many cases, the substance monitored has been chosen for convenience rather than because it represents an 'ideal' marker. The objective of studies with these surrogates is to develop a picture of how ETS varies within a particular situation, and how individuals move about both at home and during a typical working day. From such information, it should be possible to deduce the level of exposure to ETS. Once this is determined, it may be possible to relate exposure to uptake (i.e. dose), which can then be related eventually to any health effects. At present, however, we are a long way away from this possibility. As is often the case in biology, the sensitivity of methods of chemical analysis lags a long way behind our natural detectors: it is possible to smell ETS in locations where the analyst struggles to obtain readings. Whatever surrogate for ETS is selected, the chemist has to work at the limit of detectability, whereas the human nose can sense ETS readily.

A third and perhaps more relevant (in health terms) method for assessing exposure to ETS is to measure the result of the exposure. This can be achieved by measuring how much of a chosen surrogate is present within a person after a period of exposure. Several surrogates have been studied (see Table 12.3), and of these nicotine would seem to be ideal, because it is present in comparatively high concentration in tobacco smoke, and because the only other sources (exposure to nicotine used as a pesticide in warehouses and trace amounts in tomatoes and one or two other vegetables) are unlikely to contribute much to levels found in humans.

Nicotine presents problems as a surrogate, however, because it is not representative of all the components of ETS, and does not persist long in man (its half-life is about 2 hours) as most of a dose is metabolised. Figure 12.1 shows the difficulty in using nicotine as the marker substance for ETS in room air: the peak due to nicotine is small and located amongst many others. Furthermore, nicotine is found in the gas phase of ETS, and with respect to possible health effects, the particulate phase may be more impor-

Table 12.3. *Uptake of components of ETS by man*

Component	Body fluid	Change in concentration with ETS exposure	Smoker
Carbon monoxide	Blood	about 1% rise in COHb	7%
Nicotine (ng/ml)	Plasma	from 1 to 3	14.8
	Urine	from 10 to 100	1750
	Saliva	from 2 to 45	670
Cotinine (ng/ml)	Plasma	from 0.8 to 1.8	275
	Urine	from 1.5 to 9.5	1400
	Saliva	from 0.7 to 2.6	310
Thiocyanate	All values are unreliable because of contribution from the diet.		
Mutagenic activity	Urine	Only demonstratable after concentration of the urine (see the text)	

Note the very low levels of the markers due to exposure to ETS relative to those found in smokers (they are usually less than 1%). Only nicotine and its metabolite cotinine are specific (or almost specific: see the text) to tobacco.

tant[3]. Because of the short half-life, the most used indicator of exposure to ETS is not nicotine itself, but its metabolite, cotinine (with a half-life of about 15 hours in males and 10 hours in females (see Fig.12.2), or, more appropriately, nicotine plus cotinine[4]. Although the determination of nicotine and cotinine levels in subjects exposed to ETS may seem ideal, there are considerable technical problems in this type of investigation. The levels of each are so low that great care must be taken in handling the samples so as not to add to the apparent levels by contamination, especially if the chemist is a smoker. Most studies use gas chromatography for the detection of nicotine and cotinine, but the currently much more expensive and more sensitive technique of radio-immunoassay would seem to be worthwhile for future investigations.

Despite various difficulties in measuring ETS using surrogates, there is substantial literature quantifying it in different situations. Most urban populations are exposed to ETS for at least some part of the day (exposure may

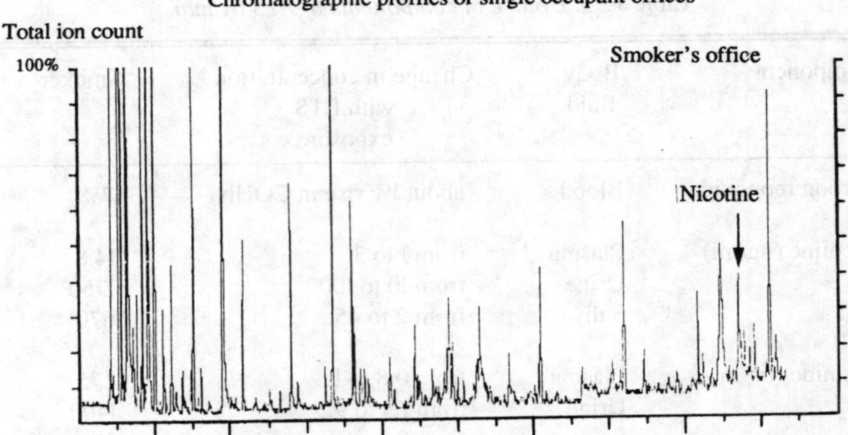

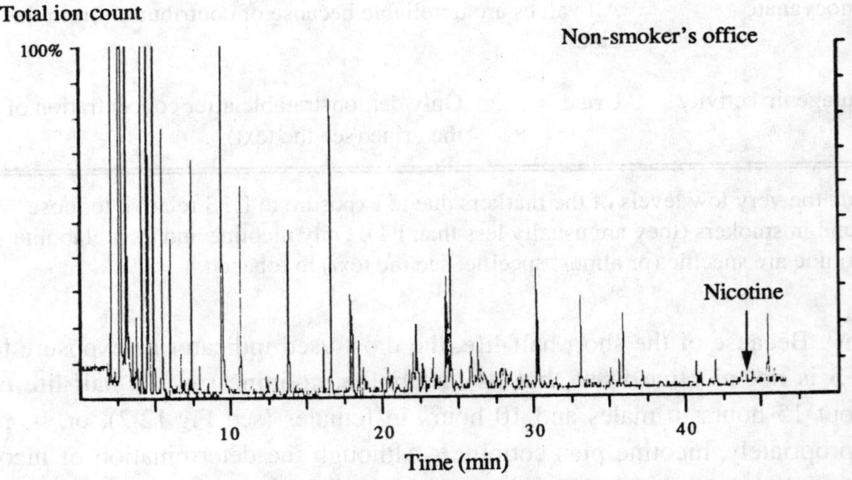

Fig. 12.1. Gas chromatogram of room air in offices occupied by a smoker (above) or a non-smoker (below). Note the presence of many substances detected by the atomic absorption spectrometer, and the small peak produced by nicotine, even in the office where smoking has occurred. The determinations were made by Dr C. Proctor. (I am very grateful to Dr C. Proctor for providing this previously unpublished figure.)

be less in certain rural situations), and most of that dose of ETS is received inside confined spaces (e.g. buildings and cars), with little contribution coming from the time spent out of doors. Using nicotine in air as the marker for ETS, measurements (in μg/m^3) range from 1.0 in a public transport waiting room to 10.3 in a cocktail lounge, to 65 in an open car occupied by three smokers, and, under the same conditions but in an unventilated saloon

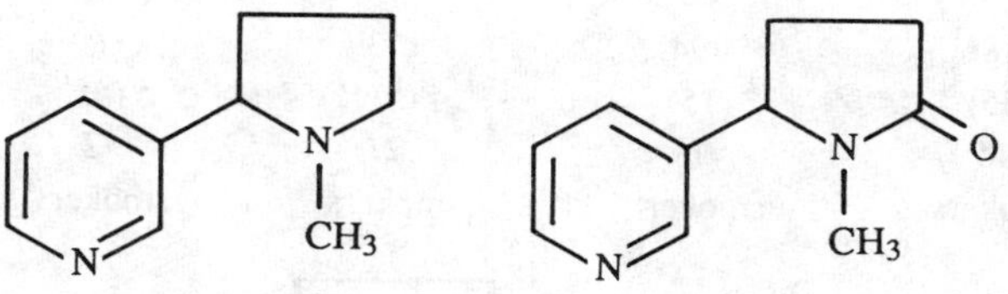

Nicotine

Cotinine

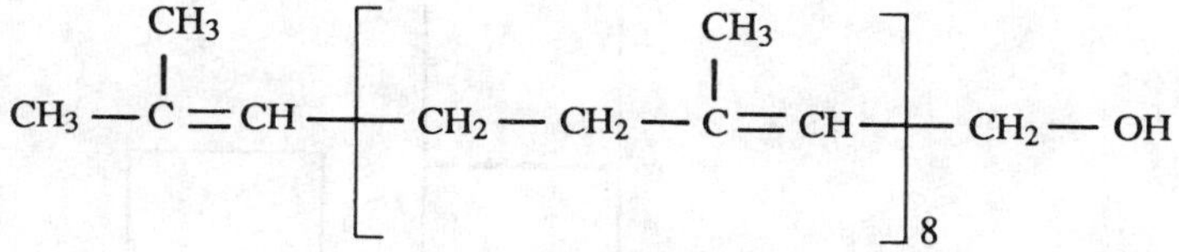

Solanesol

Fig. 12.2. The chemical structure of substances used as surrogates for ETS.

car, an exceptionally high level of 1010 (see O'Neill *et al.*, 1987). As would be expected, the highest levels recorded have been obtained from cars, conference rooms, cafeterias and the like, whilst the lowest levels (more than ten times lower) were found in laboratories and hospital reception areas. These concentrations result in blood levels in non-smokers many times smaller than those found in smokers.

Measurements of carbon monoxide as the surrogate for ETS are affected by the numerous other sources of this gas. These concentrations result in blood levels of surrogates for ETS at least 100 times lower than those found in smokers. Even accounting for considerable variation from site to site, smoking does not cause an enormous increase in the concentration of carbon monoxide: there is only about a 50% rise where smoking has occurred relative to the same place without smoking (typically from 1–2 ppm to 3 ppm). When nitrogen oxides are measured, there is very little, if any, increase in their concentration indoors after smoking relative to the levels detected outside. With sulphur dioxide, the levels are raised by smoking, but usually only two-fold.

The concentration of respirable particles, on the other hand, increases up to ten-fold with smoking, especially in confined spaces: in a bar with grill, from 63 μg/m^3 to 589 μg/m^3 (the high level in the absence of smoking indicates that some may have come from the cooking), and in an inn from

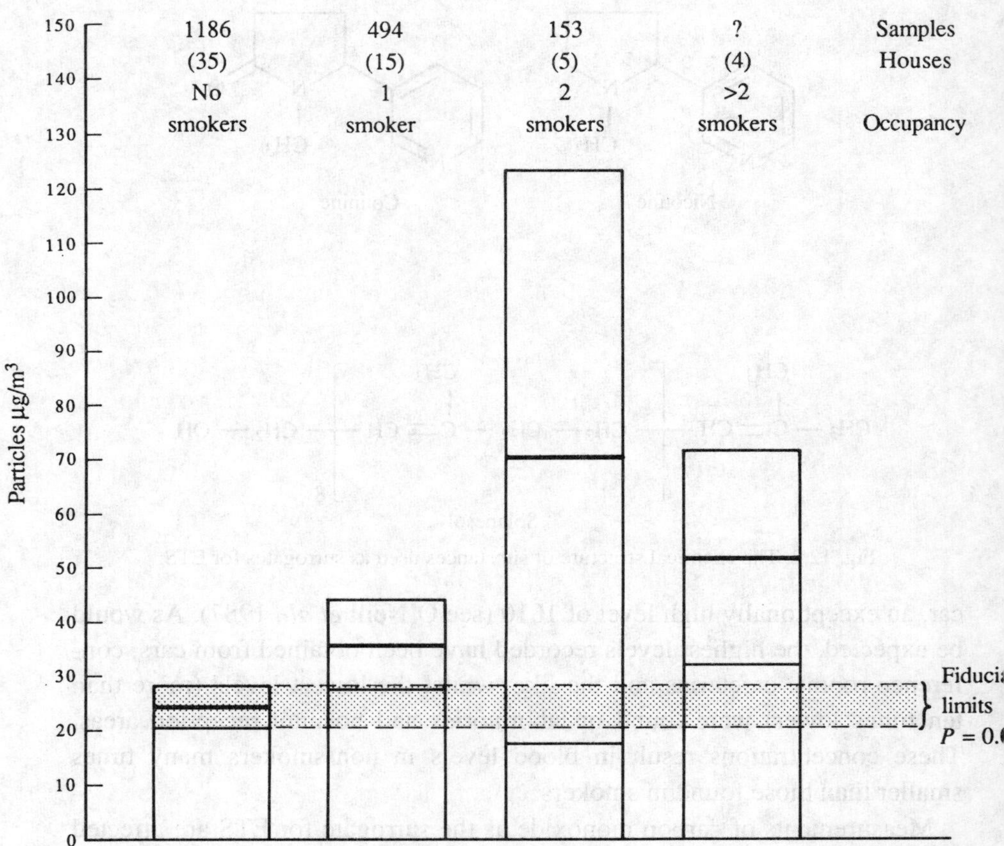

Fig. 12.3. The increase in respirable particles in houses according to the number of smokers present. Note that even in the absence of smokers there are appreciable levels of particles, indicating several sources of this pollutant. Only with two or more smokers present is the level of particles significantly higher than in a non-smoking household. The figure has been constructed from the data in Spengler *et al.* (1981).

22 μg/m³ to 239 μg/m³. The concentration of respirable particles also increases in the home as the number of smokers increases (Fig.12.3), although with two or more smokers present, the levels are only double those in a non-smoking household.

Carcinogens and ETS

Although nicotine is most frequently monitored as a marker for ETS, it is probably inappropriate to base assessments of any lung cancer risks from

Table 12.4. *Suspected carcinogens in sidestream smoke*

Carcinogen	Quantity (μg/cigarette)	Phase
Formaldehyde	5000	G
Benzene	480	G
Nitrosonornicotine	9	P
Nitrosodimethylamine	4	G
3-Pyridyl-3,3-(*N*-methyl-*N*-nitrosoamino)-propylketone	4	P
2-Toluidine	3	P
Nickel	2.4	P
Nitrosodiethylamine	1	G
Nitrosopyrrolidine	0.9	G
Cadmium	0.72	P
Benz[*a*]anthracene	0.28	P
4-Aminobiphenyl	0.14	P
Benzo[*a*]pyrene	0.14	P
Hydrazine	0.09	G
Nitrosodiethanolamine	0.08	P
2-Naphthylamine	0.05	P
Polonium 210	0.4pCi	P

The values above are for the maximum quoted yields from SS.
Note that the possible carcinogens are present in both the particulate and gas phases of SS. In cigarette smoke (MS), the main toxic components are in the particulate phase, but this may not be the case for ETS. SS is greatly diluted with room air and aged before becoming ETS. Most of the suspected carcinogens have been shown to be active in animal tests only. The maximum level of exposure to ETS in animals or man is limited by the non-carcinogenic toxic action of carbon monoxide.
The information in this table has been adapted from several sources.

ETS on such measurements because it is unlikely that nicotine is a carcinogen (but, see the two contrasting views of Hoffmann and of Roe in Wald & Froggatt, 1989). A carcinogen is a substance which, administered by any route, increases the incidence of malignant neoplasms in animal or man. There is a second class of toxic substances, the co-carcinogens, that do not cause tumours themselves, but rather increase the effect of a carcinogen. At present, these two classes of toxic chemical can be distinguished only in experiments where animal skin is painted with the agents: carcinogens cause tumours in such a test, whilst co-carcinogens potentiate a carcinogen. Were nicotine to be metabolised to a carcinogen, one would expect to detect some effect in pipe and cigar smokers, yet in most epidemiological studies from Britain and USA, the association between lung cancer and smoking pipes or cigars is substantially less than the association between lung cancer and smoking cigarettes.

The suspected carcinogens in tobacco come from the oxidation of organic molecules during the process of combustion: Table 12.4 contains the amounts per cigarette of these agents in SS, from which ETS derives. It is notable that in all cases, the amounts of suspected carcinogens listed in Table 12.4 are higher in SS than in MS, in some cases, e.g. *N*-nitrosodimethylamine, 10–200 times higher. Over the last 25 years, filters have been added to cigarettes to reduce the dose of tar and nicotine. These reductions have resulted in a changed composition of MS, but have affected SS much less.

The higher emitted amounts per cigarette of carcinogens in SS relative to MS is an interesting laboratory finding, but it is not clear if this is important in real situations. In one set of *in vitro* experiments in which cells grown in tissue culture were exposed to fresh SS, 86% of the cells died; if the SS was aged for only 8.7s, however, this level of cytotoxicity reduced to 5% (Sonnenfeld *et al.*, 1985). This result emphasises the difference between SS and ETS: ETS is subject to ageing, which, under virtually all circumstances, would be greater than the 8.7 seconds in the experiment described above. The degree of dilution of SS in room air to become ETS would also be expected to reduce cytotoxicity. Similar studies using ETS, rather than SS, have not been reported yet.

Is ETS a health risk?

In an ideal world, this question could be answered simply by direct evidence. Unfortunately, this is not possible. When one considers the limited amount of direct evidence, albeit limited, that exists, it suggests that ETS does not present a health problem. For example, one would expect to observe pre-neoplastic lesions in passive smokers if ETS were to affect healthy bronchial mucous, but no such lesions have been observed. The predominant form of lung cancer detected in the epidemiological studies on the possible effects of ETS is adenocarcinoma, which is found distal to the characteristic bronchial carcinoma seen in cigarette smokers. How the puta-

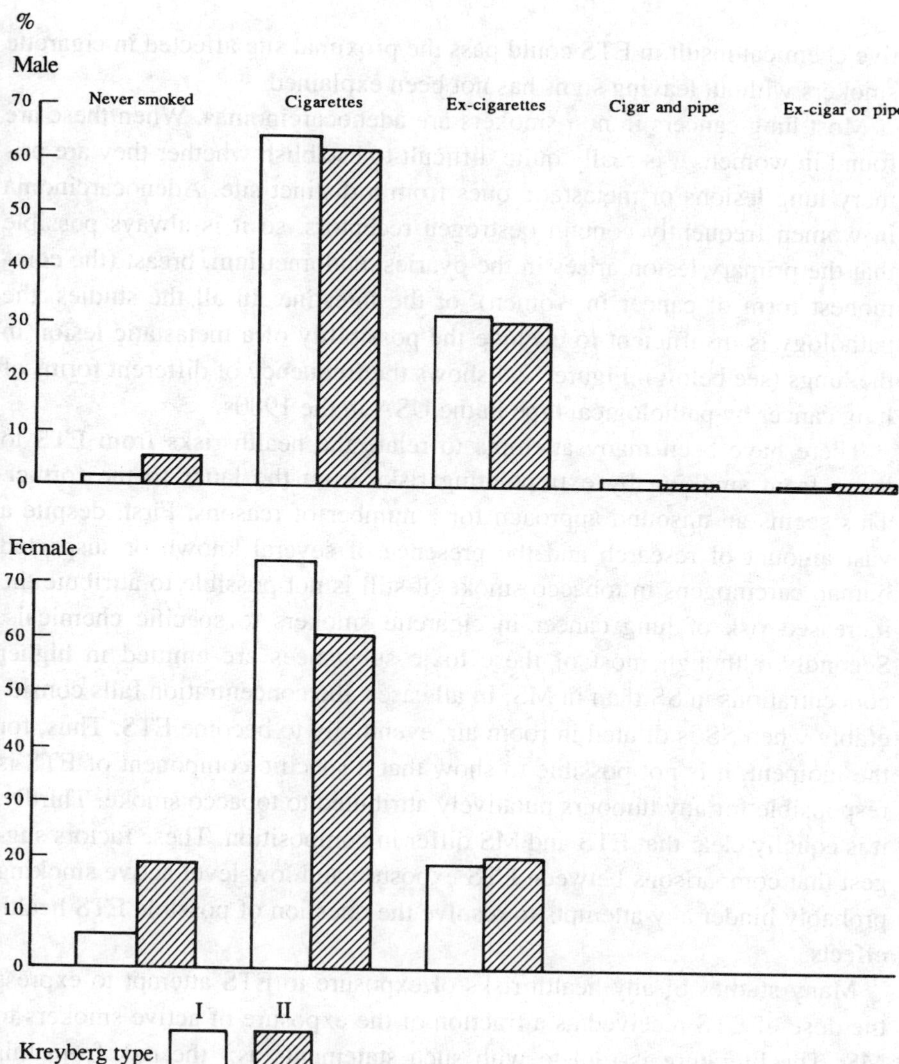

Fig 12.4. Incidence of different forms of lung cancer in the USA in the 1980s. Kreyberg I consists of squamous, large cell, small cell and oat cell carcinomas. Kreyberg II consists mainly of adenocarcinomas. Note Kreyberg I cases are rare in patients who have never smoked. The adenocarcinomas (Kreyberg II) occur in a higher proportion of women who never smoked than is the case for men. A proportion of the adenocarcinomas in women may have been metastatic lesions from endometrial cancer, as this type of cell often contains oestogen receptors.

In both men and women, the ratio of Kreyberg II to I increases with age, being twice as common in those over 65 as they are in those under 45 years old.

This figure has been constructed from data contained in Wynder & Covey (1987), *European Journal of Cancer, and Clinical Oncology*, **23**, 1491-6.

tive chemical insult in ETS could pass the proximal site affected in cigarette smokers without leaving signs has not been explained.

Most lung cancers in non-smokers are adenocarcinomas. When these are found in women, it is really quite difficult to establish whether they are primary lung lesions or metastatic ones from a distinct site. Adenocarcinoma in women frequently contain oestrogen receptors, so it is always possible that the primary lesion arises in the ovaries, endometrium, breast (the commonest form of cancer in women), or the intestine. In all the studies, the pathology is insufficient to exclude the possibility of a metastatic lesion in the lungs (see below). Figure 12.4 shows the frequency of different forms of lung cancer by pathological type in the USA in the 1980s.

There have been many attempts to relate the health risks from ETS to those from smoking by extrapolating risks from the latter to the former. This seems an unsound approach for a number of reasons. First, despite a vast amount of research and the presence of several known or suspected human carcinogens in tobacco smoke, it still is not possible to attribute the increased risk of lung cancer in cigarette smokers to specific chemicals. Secondly, although most of these toxic substances are emitted in higher concentrations in SS than in MS. In all cases, the concentration falls considerably when SS is diluted in room air, eventually to become ETS. Thus, for the moment, it is not possible to show that a specific component of ETS is responsible for any tumours putatively attributed to tobacco smoke. Thirdly, it is equally clear that ETS and MS differ in composition. These factors suggest that comparisons between ETS exposure and low-level active smoking probably hinder any attempts to resolve the question of possible ETS health effects.

Many studies of any health risks of exposure to ETS attempt to express the dose of ETS received as a fraction of the exposure of active smokers to MS. The literature is replete with such statements as: ‘the risk from this level of exposure [to ETS] is equivalent to smoking one cigarette a day’. Are such calculations useful? As shown above, ETS is a very different mixture of chemicals from the smoke inhaled by a smoker. In addition, it is taken up in a different manner. Regardless of evolutionary changes in the construction and composition of cigarettes and often related changes in individual styles of smoking, virtually all smokers take smoke into the lungs via the mouth, hold it there for a time and then exhale through the mouth and sometimes the nose. In the case of ETS, however, the non-smoker takes ambient diluted smoke via the nose. Thus a passive smoker not only absorbs a different composition of smoke, but also does so by a quite different route[5]. It is therefore questionable to attempt, as many authors have done, to estimate doses of ETS by comparing active and passive smoking.

Those estimates which suggest that the adverse health effects of ETS are as much as half those of smoking (as some do, see below), or equivalent to smoking seven cigarettes per day, would appear to be based on distorted calculations. From a scientific standpoint, uptake of ETS, when measured by the levels of nicotine or cotinine in body fluids, would lead one to conclude that an appropriate level would not exceed one cigarette per day, and might be well below that.

Just as there has been little helpful direct toxicological evidence with respect to humans, animal experiments similarly have been unhelpful. Attempts to induce lung cancer in animals after exposure to ETS under the controlled conditions of a toxicity test are expensive and fraught with technical problems, and thus, not surprisingly, there have been very few reports of the toxicological effects of SS and ETS. Animal toxicological experiments, when done, involve giving a very high dose of potential intoxicant (in this case SS) on the perhaps improbable grounds that this will reveal toxicity more readily than would a lower dose. This cannot be achieved easily with smoke from combustion of tobacco because the carbon monoxide generated by burning will kill the animals. Nevertheless, with a very high concentration of SS (45 cigarettes per hour, 10 hours exposure per day, 5 days per week for 3 months), Adlkofer and colleagues (see in Perry & Kirk, 1988) failed to produce pulmonary lesions that were visible under the light microscope in either hamsters or rats. Some variable changes were seen in the respiratory tract when the electron microscope was used, but this damage to the ciliated cells of the trachea and lung parenchyma was reversible, not being detectable 90 days after the period of exposure. Thus, SS applied for a long time in a dose much higher than man would ever encounter (depending upon the surrogate used in the tests, it was up to 100 times the maximal human exposure level) failed to produce clear-cut signs of a carcinogenic or pre-carcinogenic action. Thus it seems improbable that ETS, with its different and diluted chemical composition, would either.

Epidemiology

Over the years, statistical techniques have been developed for evaluating the vulnerability of classes of individuals to diseases. Because the first steps in the evolution of these techniques occurred with infectious diseases, this medical discipline is called epidemiology, by reference to the epidemic nature of the infections being investigated.

Epidemiology attempts to relate the occurrence of disease in a specific population of individuals to some common factor. Epidemiological studies generally take one of two forms: retrospective or prospective studies. Retrospective studies begin with knowledge of the disease as the outcome, and then look back at the lifestyle of the affected patients; measuring and attempt to link the perceived common factor among that population with the disease. Prospective epidemiological studies, on the other hand, select a defined population and measure the fate of these individuals with the passage of time. The great advantage of epidemiology is that the study measures the incidence of disease in man, as opposed to animals, thus avoiding the need for projection from one species to another. The great problem of epidemiology is that it is often difficult or impossible to attribute observed effects to a single factor, and thus the studies may fail to answer the question asked.

Control of epidemiology

Whilst fairly careful control over animal experiments may be achieved by regulating genetic differences between groups, doses of drugs, diet, and environmental conditions (e.g. level of stress from noise, duration and intensity of light and dark), such control is not easily achieved in experiments on man, and can never approach that possible in animal experiments. In selecting the type of study, epidemiology or animal experiments, the investigator thus has to choose between relevance and control.

For example, in the case of retrospective epidemiological studies, if the investigator has identified, say, 100 individuals with a disease, he will seek to determine whether any characteristic of their life-experience or make-up is associated with that disease[6]. In all cases of contemporary interest most individuals will be exposed to the characteristic under investigation, and most of those, even those with a high dose, will not succumb (e.g. most heavy cigarette smokers do not develop lung cancer). Individuals vary so much in their sensitivity that a few will be affected by the low dose, and yet most receiving the high dose will be resistant. This is the problem confronting the epidemiologist. From this individual variation he must seek to associate a characteristic with a consequence. As epidemiology is almost always observational in nature, usually it is not possible to force the argument to a conclusion by making an intervention, as is the case in many animal experiments.

One technique employed in attempts to associate some characteristic from the lifestyle of the patients with disease can be achieved by means of case-control studies. Here a disease of comparable severity to the one under investigation is identified, and from this a group of patients is assembled to act as a control. Exposure to the suspected characteristic is next sought in the two groups. For example, researchers might attempt to determine the proportion of cigarette smokers in 100 patients with cancer of the lung, and then compare this with the proportion of cigarette smokers among patients having some other neoplasm.

There would still be potential confounding factors in such a case-control study. The cigarette smokers in both groups could have been drawn predominantly from the working class, and thus many have been exposed to unsuspected industrial chemicals in their manual jobs. If most of the non-smokers were drawn from the professional ranks they would not be exposed to industrial chemicals to the same extent: perhaps too, they may have had freer access to medical services. To avoid such obvious sources of bias it is necessary that the test and control groups exhibit similar demographic profiles. Among the comparisons normally made in this respect would be age, sex, level of educational attainment, income, place of abode, and so on. If such demographically matched groups cannot be obtained at the outset, the epidemiologist may resort to matching patients at the time of statistical analysis in order to balance the groups.

Characteristics that are strongly associated with a condition present few problems: the association stands out from the data quite clearly. But when weaker associations are sought, quite elaborate standardisations of data may be a prerequisite to reaching an unambiguous answer. Although such complex and sometimes unexplained calculations are often made, there is no audit of the appropriateness of the standardisation of epidemiological data. A quite different approach is taken with respect to the licensing of new drugs and products[7]. Surely epidemiology should be subject to procedures equivalent to those of 'good laboratory practice' that apply to drugs and similar products. Although it would seem that epidemiology complies with the procedures normal for academic disciplines, with referees and editors evaluating papers for the validity of the techniques used, including standardisation of data, and the quality of the investigation, many studies using questionable techniques have been published, and their conclusions have in turn encouraged changes in social behaviour. As noted above, the direct evidence fails to establish any health-risk. The question here is whether the epidemiology shows that ETS is damaging to health?

The epidemiology of ETS

Lung cancer

There have been many epidemiological studies that have attempted to measure an association between exposure to ETS and lung cancer in non-smokers. With five notable exceptions, the 24 studies reviewed by the author failed to achieve a statistically significant association. The thoroughness of the procedures to obtain information varied with the different groups investigating this problem. The way the variation could have influenced the results obtained can best be appreciated by considering one careful study in detail.

The Garfinkel et al. *case-control study of 1985*

One particularly careful epidemiological case-control study, performed by Garfinkel *et al.*, in 1985, provides a clear sense of the inherent complexity of the design of this type of investigation. The basic design of the Garfinkel *et al.* study was to identify two populations of non-smoking patients, one with established lung cancer and the other a suitable control disease, and then to compare the two groups with respect to the smoking behaviour of their spouses. The presumption behind this comparison was that if the spouse smoked, the patient would have been exposed to ETS in the home, whereas less exposure would have occurred in an entirely non-smoking household. Realistically, only women can be used for the cases of non-smoking lung cancer because so many men smoke or have smoked in the past. The control group selected by Garfinkel *et al.*, women with cancers of the colon or rectum, was chosen on the ground that these two diseases represented conditions of similar severity, and because it has been established that colon–rectum cancers are not associated with cigarette smoking. In an effort to improve the precision of the study, three control patients were matched to each lung cancer case.

There was some difficulty in identifying the cases of lung cancer because of the rarity of this disease in non-smokers. The basic pool from which Garfinkel *et al.* drew their population consisted of 1175 cases of lung cancer in women reported in four large hospitals. The hospital records showed that 892 (76%) of the cases were smokers. When the histological evidence of lung cancer in these remaining 283 cases was reviewed by a pathologist on a blind basis (i.e. from the evidence on the slides, without any other clinical findings)[8], in 36 cases there was no histological evidence of lung cancer. Because the smoking behaviour of the patients in some of the remaining cases was not clear from the hospital records, a questionnaire was administered to the surviving patients, their spouses, or, if neither of these was available, to other ostensibly reliable respondents. After these review procedures, the initially identified group of 283 non-smokers with lung cancer was reduced to only 134 cases (i.e. to less than half the initially identified cases), which represented just three cases per hospital per year. Most of the women were elderly, more than half being over 70, and more than one-fifth over 80, at the time of diagnosis.

The smoking status of the husbands of the lung cancer and colon–rectum cancer control patients was then determined on a blind basis by seven interviewers using a structured questionnaire. If the patient was alive, information was obtained from her; otherwise, the patient's spouse or children were interviewed. For the one-quarter of the patients for whom personal or family responses were not available, another person who had known the patient for at least 25 years acted as informant. Because the majority of the patients were elderly (over 70), and their children, in their 40s, would have lived away from home for a number of years, the establishment of the smoking status of the patients' spouses is one of the least reliable sets of data subjected to analysis, even though the Garfinkel *et al.* study was a carefully executed attempt to address the question of the possible relationship between exposure to ETS and lung cancer.

The underlying assumption of the Garfinkel *et al.* study – that if a spouse smokes the subject will be exposed to ETS in the home, whereas in a marriage between non-smokers, no such exposure would occur – is recognised by epidemiologists to provide a poor measure of previous ETS exposure because virtually all people are exposed outside the home. For this reason, Garfinkel *et al.* attempted to measure any effects due to such exposure to ETS. The questionnaires were complex, and sought information relating to the extent of exposure to ETS in the home over the last 5 and 25 years,

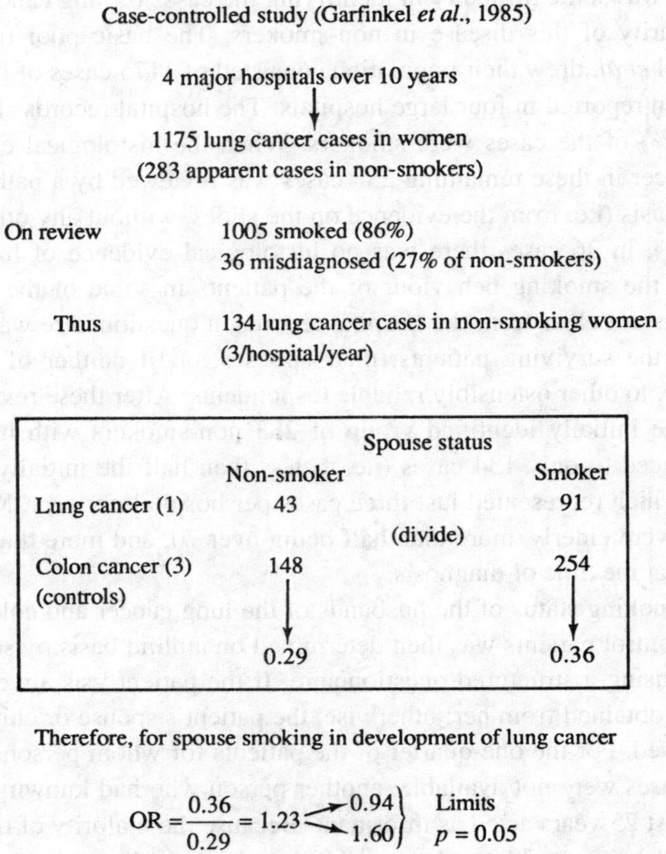

Case-controlled study (Garfinkel *et al.*, 1985)

4 major hospitals over 10 years

↓

1175 lung cancer cases in women
(283 apparent cases in non-smokers)

On review — 1005 smoked (86%)
36 misdiagnosed (27% of non-smokers)

Thus — 134 lung cancer cases in non-smoking women
(3/hospital/year)

	Spouse status	
	Non-smoker	Smoker
Lung cancer (1)	43	91
	(divide)	
Colon cancer (3) (controls)	148	254
	↓ 0.29	↓ 0.36

Therefore, for spouse smoking in development of lung cancer

$$\mathrm{OR} = \frac{0.36}{0.29} = 1.23 \begin{matrix} \nearrow 0.94 \\ \searrow 1.60 \end{matrix} \Big\} \text{ Limits } p = 0.05$$

Fig. 12.5. Design of a case-control epidemiological study of lung cancer and exposure to ETS. The exemplary case-control study of Garfinkel *et al.* (1985) is used in this illustration.

exposure outside the home over the same time periods, hours of exposure per day, and the like. But, as is the case with all such questionnaires, the reliability of the responses depends upon the memory of the respondent, or the knowledge and memory of the surrogate respondent. Thus, although great care was taken in establishing the smoking behaviour of each participant and in confirming the diagnosis of the disease, it was impossible to exert a comparable degree of control over the ETS exposure.

The effect of spousal smoking on the two diseases was obtained from a 2×2 contingency table (Fig.12.5), and was expressed as an odds ratio (OR), in this case 1.23, where the lung cancer rates of those with non-smoking spouses was taken as 1.0. Thus, exposure to ETS from the husband's smoking resulted in the non-smoking wife's having 1.23 times the risk of developing lung cancer relative to those with a non-smoking husband. However, the OR is of little importance itself, because of the inherent variability between people. This variability can be measured in this type of investigation, using standard statistical procedures. Following such calculations, it is possible to establish confidence limits for the OR, in this case at $P = 0.05$, 0.94–1.60[9]. This means that if such a test were repeated 100 times, the OR would lie between 0.94 and 1.60 in 95 instances. Because these confidence limits enclose the critical value for the OR of 1 (i.e. zero effect), there is no significant effect (at $P = 0.05$) of the husband's smoking on the development of the non-smoking wife's lung cancer.

Because of the different sub-groups considered, Garfinkel *et al.* made 88 comparisons, 7 of which yielded ORs significantly greater than 1.0 (P = 0.05), and 7 others an OR significantly less than 1.0 (P = 0.05). In other words, a significantly increased and decreased risk of developing cancer was shown in an equal number of situations. Of course, as the initial 134 cases of lung cancer are divided into different sub-categories the number of cases in each analysis is reduced, and because there are so few cases considered, each one exerts a greater influence, such that any errors introduced lead to greatly distorted ORs. Garfinkel *et al.* found, for example, a statistically significant increased OR for lung cancer in the sub-group where the husband was a heavy smoker. The evidence of spousal smoking behaviour for the majority of these patients was obtained, however, from a proxy respondent rather than from either the patients themselves or their wives, which reduces the reliability of the responses used in the analysis. Indeed, there was no evidence at all of a relationship between spousal smoking and lung cancer if only those cases are considered in which the patients themselves, or their spouses, supplied the information.

Misclassification: a possible source of error in epidemiological studies

The case-control study by Garfinkel *et al.* is a most revealing investigation, and can serve as a guide to the thoroughness of other epidemiological studies on ETS health risks. The three elements of critical evaluation found in it (confirmation of the diagnosis of lung cancer by a pathologist; care in detecting any smokers or ex-smokers misclassified as non-smokers; and some attempt to evaluate exposure to ETS beyond simply whether the husband smoked) testify to the study's thoroughness, and yet, as was discussed above, the quality of the data subject to analysis is uneven. The careful design of the Garfinkel *et al.* study provides some sense of the effect of misclassifying women smokers as non-smokers. Although the interviewers in the Garfinkel *et al.* study were experienced and conducted the interviews on a blind basis, it is not clear whether they could detect 'untruthful' answers[10]. The possibility remains that some of the study's 'non-smoking' wives with lung cancer were in fact smokers or ex-smokers. Furthermore, smokers tend to marry smokers, and non-smokers non-smokers, so such misclassification must be suspected as any smoking by an alleged 'non-smoker' would not be immediately detectable in a house that smelt of smoke and contained ashtrays, lighters and cigarettes.

When the initially wrongly classified cases (those that were detected in review) were included in the analysis, the OR for husband smoking was 1.61. When they were excluded, the OR fell to 1.23 (with an upper confidence limit of 1.60, which, of course, is less than 1.61). It is probable that some misclassification error remained despite stringent attempts to control this, given the fact that proxy respondents were used to obtain information for a quarter of the subjects. Had one been able to correct for this too, the OR would possibly have been reduced further. Clearly, the opportunity for misclassification of smokers or ex-smokers as non-smokers is an important potential weakness in all studies of this kind.

What evidence is there that smokers or ex-smokers are sometimes misclassified as non-smokers in case-control studies? First, it has been shown repeatedly that cigarette sales are greater than any estimates derived from smokers' admitted consumption. Secondly, there is persuasive evidence of the unreliability of memory based on responses to identically phrased questions about smoking behaviour that were posed on separate occasions. This problem has been reviewed by Lee[10] (1988), who reports disagreements between consecutive responses in various studies as 0–39.9%. When a marker for smoking (usually nicotine or cotinine levels in various body fluids) was measured in previously smoking subjects, the proportion of false claims was 0–38.8% in different investigations. Moreover, if questions of smoking behaviour were posed in a medical context (i.e. after a number of health-related questions), the proportion of denials of smoking by smokers or ex-smokers increased. Thus, although a questionnaire may be specially designed to detect the misclassification of smokers as non-smokers, it is probable that some degree of false-reporting occurs. Lee has also attempted to measure the influence of misclassification of smokers or ex-smokers as non-smokers on the results of epidemiological studies of the health effects of ETS, and has found that the bias is systematic, elevating the OR. A 5% level of misclassification could inflate the OR seen in the studies of ETS and lung cancer to completely explain the results for male populations and for western women. A higher level would be needed to explain levels for Asian women, but there is some anecdotal evidence that suggests that the misclassification level among Asian women is remarkably high.

Meta-analysis

The first suggestions that ETS may cause lung cancer came from two studies published in 1981: a retrospective case-control study done in Greece by Trichopoulos *et al.*, and a prospective study in Japan directed by Hirayama. Since then, there have been several similarly-designed studies, most of which have failed to detect a statistically significant increase in risk of lung cancer from exposure to ETS. A more recent publication by Wald *et al.* (1986), however, reporting ETS to pose a significant hazard, has proven to be quite important, at least with respect to its political influence in Britain. This report, entitled 'Does breathing other people's smoke cause lung cancer?', did not present new results, but instead combined those from previous epidemiological studies selected by Wald *et al.*, and then subjected the pooled results to a statistical evaluation called meta-analysis. Based on the increased number of lung cancer cases in non-smokers, Wald *et al.* reported a significant hazard from exposure to ETS. To evaluate the correctness or otherwise of the meta-analysis performed by Wald *et al.* it is necessary to consider critically the studies upon which it is based, focusing particular attention on the design of the studies, and any influences that may have confounded the results. In most of the case-control studies, the critical control measures that had been exercised by Garfinkel *et al.*[11] were either missing or perfunctory.

Wald *et al.* (1986) reviewed 13 published case-control studies that considered lung cancer risk from exposure to ETS, including that by Garfinkel *et al.* described above, and noted that: 'few of the studies have shown a significant risk of lung cancer associated with such smoke, but this cannot be taken as negative evidence because most of the studies were too small to detect the small excess risk that would be expected'. Wald *et al.* excluded three of the 13 studies from consideration either because data on the non-smokers were insufficient, or because there was no control group other than

Table 12.5 *The case-control studies of exposure to ETS and lung cancer considered by Wald* et al. *(1986)*

Author	Country	Cases	% conf	Controls	Proxy	Blind	OR	LL	UL
Chan 82	HK	84	82	H	No	No	0.75	0.43	1.31
Correa 83	USA	22	97	H	Yes	Yes	2.03	0.81	5.08
Trichopoulos 83	Greece	62	18	H	No	No	2.11	1.17	3.78
Buffler 84	USA	41	100	P	Yes	No	0.80	0.32	1.99
Kabat 84	USA	24	100	H	No	No	0.79	0.25	2.48
Garfinkel 85	USA	134	100	H	Yes	Yes	1.23	0.81	1.86
Akiba 86	Japan	94	53	P	Yes	No	1.48	0.87	2.52
Lee 86	England	32	?	H	No	No	1.03	0.41	2.58
Koo 85	HK	86	16	P	No	No	1.54	0.89	2.67
Pershagen 87	Sweden	67	99	R	Yes	No	1.27	0.75	2.18

The important paper by Wald *et al*. (1986) subjected ten acceptable (to them) case-control studies on the effects of ETS exposure (assessed by spouse smoking behaviour) and lung cancer in non-smokers to meta-analysis to obtain a 'global' OR of 1.27 (confidence limits, $P = 0.05$, 1.05–1.53). In the category of 'exposed to ETS', there were 402 lung cancer cases in women and only 23 in men, so only the data relating to women are included in Table 12.5, although Wald *et al*. included the male patients in his meta-analysis. In each case, only the first author and year of publication is used to identify the individual study: see the Reference section for full details.

Abbreviations used:

% conf: % lung cancer cases with histopathological confirmation.

Proxy: whether or not the smoking behaviour was determined in some cases from an informant other than the patient or her husband.

Blind: whether or not the interview that determined smoking behaviour was performed on a blind basis.

OR: odds ratio

LL: lower limit of the OR at $P = 0.05$ (significant effect if LL > 1.0)

UL: upper limit of OR

HK: Hong Kong

H: Hospital-based controls

P: population-based controls

R: registry-based controls

population estimates. Wald *et al.* then combined the remaining ten case-control studies he considered to be acceptable and determined the rate of lung cancer cases associated with spousal smoking, comparing this rate with that from the pooled controls (six deriving the controls from hospital patients, and four from the local population: Table 12.5). The meta-analysis 'weighted' each study, according to its estimate of risk. The relative risk (= OR) for lung cancer due to exposure to ETS was found to be 1.27 (fiducial limits, $P = 0.05$, 1.05–1.53); thus by increasing the population size, the authors found a statistically significant effect. Only one of the ten studies analysed (that by Trichopoulos *et al.*, 1981) showed a statistically significant OR for lung cancer in non-smokers exposed to ETS from their smoking spouse.

Trichopoulos *et al.*'s case-control study (1981 and re-reported in 1983) is worth considering in detail, not only because it was the first report on this subject and started the whole ETS lung cancer controversy, but also because its results contribute substantially to Wald *et al.*'s meta-analysis. The design of Trichopoulos *et al.*'s study was much like that of Garfinkel *et al.*, except that it was executed without the degree of care that characterised the American study. Indeed Uberla has described the Trichopoulos *et al.* work as '.... a textbook example for some of the errors which must be avoided if a case-control study is to be valid' Uberla (1987).

In more than one-third of the lung cancer cases Trichopoulos *et al.* considered, there was no histological confirmation of the diagnosis. Patients with adenocarcinoma (the most common histological type of lesion in female non-smokers with lung cancer) and alveolar carcinoma were omitted from the study (although how this was achieved without full histology is unclear), and the smoking behaviour of the cases and controls was determined by one doctor who knew the intention of the study. In addition, lung cancer cases came from three hospitals, whereas the controls (orthopedic disorders) were selected from another hospital, thus raising the possibility of a systematic bias. The cases and controls were not matched demographically, at least not in the first report, and no attempt was made to control for confounding factors, such as diet, and exposure to airborne pollution at work or in the home. Moreover, the ORs were calculated incorrectly in the 1983 paper: the quoted values for the non-smokers with a spouse who smoked up to 20 cigarettes per day was 2.4 (it should have been 1.95), and for more than 20 a day 3.4 (which should have been 2.54).

It is clear that many of the inadequacies of the Trichopoulos *et al.* and other individual studies have been incorporated in the meta-analysis. In only a minority of the ten individual studies was there satisfactory histopathological confirmation that the lung cancer cases were primary lesions and not metastatic from another site. Moreover, no attempt could be made to quantify exposure to ETS in the home or allow for exposure to ETS outside the home, and thus the measurements were more of the influence of an attribute than of a toxic substance with a dose–response relationship. With regard to the potential for misclassification of smokers and ex-smokers as non-smokers, although Wald *et al.* considered this, and made 'reasonable' allowances for the influence of such a confounder in their analysis. None the less, they assumed a lower level of misclassification than either Lee or Garfinkel *et al.* did in their studies. Wald *et al.* considered the increased risk of lung cancer run by a misclassified smoker to be eight times that of a non-smoker, as compared to the factor of 10 determined by Lee. If one were to apply Lee or Garfinkel *et al*'s. alternative allowances, the lower confidence limit of 1.05 calculated by Wald *et al.* would have been reduced perhaps to below 1.0, so no significant effect would have been detected.

Wald *et al.* also produced a second set of results based on three prospective epidemiological studies of the risk of lung cancer from ETS[12]. These results indicated that the risk ratio (RR) for the effect of spousal smoking on non-smoking partners was 1.44 (fiducial limits, P = 0.05, 1.20–1.72). Again, only one individual study, that by Hirayama (1981) describing results from Japan, demonstrated an RR significantly greater than 1.0. Hirayama's study was not designed to examine the influence of ETS on lung cancer in non-smokers, but rather was adapted for this purpose after the cohort was established and under observation for a number of years. The study is marked by inadequate pathological confirmation of lung cancer[13] and by the fact that cause of death was taken from death certificates, which is known to be an unreliable procedure.

Hirayama's reports also contain some unusual findings regarding the risk from ETS measured relative to that from smoking cigarettes. Generally, exposure to ETS is quoted as carrying the same risk as smoking 0.1–1 cigarettes per day; in other words, the OR is a small fraction of that associated with smoking. Hirayama's study found a much higher risk from ETS: '... the results clearly indicate that the effect of passive smoking is about one-half to one-third that of direct smoking in terms of mortality ratio or relative risk'. Curiously, the positive case-control report of Trichopoulos *et al.* echoes this quotation: '... strange that the relative risk associated with passive smoking in this study (2.4 for all categories of smokers combined) is only slightly lower than the figure of 2.9 associated with active smoking by the women themselves'. Both sets of findings should be considered in the light of Garfinkel *et al*'s case-control study in which pathologically-confirmed lung cancer in the absence of smoking was so difficult to find. Whereas Garfinkel *et al.* found most of their cases from the elderly, half of Hirayama's non-smoking lung cancer cases occurred in patients under 60 years old. A distorting influence in Hirayama's study may have been the unrepresentative nature of his population. Hirayama decided to relate lung cancer in the non-smoking wives to the age of their husbands, rather than to that of the patients. Moreover, the composition of his cohort, particularly its age profile, provides evidence of inhomogeneity compared with that of Japan's population as a whole (Schneider, 1989)[14].

Other epidemiologists have questioned Hirayama's findings. Uberla (1987) has shown that if one excludes the women married to industrial workers from the comparison, the relative risk due to exposure to ETS is no longer statistically significant in Hirayama's cohort. Schneider has attempted to calculate values 'corrected' for inhomogeneity of the cohort with respect to age; these values reduce the RR, but he concludes that '... the arbitrariness and relativity of an overall risk measure cannot be avoided...'. Unrepresentative populations giving rise to unusual findings may also be the explanation behind the curious finding of Koo *et al.* (1985), who showed that lung cancer in Hong Kong Chinese smokers was first detected in patients at a greater age (64.8 years) than was found for never-smokers (57.8 years, $P<0.001$). One is left with the distinct impression that there must be some, as yet, unidentified factor that is highly correlated with the lung cancer in non-smokers in Japan and China.

Validity of meta-analysis

On one level, the validity of Wald *et al.*'s meta-analysis must depend upon the validity of combining data relating to different ethnic groups. If one includes the three prospective studies with the ten case-control ones considered by Wald *et al.*, five of the thirteen related to the United States, four to Europe, and four to Asia. Yet Asia is distinct, environmentally and otherwise. For example, it is likely that in Asia the relative smallness of living rooms results in higher concentration of ETS than would be found in the West. The fumes coming from open fires and kerosene stoves that are often used for cooking in Asia may also have some effect. There are also certain still-unidentified factors relating to lung cancer in China. For example, three-quarters of lung cancer deaths in Shanghai, and two-thirds in Hong Kong, are not associated with smoking. Although only 1 in 100 women is reputed to smoke in Xuan Wei county of Yunnan Province, the rate of lung cancer in women is one of the highest in all China. Koo *et al.* (1983) examined lung cancer incidence in non-smoking Hong Kong Chinese, and added further doubt about the representative nature of findings from these races. The Danjia/Helao people live on open-decked boats, spending most of their time free of indoor pollution. As the boats are small and made of wood, the husbands generally smoke out on the deck, so as not to run a high risk of fire in the confined cabin. Yet this population yielded the highest RR for female lung cancer of all the ethnic groups in Hong Kong. It is unlikely that ETS was responsible for these cancers.

In any statistical analysis, the quality of the data used determines the reliability of the conclusion reached. As can be seen in summary Table (12.5), evidence of smoking behaviour was obtained only from the response to questions asked of those in the trial, leaving open the possibility of misclassification of smokers and ex-smokers as non-smokers. The majority of studies also included no histopathological confirmation that the cases involved primary lung cancer. If careful work eliminated more than half the ostensibly-usable cases in the Garfinkel *et al.* study, one must question the reliability from meta-analysis in the Wald *et al.* study, which included studies where this degree of care was the exception rather than the rule[15].

Is ETS associated with lung cancer in non-smokers?

The discussion above leaves one highly doubtful that ETS has been shown by the available studies to be associated with lung cancer. Even the better studies have flaws, and the vast majority are so flawed that they cannot be relied upon with any confidence. Just one problem, misclassification, may account for the results of many of the studies. And even with no explanation, fewer than 20% of the studies with elevated ORs have results that are statistically significant. The data set is not impressive, and this author would conclude that the association has not been established.

As this book goes to press, the United States Environmental Protection Agency is considering a draft risk assessment that concludes that ETS causes some 3700 lung cancer deaths in the United States each year. The assessment is based on a meta-analysis of the available studies. This meta-analysis has been heavily criticised for many of the same reasons that the Wald *et al.* study has been criticised in this chapter. But some interesting new analyses of the data have been undertaken. For example, if only US studies are considered, meta-analysis yields a non-significant OR close to one. Asian results are higher, and the reason for the difference is unclear. But if results are so markedly different for different ethnic groups it seems clear that such groups should be analysed separately.

Another interesting point concerns the reliance on data from studies of non-smoking women in the home to justify public smoking restrictions. A better set of data would be those from ETS exposure at work. Of the seven studies with eleven endpoints, six have ORs below one and four are above one. Of these last four only one is statistically significant (a second study by the same authors found an OR below one). One wonders if these data do not more closely reflect the correct picture, for among other possible explanations, the problem of misclassification might be less among working populations.

Cancer at sites other than the lung

The great difficulty in determining any supposed increased risk of lung cancer from ETS has been considered in detail above. When one considers the risk of ETS-induced cancer at other sites, the quality of information is even less acceptable. Most of the precautions in design necessary for good epidemiological studies are missing from those studies that attempt to correlate exposure to ETS and cancer at other sites. The manner in which such studies have determined exposure of non-smokers to ETS is irregular, and there has been little control for such confounding factors as exposure to industrial chemicals, consumption of alcohol, diet, and sexual behaviour, all of which have been implicated by various authors as possible cancer-causing events for different sites in the body. Moreover, precautions necessary to avoid misclassification of smokers or ex-smokers as non-smokers in case-control studies have generally been absent. Accordingly, little can be concluded from such studies.

Irritation and annoyance

While the epidemiological evidence demonstrating serious health risks from exposure to ETS is flawed and sketchy at best, in certain situations some individuals will experience irritation and become annoyed by accumulated ETS. A proportion of the public opposition to exposure to ETS stems from its potential for causing such irritation and annoyance.

If smoking in a room has been heavy, some sensitive individuals react with bronchospasm and irritation of the eyes and mucous membranes. The odour and attendant annoyance at the presence of ETS is caused by components of the gas phase, whereas irritation arises from particles. An indoor air filtration system can remove the particles, but will not change the gas phase. The introduction of fresh air is necessary to dilute the persistent smell. The latter has, from a practical standpoint, been made more difficult to achieve by the tendency since the oil crisis in 1974 to design buildings and their services with emphasis on conservation of heat. Even with the use of heat exchangers, in many cases not enough fresh air enters buildings to dilute the smell of ETS.

Irritation can be measured as the frequency of blinking of the eyes, whereas odour depends upon the subjective reaction of individuals. Both odour and irritation contribute to the annoyance of occupants of buildings. Once annoyance has been established, it persists even though the concentration of ETS declines. Experiments have been performed on volunteers in exposure chambers to determine the characteristics of annoyance, and these have shown that the relative humidity is a major factor in an individual's reaction to ETS. Short-term exposure to ETS at a relative humidity of 65% resulted in less of a perception of both odour and irritation than was the case at 30% relative humidity. In addition, research in normal buildings has found that the kind of activity occupants are involved in during the course of exposure of ETS will be an important influence on the degree of annoyance experienced. A greater level of ETS is tolerated in a pleasurable, mind-preoccupying situation (e.g. playing cards for money) than if the time is spent on mundane, boring tasks (like work!).

Pulmonary function

Another alleged negative health effect of exposure to ETS is reduced pulmonary function. In a simplistic way, the physiological processes of our bodies can be viewed as consumption of nutrients by oxidation with subsequent utilisation of the resultant energy. Without a supply of oxygen, life can continue for only a short time before irreversible damage occurs. The net input of oxygen depends upon a functioning lung, so any disease that lowers pulmonary function will reduce the ability of organs to respond to any increased need for oxygen. The normal intake of air is 6–7 litres/min in the unstressed state, which can be increased to 100 litres/min or more at will, and thus humans have about a 15-fold capacity to increase ventilation. This large ventilatory reserve means that individuals can vary considerably in their rate and depth of breathing without any effect on normal bodily functions. In order to see a reduction in pulmonary capacity due to disease, the system has to be assessed at or near its maximal capacity[16]. Reduced maximal ventilatory capacity may be manifest as intolerance to exercise (lack of fitness), but the problem may lie with some other vital system, for example, the heart or circulatory system.

One difficulty in investigating whether chronic exposure to ETS damages the lungs and how FEV_1 is affected by ETS concerns the difference between acute and chronic effects. Whereas a subject can be assessed before and after exposure to a definite level of ETS to measure acute effects, no such simple comparison is available when considering chronic effects. Probably all individuals have suffered some exposure to ETS, and thus the study design of such chronic exposure depends upon determinations of lung function in those exposed to high levels relative to others at lower levels of ETS.

Acute experiments on volunteers in exposure chambers produced small reductions (3–5%) in FEV_1, even though there was no change in vital capacity. These tests were also characterised by eye irritation, often occurring before the decrease in FEV_1. Because the subjects were aware of irritation due to ETS, it is not clear how much of the small effect on FEV_1 represented a subjective response. Unfortunately, it is not possible to perform this type of experiment on a blind basis.

Similar experiments with asthmatic subjects have produced conflicting results, varying from no reduction in FEV_1 to a 21% reduction. Another way of measuring acute respiratory function is to cause a bronchoconstriction by inhalation of a drug (e.g. histamine), and observe the effect on FEV_1. When asthmatics were subjected to such procedures, exposure to ETS led to a lower dose of histamine being needed to cause a 20% reduction in FEV_1. This result may indicate that the lungs of asthmatics are more sensitive than are those of normal people to any bronchoconstrictor substances in ETS, but it is in any case well known that asthmatics are extremely sensitive to bronchoconstictor substances such as histamine. These experiments could not be performed on a blind basis, and thus anxiety, which exacerbates reactions in some asthmatics, may have contributed to the effect. For these reasons, the relevance of the above findings is not immediately obvious, although there could be a problem of a greater than normal sensitivity of asthmatics to ETS.

The chronic effects of ETS exposure on pulmonary function have also been investigated in a number of studies. Problems in estimating ETS exposure are inherent in the design of the studies (which are similar to those described above with respect to lung cancer), and exposure has usually been determined on the basis of spousal smoking behaviour. There is, accordingly, the now familiar possibility of confounding the result by the misclassification of smokers or ex-smokers as non-smokers. However, more subjects have been available for assessments of lung function than was the case for lung cancer, the latter being a rare disease in non-smokers.

FEV_1 has been measured as an index of lung function in cross-sectional studies, which assemble three categories of subject: smokers and ex-smokers, non-smokers living with smokers, and non-smokers living with non-smokers (who are thus considered to be free from ETS exposure in the home). In a French co-operative study of this kind, 7800 adults were examined and 1675 were found to be non-smokers. There was evidence of a lower lung function in both men and women from living with a smoker. However, the differences detected were small and inconsistent with respect to age[17]. Other factors, such as social class, level of educational achievement, exposure to air pollution other than ETS, and the like did not provide an explanation for the small effect seen.

The results in the French study have been confirmed generally in studies from other countries, and the total suggests that ETS in the home is associated with a small reduction in lung function relative to those not so exposed. The medical importance of this small reduction in lung function is not known, however, since, in general, the results were all in the 'normal' range. In any case the magnitude of the effect is much less than the reduction seen in smokers, and in all studies there remains the problem of misclassification of smokers and ex-smokers as non-smokers.

Acute respiratory illness does not appear to be a by-product of exposure to ETS. The Surgeon General has stated that there are no published reports describing the effect of exposure to ETS and acute respiratory illness in adults and attempts to correlate such exposure to chronic signs of respiratory disease (cough, wheezing and production of phlegm) using spousal smoking as a surrogate for exposure to ETS have failed to show any relationship. Any effect, if indeed there is one at all, is small, and could be due to undetected previous smoking by a small proportion of the subjects. In summary then, there are no important medical implications with respect to ETS and chronic pulmonary disease in adults.

Respiratory effects in children

There are many reports of the effects of ETS exposure on the respiratory systems of children. With young children (under about 8 years old) there probably has been no contribution to any effects seen from undeclared smoking by the children. The possibility of such confounding increases as the children studied grow older, because of a general tendency to deny such smoking in the age-group from about 8 years of age upwards[18]. Accordingly, any longitudinal study that follows the physical development of a cohort of children may be subject to a systematic bias in its later stages. An advantage of conducting studies on children, however, is that there is no workplace exposure to airborne substances that could confound the attempts to measure the consequences of exposure to ETS.

Most of the studies on the effects of ETS exposure on children have been carried out in the United States. Studies in this area have consistently found a small decrease in FEV_1, usually of the order of 0.5% per year. Maternal smoking has been found to be a major influence, perhaps as one would expect since young children generally have somewhat less contact with their fathers. One study from the Peoples' Republic of China, however, runs counter to this trend. Chen and Li (1986) studied 8–16 year old children of both sexes and found a reduction in FEV_1 that was correlated with paternal smoking behaviour (none of the mothers in this study was a smoker). This study should be considered with caution, because the relationship between smoking and lung disease in China appears to be quite different from that in the West (see above: the section on lung cancer).

The mechanism by which the small decrease in children's FEV_1 is brought about is not known. The simplistic conclusion that exposure to ETS impairs the growth of lungs should not be accepted without consideration of the rate of respiratory infections (see below), variations in diet, and other potentially confounding influences. Also, any effect could possibly be due to exposure *in utero* to maternal active smoking rather than exposure to ETS after birth. Whatever the mechanism, the observed effect is small, disappears by the time maturity is reached, and is unlikely to have any adverse effect on health.

There are several reports that the incidence of respiratory tract infections in the children of smokers is higher than that found in children from non-smoking households. This has not been a consistent result, however, and other studies have failed to find such an effect. Very real problems exist in determining the number of respiratory tract infections in children. For example, because smoking parents are more likely to have respiratory tract infections than are non-smoking ones, any effect seen in the children may simply reflect infections acquired from the parents. Moreover, there is strong potential for both under- and over-reporting because no consistent standard has been adopted by parents in reporting the signs and symptoms of their children's respiratory tract infections. It is perhaps important too that children had fewer signs and symptoms when the father, as opposed to the mother, completed the questionnaire. In all research on the long-term health effects of ETS, the problem of inherently subjective responses from patients or subjects with respect to outcome and exposure makes it difficult, if not impossible, to reach unambiguous answers. Nowhere is this more true than in measuring the rate of respiratory tract infections in children. One is left with a feeling of nihilism: clear-cut answers are not available.

Another childhood ailment that some have associated with exposure to ETS is infection of the ear, which at times exhibits effusions. Such conditions can be extremely painful, and in many cases, there is a close correlation between infections of the ear and upper respiratory tract, with both viruses and bacteria being the possible pathogenic organisms. Said *et al.* (1978) were able to relate candidates for tonsilectomy or adenoidectomy (often prescribed in an attempt to remove the site of infection) with parental smoking in the home. In a more extensive investigation, Iverson *et al.* (1985) attempted to associate middle ear effusions in children with various environmental influences, finding that parental smoking was the only significant factor (OR = 1.6, confidence limits at $P = 0.05$, 1.0–2.6). Similar positive correlations have been reported in other studies. All such studies used parental smoking as a surrogate for exposure to ETS, and determined exposure from questionnaires administered on a non-blind basis, which renders the data somewhat suspect. Moreover, in view of the fact that smokers have more respiratory tract infections than do non-smokers, and can in turn infect their children, the statistically significant elevated rate of ear infections in the children of smokers may simply reflect the passing on of respiratory infection from parent to child. Accordingly, the data from these studies are subject to the same limitations that apply to that of upper respiratory tract infections considered above.

Cardiovascular disease

Diseases of the cardiovascular system represent the major cause of premature death in the Western world. Any disturbance in the circulation can lead to damage to a variety of organs[19]. In most cases, death is caused either by heart failure or by hypoxic destruction of part of some other vital organ, such as the brain. In addition to cases where sudden death is the outcome, cardiovascular disease exerts a high cost with respect to increased morbidity of the population at large.

Cigarette smoking is associated with increased incidence of coronary heat disease: 40 cigarettes per day increases the RR for this condition between two- and three-fold. It is generally held that unlike the case with lung cancer, where the increased risk persists for a long time after ceasing to smoke, the increased chance of getting coronary heart disease returns halfway back to the non-smoking level within a year of giving up the habit (for an alternative view see Cook & Shaper, 1989). This suggests that the chemical insult contained in cigarette smoke differs for the two diseases. Those components of smoke suspected of being related to cardiovascular morbidity are nicotine, carbon monoxide, and nitrogen oxides. Each of these substances can be shown to exert marked pharmacological effects on blood vessels and the heart. However suggestive this may be of a toxicological mechanism, the cause of cardiovascular morbidity in man remains unknown.

Any assessment of the effects of ETS on the cardiovascular system must distinguish between healthy subjects and those with some degree of established cardiovascular disease. In healthy volunteers exposed to ETS, the cardiovascular system experiences no change in normal reactivity at rest or during the stress of exercise, even though there is evidence of uptake of components of smoke. In subjects with pre-existing cardiovascular disease, there seems to be little reaction to ETS, except in patients with angina pectoris[20]. Exposing these sensitive, anginal, patients to ETS led to an increase in arterial blood pressure and heart rate, indicating a more rapid than usual onset of angina. This result could be anticipated because the combination of carbon monoxide with the haemoglobin of the blood to form carboxyhaemoglobin would normally reduce the capacity of the patient to transfer oxygen from the circulation to the myocardium. As with lung function in growing children, ETS can only be seen to exert an effect when the system under consideration is performing at its maximal capacity due to stress. In angina, it is the pathology of the disease that reduces the normal overcapacity of the system to such an extent that signs and symptoms are seen. Yet, because the effect of ETS in tests on anginal patients was not determined on a blind basis (it is not possible to hide the odour), anxiety, which can also initiate anginal attacks, may have constituted a subjective component of the response.

What are the long-term effects of ETS with respect to development of cardiovascular diseases? Again, investigators of this question use the study design employed elsewhere to determine the RR for lung cancer, namely spousal smoking behaviour as a surrogate for long term exposure to ETS. There are reports from several studies, but each is flawed in some way (see Weetman Munby (1990) Hulka (1986)). For example, Garland *et al.* (1985) found an elevated risk of death from cardiovascular disease in non-smoking women as a result of the smoking of their husbands (RR = 2.7), but in that study the unexposed non-smoking group contained only two women who died from heart disease. Furthermore, the husbands who smoked were grouped with ex-smokers, which seems inappropriate. Moreover, the statistical analyses used in the study were probably inappropriate given the small size of the populations considered.

Much the same criticism can be applied to the reports from Sweden (Svendsen *et al.*, 1985), which found a two-fold (but not statistically significant) increased risk with respect to coronary heart disease. The only study yielding a statistically significant RR for ischaemic heart disease (1.3, $P < 0.01$) in non-smoking women from the smoking behaviour (more than 19 cigarettes per day) of their husbands is that by Hirayama (1984). When Hirayama first reported the results from his cohort study in 1981, he found no association between spousal smoking and coronary heart disease. Later, in 1984 (i.e. after 17 years as opposed to 14), he found a significant association. This change implies an implausibly strong association in the last 3 years of the follow-up. The serious flaws in the methodology and design of this Japanese prospective epidemiological study described above in relation to lung cancer studies apply equally to the results relating to cardiovascular disease.

Although several attempts have been made to evaluate increased risks of cardiovascular diseases from exposure to ETS, compelling evidence of an association between these factors is lacking. This is not entirely surprising, because epidemiological techniques are not particularly suited to this task for at least two reasons. First, the diseases are extremely common, and associating a factor with a disease is most easily achieved when only a few people are affected. Secondly, much evidence indicates that cardiovascular diseases are multifactorial, i.e. can be caused by several risk factors (e.g. diet, heredity, and the like) acting together.

Effects of ETS on the foetus

A final area of concern to the public at large is whether exposure to ETS in pregnancy has any effect on the developing foetus. According to the fourth report of the Independent Scientific Committee on Smoking and Health (Froggatt, 1988), maternal cigarette smoking may produce three well-established effects on the new-born, namely, (1) reduced birth weight; (2) increased perinatal mortality; and (3) retarded physical and mental development of the child in early life. The same report also states that a pregnant woman's exposure to ETS as a result of her husband's smoking exerts deleterious effects *in utero*. Certain studies have suggested birth weight was reduced by 24 g (i.e. by at least 10% of the effect associated with maternal smoking) and that the baby ran an increased risk of being of low birth weight (< 2500 g).

Not all studies support these conclusions, however, and difficulty in controlling confounding factors could account for the different findings. Perhaps the clearest study comes from Shanghai in China, where women smoke very little and so the risk of misclassification is low. In that study, Chen *et al.* (1989) were able to assess the effects of paternal smoking (as a surrogate for ETS exposure) on mean birthweight and on frequency of babies with a low birthweight for 1058 babies born to non-smoking mothers. In neither comparison was there a statistically significant effect of paternal smoking, even after allowing for possible confounders by multiple linear and logistic regression analyses. The extent to which the results of the Chinese study are applicable to Western populations is not clear.

In another study, this time from the USA (Martin & Bracken, 1986), nearly 4000 births were considered, and a reduced birth weight (by 75 g on average) and a 2.7-fold increased risk of a baby being born with a low birth weight (< 2500 g) were observed. As this was a prospective study, it should have been less susceptible to 'recall bias' and thus to possible confounding influences than many of the retrospective studies have been. It was possible in this study to attempt to determine the magnitude of the various influences by performing a multiple regression analysis. Of 15 factors considered, a high percentage of the reduction in babies' birth weight was accounted for by gestational age, whether or not the child was the mother's first, and the mother's ethnic group (whether she was non-white). After multiple regression analysis, the effect of exposure to ETS from the father's smoking ceased to be statistically significant.

The interpretation of published findings on the effects of ETS on the unborn presents many problems. First, a foetus will receive the chemical insult from the effect of maternal exposure to ETS via the same route that it receives maternal active smoking, viz by placental transfer. There may be significant confounding factors operating, such as consumption of alcohol and caffeine (in beverages including tea and coffee), which have been claimed to produce the same three effects attributed above to active smoking. These factors were considered in the Shanghai study. Finally, it is also possible that any such effects could be the result of an action on the father's sperm caused by active smoking, rather than an effect of ETS.

The studies of the effects of ETS on the foetus are accordingly complex to analyse. In foetal development, as in the development of cardiovascular disease, multifactorial influences can lead to a single effect. Researchers simply cannot be sure that a single influence, such as exposure to ETS, is responsible for a given medical outcome. While it is clear that pregnant women are ill advised to smoke, it is not clearly established whether pregnant women should avoid the smoke of others. Moreover, publicising advice to avoid ETS during pregnancy does have one clear effect: to induce anxiety and guilt in mothers-to-be whenever they encounter ETS. Is it possible that such anxiety is itself harmful to the foetus?

Conclusion

Are the conclusions of some that ETS is damaging to health correct? It is clear that in some cases conclusions have been reached (especially by expert committees) that are not supported by the data in original published studies. This is particularly true for an alleged association between ETS and lung cancer in non-smokers. The evidence, at its strongest, is merely suggestive of an association, not a casual connection. Yet such evidence has been given far more weight than is appropriate, largely because of questionable analogies to smoking. The reasoning goes as follows: carcinogens and co-carcinogens are present in tobacco smoke and persist in ETS, and thus non-smokers must experience some level of exposure to these potentially toxic chemicals. Therefore, ETS must exert some adverse health effect, albeit to a lesser degree than smoking. Yet this analogy to active smoking is quite inexact. The mixture of chemicals and the route of absorption of ETS are different from those occurring in cigarette smokers. Lung cancer in non-smokers is a rare disease (10 per 100 000 per year) and follows a different pathological sequence from that seen in cigarette smokers. Accordingly, it is hazardous to draw conclusions about ETS and lung cancer from the evidence of health effects in cigarette smokers.

Close examination of the available evidence in this chapter has identified the following conditions as possibly associated with ETS:

1. Irritation of the eyes, and to a lesser extent of the mucus membranes of the nose and mouth.
2. Bronchiospasm in asthmatics.
3. Slowed incidence of respiratory tract development and increased incidence of ear infections in children.

These effects sometimes manifest themselves only in sensitive individuals or persons with pre-existing conditions. Moreover, as can be seen from the discussion above, it is not at all clear that these effects are caused by exposure to ETS.

Notes

1. Other vegetable matter produces a similar spectrum of products from combustion, with only nicotine and its derivatives being peculiar to tobacco.
2. Not all the individual compounds persist in smoke for the same amount of time; some can be detected for only a few milliseconds after combustion, whilst others can be found several hours later.
3. In MS nicotine is distributed such that nearly all of it is in the particulate phase adsorbed on the surface of the particles. This further emphasises the difference between ETS and MS.
4. There is considerable person-to-person variation in cotinine half-lives, with the highest values usually being found in non-smokers. It has recently been proposed that solanesol (Fig.12.2) is a better surrogate for ETS as it is specific to tobacco smoke and is found wholly in the particulate phase.

5. Of course, a smoker is a passive smoker too, who receives from his own cigarette the highest possible dose of ETS. He receives it, however, in the same manner as the passive smoker.
6. It is most unlikely that any such characteristic will be specific to those affected; were it so, associating the characteristic with the disease would be simplicity itself, and no epidemiological study would be necessary.
7. In the case of new drugs and agrochemicals, for example, all data have to be available to the licensing authorities, and the difficult, possibly atypical, results are considered by a process of open argument between the scientists and the regulators. Furthermore, the experiments of the animal scientists attempting to develop and market a new product are subject to unannounced audits. All the animals and records can be checked in mid-experiment, and the full records must be available at the end of the study.
8. Confirmation of lung cancer by a pathologist examining slides on a blind basis represents very well-established, probably almost error-free, data.
9. P is the probability of obtaining this result by chance: when $P>0.05$ in most biological and medical investigations, it is normal to reject the null hypothesis that the results arose by chance, and accept the alternative hypothesis that some factor is responsible for the results obtained.
10. It is now so widely believed that smoking has an adverse effect on health that most smokers feel guilty about their smoking behaviour and about how they may have damaged their health. Often people try to reduce the embarrassment of potentially damaging the health of themselves or others by claiming to smoke less than is the case, or by having a 'lapse in memory' concerning periods of smoking.
11. i.e. confirmation of lung cancer diagnosis by a pathologist on a blind basis; care in avoiding misclassification of current or ex-smokers as non-smokers; and structured attempts to measure and evaluate exposure to ETS beyond simple reference to whether a spouse or family member smoked.
12. The design of these studies was typical: a cohort of subjects was identified, assessed for social behaviour, and followed for a number of years, and the health outcome recorded. Subsequent analysis then associated spousal smoking with a lung cancer in non-smokers.
13. Indeed, it was estimated in one sub-sample of cases that only 10.5% of the diagnoses were examined by a pathologist, and then not on a blind basis.
14. In Hirayama's study, only 2% of the cohort was over 70, whereas in Japan this age-group was represented in a much higher proportion of the population (12%).
15. The Wald *et al.* meta-analysis of ten case-control and three prospective epidemiological studies derived a global OR of 1.35 (fiducial limits, $P = 0.05$, 1.19–1.54) using a well-known statistical method for calculating the OR and its confidence limits developed by Mantel (1987). Yet Mantel considers his technique unsuitable for low risk situations (i.e. for ORs less than 2.0).
16. Tests of ventilatory function most often involve having a subject inspire and expire maximally into a recording spirometer, which measures the volume of air against time. Several values indicative of ventilatory function can be determined in such a test (Fig. 12.6). Normally, more than 80% of the vital capacity (volume of air expelled after maximal inspiration and expiration) can be exhaled in 1 s; this is called FEV_1 which is the forced expiratory volume in 1 s. FEV_1 depends upon sex (higher in males), age (decreases with age after maturity), body size usually indicated by height, ethnic origin

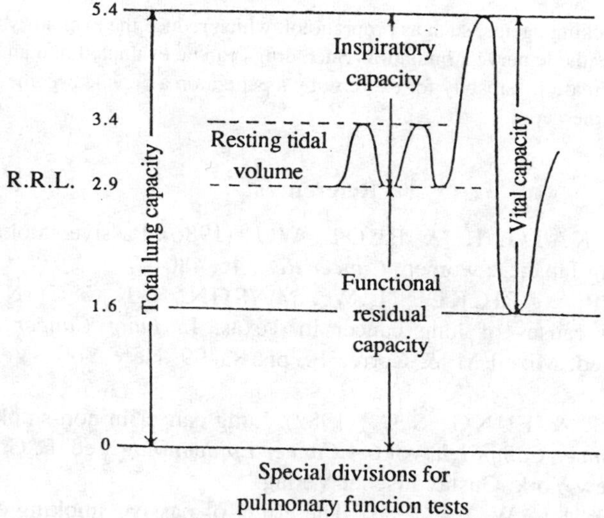

Fig. 12.6. Assessment of respiratory function. The subject co-operates by breathing into a recording spirometer, which measures volume of air against time. After a maximal inspiration, the air is expelled from the lungs as rapidly as possible. Vital capacity is the volume of this maximal expiratory effort, and FEV_1 is the volume of air expelled in 1 second.

and, of course, the presence of any respiratory disease. For a healthy Caucasian male of 35–49 years of age the mean value of FEV_1 is 3.4 litres (range 2.1–4.1).

17. Passive smokers were worse off at ages 30–34, 40–44, 45–49 and 50–54, but were better off at ages 25–29, 35–39 and 55–59.
18. Smoking in children is generally disapproved of, and thus smoking by relatively young children is frequently denied.
19. Perhaps the best way to view the inter-relationship between the different diseases is to envisage a mosaic of interdependent physiological parts, where disturbances of one element of the system will relate to others, e.g. a reduction in the effective perfusion of the brain with blood due to fatty deposits on the intima (linner lining) of the blood vessels will result in mental confusion. Essential hypertension (elevated arterial blood pressure of unknown cause) does not kill so many itself, but is the cause of much heart failure (due to the continual increased load on the heart) and cerebral infarction (hypoxic death of brain cells), following disturbances in blood flow.
20. Angina pectoris is a blanket term covering a syndrome having more than one pathological mechanism. It is characterised by a feeling of pressure and pain in the chest and (sometimes down the left arm), the pain being prominent during exercise or anxiety. In all cases of angina the fundamental problem is believed to be an inability to transfer enough oxygen from the coronary circulation to the heart muscle, this becoming especially apparent under conditions where there is increased oxygen demand, such as in exercise. The physiological process that increases the coronary blood flow to meet the increased oxygen requirements of the heart in exercise is an increase in sympathetic nerve activity, a process also activated in anxiety; an attack of angina can also be initiated by anxiety. Prominent amongst the drugs to treat angina pectoris are the beta-

adrenoceptor blocking agents, such as propranolol, which reduce the excitatory response of the heart to sympathetic nerve stimulation. Such drugs can be evaluated in man by measuring an increase in capacity for exercise by a patient on a bicycle ergometer before pain prevents further work.

References

AKIBA, S., KATO, H. & BLOT, W.J. (1986). Passive smoking and lung cancer among Japanese women. *Cancer Res.*, **46**, 4804–7.

BUFFLER, P.A., PICKLE, L.W., MASON, T.J. & CONTANT, C. (1984). The causes of lung cancer in Texas. In *Lung Cancer Causes and Prevention*, ed. Mizell M. & Correa P., pp. 83–99. New York: Verlag-Chemie Int. Inc.

CHAN, W.C. & FUNG, S.C. (1982). Lung cancer in non-smokers in Hong Kong. In *Cancer Campaign*. vol 6 – *Cancer Epidemiology*, ed. E. Grundman, pp. 199–202. New York: Gustav Fischer Verlag.

CHEN, Y. & LI, W.X. (1986). The effect of passive smoking on children's pulmonary function in Shanghai. *American Journal of Public Health*, **76**, 515–18.

CHEN, Y., PEDERSON, L.L. & LEFCOE, N.M. (1989). Passive smoking and low birthweight. *Lancet*, **ii**, 54–5.

COOK, D.G. & SHAPER, A.G. (1989). Stopping smoking and risk of ischaemic heart disease. *Lancet*, **i**, 895.

CORREA, P., PICKLE, L.W., FONTAM, E., LIN, Y. & HAENSZEL, W. (1983). Passive smoking and lung cancer. *Lancet*, **ii**, 595–7.

FROGGATT, P. (Chairman) (1988). *Fourth Report of the Independent Scientific Committee on Smoking and Health*. London: Her Majesty's Stationery Office.

GARFINKEL, L., AUERBACH, O. & JOUBERT, L. (1985). Involuntary smoking and lung cancer: a case-control study. *Journal of the National Cancer Institute*, **75**, 463–9.

GARLAND, C., BARRETT-CONNOR, E., SUAREZ, L., CRIQUI, M.H. & WINGARD, D.L. (1985). Effects of passive smoking on ischemic heart disease mortality of non-smokers: a prospective study. *American Journal of Epidemiology*, **121**, 645–50.

HIRAYAMA, T. (1981). Non-smoking wives of heavy smokers have a higher risk of lung cancer: a study from Japan. *British Medical Journal*, **282**, 183–5.

HIRAYAMA, T. (1984). Cancer mortality in non-smoking women with smoking husbands based on a large-scale cohort study in Japan. *Preventive Medicine*, **13**, 680–90.

HIRAYAMA, T. (1985). Passive smoking – a new target of epidemiology. *Tokai Journal of Experimental and Clinical Medicine*, **10**, 287–93.

HULKA, B.S. (Chairman) (1986). *Environmental Tobacco Smoke. Measuring Exposures and Assessing Health Effects*. Washington: National Academy Press.

IVERSON, M., BIRCH, L., LUNDQVIST, G.R. & ELBROND, O.

(1985). Middle ear effusion in children and indoor environment: an epidemiological study. *Archives Environmental Health*, **40**, 74–9.

KABAT, G.C. & WYNDER, E.L. (1984). Lung cancer in non-smokers. *Cancer*, **53**, 1214–21.

KOO, L.C., HO, J.H-C. & LEE, N. (1985). An analysis of some risk factors for lung cancer in Hong Kong. *International Journal of Cancer*, **35**, 149–55.

KOO, L.C., HO, J.H-C. & SAW, D. (1983). Active and passive smoking among female lung cancer patients and controls in Hong Kong. *Journal of Experimental Cancer Research*, **4**, 367–75.

LEE, P.N. (1988). *Misclassification of Smoking Habits and Passive Smoking: a Review of the Evidence*. Berlin: Springer-Verlag.

LEE, P.N., CHAMBERLAIN, J. & ALDERSON, M.R. (1986). Relationship of passive smoking to risk of lung cancer and other smoking-related diseases. *British Journal of Cancer*, **54**, 97–105.

MANTEL, N. (1987). Lung cancer and passive smoking. *British Medical Journal*, **294**, 440.

MARTIN, T.R. & BRACKEN, M.B. (1986). Association of low birth weight with passive smoke exposure in pregnancy. *American Journal of Epidemiology*, **124**, 633–42.

O'NEILL, I.K., BRUNNEMANN, K.D., DODET, B. & HOFFMANN, D. (eds) (1987). *Environmental Carcinogens Methods of Analysis and Exposure Measurement*. vol. 9 – *Passive Smoking*. Lyon: International Agency for Research on Cancer.

PERRY, R. & KIRK, P.W. (eds) (1988). *Indoor and Ambient Air Quality*. London: Selper Ltd.

PERSHAGEN, G., HRUBEC, Z. & SVENSSON, C. (1987). Passive smoking and lung cancer in Swedish women. *American Journal of Epidemiology*, **125**, 17–24.

SAID, G., ZALOKAR, J., LELLOUCH, J. & PATOIS, E. (1978). Parental smoking related to adenoidectomy and tonsillectomy in children. *Journal of Epidemiology and Community Health*, **32**, 97–101.

SCHNEIDER, B. (1989). Risk assessment for inhomogeneous subgroups. In the symposium *Assessment of Inhalation Hazards: Integration and Extrapolation Using Diverse Data*, at the Hannover Medical School, Federal Republic of Germany, February 1989.

SONNENFELD, G., GRIFFITH, R.B. & HUDGENS, R.W. (1985). The effect of smoke generation and manipulation variables on the cytotoxicity of mainstream and sidestream cigarette smoke to monolayer cultures of L–929 cells. *Archives in Toxicology*, **58**, 120–2.

SPENGLER, J.D., DOCKERY, D.W., TURNER, W.A., WILSON, J.M. & FERRIS, B.G. Jr. (1981). Long term measurements of respirable sulfates and particles inside and outside homes. *Atmospheric Environment*, **5**, 23–30.

SURGEON GENERAL, A REPORT OF (1986). *The Health Consequences of Involuntary Smoking,* Washington: US Government Printing Office.

SVENDSEN, K.H., KULLER, L.H., MARTIN, M.J. & OCKENE, J.K. (1985). Effects of passive smoking in the Multiple Risk Factor Intervention Trial. *American Journal of Epidemiology*, **126**, 783–95.

TRICHOPOULOS, D., KALANDIDI, A. & SPARROS, L. (1983). Lung cancer and passive smoking: conclusion of Greek study. *Lancet*, **ii**, 677–8.

TRICHOPOULOS, D., KALANDIDI, A., SPARROS, L. & MACMAHON, B. (1981). Lung cancer and passive smoking. *International Journal of Cancer*, **27**, 1–4.

UBERLA, K. (1987). Lung cancer from passive smoking: hypothesis or convincing evidence? *International Archives Occupational Environmental Health*, **59**, 421–37.

WALD, N. & FROGGATT, P. (eds) (1989). *Nicotine, Smoking and the Low Tar Programme*. Oxford: Oxford University Press.

WALD, N.J., NANCHAHAL, K., THOMPSON, S.G. & CUCKLE, H.S. (1986). Does breathing other people's tobacco smoke cause lung cancer? *British Medical Journal*, **293**, 1217–21.

WEETMAN, D.F. & MUNBY, J. (1990). Environmental tobacco smoke and cardiovascular disease. In *Indoor Air Quality and Ventilation*, Lunau, F. W. & Reynolds, G. L., eds, pp. 211–16. London: Selper Ltd.

WYNDER, E.L. & COVEY, L.S. (1987). Epidemiologic patterns in lung cancer by histologic type. *European Journal of Cancer and Oncology*, **23**, 1491–6.

YAMASAKI, E. & AMES, B.N. (1977). Concentration of mutagens from urine by adsorption with the nonpolar resin XAD-2: cigarette smokers have mutagenic urine. *Proceedings of the National Academy of Sciences, USA*, **74**, 3555–9.

Abbreviations and definitions used in this chapter

ETS

Environmental tobacco smoke. The smoke found in air as a specific result of tobacco smoking. ETS contains both the exhaled smoke and that emanating directly from cigarettes, cigars and pipes that are being smoked.

MS

Mainstream smoke. The smoke taken into the mouth from a cigarette (etc) during the process of smoking.

SS

Sidestream smoke. The smoke emanating directly from a cigarette etc between the inhalations of MS.

Respirable particles

Particles that can be inhaled arising from combustion of tobacco and other organic material. Their deposition in the respiratory tract or lung depends upon the size of the particles.

OR

Odds ratio. A numerical estimation of relative risk. It is the value obtained in a 2×2 contingency test, in which one level of risk (usually for unexposed individuals) is fixed at unity, and that relating to those exposed is expressed as a relative function of this. As with all ratios, there is no unit for an OR.

RR

Risk ratio (also known as relative risk). The ratio of the incidence of, or mortality to, a disease in a population exposed to the factor under consideration to the corresponding rate in a population not so exposed. Relative risk may be taken to be equal to the odds ratio from a 2×2 contingency table, provided there has been no bias in favour of including either exposed or unexposed individuals in the selection of cases or controls.

13

INDUSTRIAL ENVIRONMENTS

G. W. CROCKFORD

Introduction

Generally the phrase 'indoor environment' is held to apply only to the non-industrial situation and thus exclude the effect of air contamination due to work processes within the area. There is some logic in this as the population at risk is a selected one consisting of those fit enough to work and does not include children, the old and the infirm. Thus the application of air quality limits intended for the health protection of the whole population is unnecessary. However, the industrial environment has been extensively studied over decades by occupational health practitioners including occupational hygienists, medical and nursing practitioners, epidemiologists and toxicologists, giving a wealth of knowledge of the effects of chemical air contamination on humans. In addition, investigational and measurement techniques, developed over many years, are the basis of those now being used in indoor air investigations. It was therefore considered that a general book on indoor air quality must include some reference to the industrial environment.

The complexity of modern industrial activity and the multitude of materials and processes used has led to the recognition in recent years of many disease states associated with the work place. Some diseases are sufficiently strongly linked with the work place to have been ascribed the title of prescribed disease. There are some 50 of these diseases which include asbestosis, lead poisoning and chrome ulceration. Although it was not until 1895 that the principle of notification of some diseases was introduced to obtain information on them and their association with workplaces, such an association had been made more than 2000 years previously when Herodotus had noted the connection between inhaling dust and the premature death of miners. In the Spanish mines in the Americas the life expectancy of miners in the seventeenth century was said to be 6 to 18 months. Ramazzini, in the eighteenth century, had associated disease with the dust created in tool manufacture as well as with that found in mines and quarries. In the United Kingdom, a good detailed medical description of the disease associated with mineral dust exposure was available by the middle of the nineteenth century and by the end of the century legislation aimed at the prevention of silicosis was beginning to appear.

Earlier, the Romans had been aware of the development of mercury poisoning in cinnibar miners and the Greeks of arsenic poisoning and the development of neuropathy in craftsmen. The Greek Patron God of craftsmen is shown as being lame, presumably suffering from the effects of arsenic exposure which was a component of early bronzes. The health hazards of making bronze from copper and arsenic were so severe that the arsenic was eventually substituted by tin.

In retrospect, the connection between industrial activity and disease appears so obvious that it begs the question as to why effective action was not taken. Some action was taken but these were necessarily within the limits of contemporary knowledge and technology. During the industrial revolution new processes emerged creating industrial diseases on an unprecedented scale and economic and social factors operated which prevented even those who may have wanted to improve working conditions from doing so. Before any real improvements could be made in reducing the risks to health of those working in industry, a number of social and technical conditions had to be met, including the:

- ability to measure and quantify, if not the causative agent, at least some environmental factor that might indicate the health risk to operatives.
- information and knowledge to identify health hazard criteria and set environmental standards.
- legislative environment to set the ground rules regarding health and safety for industrial activity.

To the above must now be added another factor that in modern industrial societies is vitally important, and that is the ability to recognise hazards and potential risk to the health of employees at a very early stage, preferably before anybody has come to harm.

Behind these conditions lies the social will to set and achieve health and safety objectives. One of the biggest problems in this respect is the apparent inability of society to understand and respond to injury and deaths which are spread out in time and/or are dispersed throughout society. Occupational diseases are accidents in slow motion, some taking 30 years or more to run their course. Sometimes, but not always, occupational diseases are like motor vehicle accidents, dispersed in a community, even if those who are affected work in the same industry or factory. The inability to understand the human and social cost in a situation like this is a human failing that every one is prone to and which has made the task of the occupational health and safety profession, in achieving change in the workplace so difficult. Even when hundreds die as a result of one industrial event such as the Hawkes Nest tunnel constructed in West Virginia, USA and described as America's largest industrial disaster (Cherniack, 1986) it appears incredibly difficult for a society to lay down the ground rules that will prevent similar events happening in the future and in some way compensate those who have been affected.

Legislation

Faced with continuing ill health and deaths connected with people's employment, many industrialised countries introduced comprehensive legislation during the 1970s, aimed at ensuring the health of employees; in short, this legislation makes it illegal for an employer to put at risk the health and safety of an employee. In the UK this took the form of the Health and Safety at Work Act 1974. One action all employers had to take as a result of this Act was to prepare a written health and safety policy statement which gave the company's health and safety aims and objectives and named those who were responsible for seeing that the policy was followed and objectives achieved.

Where employees were members of a union, health and safety committees had to be set up with employees represented on them if requested. The legislation was, however, what is called 'enabling legislation' under which regulations could be produced and, although the objectives of the Act were very laudable, it lacked the necessary bite and guidance required for enforcement. Regulations have since been produced under the act to deal with specific problems such as asbestos and lead but materials and processes are too varied for regulations of this nature to be a realistic approach to the problem of industrial diseases.

Recognising this, in 1980 the EEC produced a Directive relating to the protection of workers, health, Directive 80/1107. This Directive called upon member states to introduce legislation that required the assessment of all health hazards associated with the use of substances and materials in an employee's workplace, the measurement of exposures to the materials and substances, the control of exposures, measurement of the effectiveness of control, the maintenance of control systems, the provision of information on health hazards to employees, the training of employees and where appropriate the health surveillance of employees. In order to comply with this Directive and the International Labour Office Convention 139 on 'carcinogenic substances and agents and future procedures and controls on substances hazardous to health' the UK Government introduced regulations for the Control of Substances Hazardous to Health in 1989. The regulations called for the assessment of health hazards and the risk to health wherever people are exposed to substances which might have an adverse health effect on employees and any others on the employer's site. The regulations call for an assessment of the health hazards and risks to health associated with the use of any substance on the employer's site, the control of any risks to health identified, the effective and proper use of any exposure control system, the maintenance of the control systems, the monitoring of exposures and the performance of the control systems, the provision of information and training and, if indicated, health surveillance.

Records have to be kept, maintained and made available to employees as well as the factory inspectors. The essential elements of occupational hygiene, namely recognition, measurement, assessment and control of those factors in the environment that might adversely affect the health and comfort of employees, are therefore present in the regulations. The occupational hygiene programme, as spelled out in the textbooks of the 1950s, is now law. It is illegal to place an employee in a working environment without first doing a full assessment of the risks to health, controlling any that are present and fully informing the employee about the risks and control measures. The regulations are obviously far reaching for both the employer and employee and, in an ideal world, occupational diseases from exposure to substances should become history. The COSHH Regulations replace much previous legislation but they do form part of a new constellation of legislation which includes regulations relating to noise, electricity and ionising radiation.

However good or far sighted the legislation, it has to be complied with by industry and enforced by somebody. In matters relating to exposure to substances this requires a reasonably accurate method of measurement which is simple and inexpensive and control measures which are reasonably practical for most employers. The targets set by legislation, on the one hand, and the ability of industry to meet them and that of the inspectorate to measure them, on the other, must coincide. Without this, there can be no enforcement of recognised and fair ground rules.

This problem still bedevils industry, particularly small companies who may not have the expertise and resources to translate legislation into a form applicable to the company's workshops. The companies are also so numerous that the enforcement agencies as presently structured cannot inspect on the scale required to see that legislation is being complied with. The concept of self enforcement or self regulation has therefore arisen. The idea is for industry to inspect itself with either insurance companies doing much of the inspection and audit of systems or independent consultants doing health and safety status assessments and audits and reporting directly to the chief executive or board of management. Whatever form of inspection and audit eventually develops, there will have to be considerably more of it than there is at the moment if companies both large and small are to achieve high standards of compliance. The problem of achieving adequate levels of environmental control in small companies has been recognised for many years and the World Health Organisation has drawn attention to the need for improved standards of occupational hygiene in such companies.

Health

It might be salutary to consider some of the health effects of substances and the situation in which industrially related disease develops.

The main dust diseases produced by exposure to crystalline forms of silica such as quartz, coal dusts, asbestos, heavy metals such as lead and mercury, and a number of other materials, are hopefully mainly a thing of the past with the hazards well recognised and the monitoring techniques and engineering controls well developed and readily available. There are new diseases being recognised as materials become widely used and new ones are developed, for example, angiosarcoma of the liver produced by exposure to vinyl chloride monomer but again, recognition of the disease normally leads to environmental control and its disappearance. The appearance of work-related clinical symptoms of ill health in a working population will today normally lead to the development of appropriate standards of control by industry. The aim, of course, must be to prevent symptoms of ill health ever appearing and, in all the major industrialised countries, new chemicals are required to be screened for possible adverse health effects before being used. Unfortunately there are many thousands of chemicals already in use which have not had this screening carried out.

The effects of dangerous preparations and materials are classified in the Dangerous Substances Directive and General Preparation Directive (Hewstone) into nine categories. They are:

- acute lethal effects
- non-lethal irreversible effects after a single exposure
- severe effects after repeated or prolonged exposure
- corrosive effects
- irritant effects
- sensitising effects
- carcinogenic effects
- mutagenic effects
- teratogenic effects

Obviously as a substance or preparation is diluted, what was once a material with an 'acute lethal effect' or 'corrosive effect', becomes one with non-lethal or irritant effects and with greater dilution these effects in turn disappear. A material with non-lethal but irreversible effects after a single exposure may produce clinically identifiable changes in exposed people but again if the material is diluted and the exposures are low, the changes produced may not be identifiable using current techniques. People also vary in their response to materials, some being badly affected whilst others are unharmed, or at least can continue working. For example some workers are tolerant of high blood lead levels. Sensitised individuals may react to very low exposure levels of some materials whilst others show no response at all. Common examples are pollen, fungal spores, and platinum compounds but there are many other chemicals used in industry today that produce sensitisation and practically all biological materials can produce sensitisation in some individuals, including practically all biological materials.

Considerations such as this lead into one of the main philosophical and practical problems of occupational health, namely the identification of criteria for the setting of exposure standards and then where to set the standard in relation to the criterion. For example, if a gas causes eye irritation, is the standard to be set to protect 90% of the exposed population, 99% or 99.9%?. The situation becomes more complex when subclinical effects are known to occur; for example, changes in nerve conduction velocity as with lead exposure, psychomotor performance test changes such as reaction times, or mental performance test changes such as speed and accuracy of adding numbers as happens with exposure to some solvent vapours. The changes may be so small that they cannot be identified in any one individual but only show up when large numbers of people are studied, as in the case with low level exposure to alcohol. The reality of the problem is that, however imprecise and wanting the information is, standards have to be set and enforced. They are the cornerstone for the practice of occupational hygiene and they can always be changed as more information becomes available. Another consideration that enters into the setting of standards is that it is impossible to protect all individuals due to the variability between people in their response to contaminants, some being extremely vulnerable or sensitive to particular substances. The state of health, age, body size and the taking of medication may all play a part in determining how a person responds to environmental contaminants. Body size is related to the sex of a person, the standard woman being two-thirds the weight of the standard man. Work environment standards are based on the assumption that the work force is healthy, that exposures are for limited periods of time only, normally an 8 hour shift or, where materials have acute effects, for short periods such as 10 or 15 minutes. The short-term exposure limits are important, for example where materials have an acute central nervous effect which may lead to loss of judgement and undermine people's ability to look after themselves in potentially dangerous situations.

Regarding the health status of workers, although pre-employment medical checks and 'self-selection' takes workers out of particular exposures that may adversely affect them, most workers are healthy and are not unduly sensitive to the working environment. However, there are some occupations where people with health or physical disabilities can, and in fact expect to, work successfully and effectively without their health being put at risk. Office work comes within this category. Thus the usual worker standards for chemical substances in the industrial environment cannot be used in such situations without the possibility of adverse response appearing in a very limited number of the workforce. It is open to debate what, if any, factor should be applied to the occupational exposure limits to protect the health of this very small group of workers, whether in the office or other industrial situation. On a cost/benefit basis it seems that the often suggested factor of 0.1, the same as that sometimes suggested for the general environment, is difficult to justify.

Target organs

Many materials affect and injure particular organs within the body more than others. The organs most significantly affected are therefore referred to as the target organs. Insoluble dusts will, for example, tend to affect the respiratory tract. The part most affected will be determined by the ability of the dust particle to penetrate the lungs. Dust particles, however small, are always falling out of the air due to the influence of gravity. The speed of fall is determined by the weight or force pulling it down which is counteracted to a certain extent by the air resistance opposing the fall. The mass or weight of the particle tends to be proportional to the cube of its dimensions, whilst the surface area on which the resistance of the air to movement is acting, is proportional to the square of the particle dimensions. Consequently, the larger the particle, or more accurately, the heavier it is in relation to its surface area, the faster it will fall and the higher its terminal velocity. The mass of a particle will also lead to it having inertia and resisting any change in direction as the air stream carrying it moves around obstructions and bends round corners. Those particles greater than 10 μm in size therefore get removed from the air stream. The size of particle reaching different parts of the lung therefore becomes smaller and smaller as the air travels deeper and deeper into the lungs. There is, however, one very difficult practical aspect to determining particle size: the dust particles can be all sorts of shapes and impossible to measure. It is therefore normal practice to refer to the size of particles in terms of the size of a droplet of water that falls at the same speed, that is to what is called the aerodynamic size of the particle. As this relates directly to the ability of the particle to penetrate into the lungs, it is one of the properties of the particle that will determine the health hazard of the aerosol. The large particles with an aerodynamic size greater than 7.2 μm do not normally penetrate into the alveolar region but tend to get caught on the mucocillary ladder where they are cleared from the lungs in about 8 to 24 hours. If the particle is sufficiently small to penetrate to the alveolar region, white blood cells called alveolar macrophages clear them from the lungs with a clearance time of weeks for inorganic particles.

This mechanism breaks down in two notable cases. The first is with crystalline silica which is found in many rocks including sandstone, flint and many minerals and the second is with fibres such as asbestos. In the first case, the crystalline form of silica is toxic to the alveolar macrophage which is killed. In the second case, the alveolar macrophage finds it difficult to handle the fibre and cannot clear it from the alveoli. Very long fibres can penetrate to the alveoli because a fibre falls at a speed proportional to its diameter, not its length. When sampling and analysing for these two contaminants in the context of health effects, it is the respirable fraction that is measured, not the total inhalable dust. Total dust becomes important if the dust does its damage higher up in the respiratory tract or if it is soluble in body fluids and so can diffuse into the tissue and be picked up by the blood circulation system.

Standards

With the experience of exposure to silaceous dusts, asbestos, lead, mercury and many other materials, it is easy to see that workplace standards are required to protect the health of those involved in industrial activity, whether they are working inside factories or outside. The simplest approach is to identify those environments causing ill health and the severity of the diseases developed, characterise them in terms of contaminant concentrations and length of time for which people have been exposed and develop a dose–response relationship to identify those situations where ill health does not develop although operatives are still exposed to the same materials. From the comparisons, a 'bench mark' in terms of, for example, silica dust levels expressed as time weighted averages can be identified above which disease may appear in the workforce and below which disease or identifiable subclinical changes are unlikely to occur.

This is an oversimplification of what happens in practice, but, nevertheless, the development of dose–response curves is the essence of standard setting. For many materials, human response data from industry and laboratories may not be available, so other information is also used, obtained from animal and bacterial studies and from a comparison of the material with others having similar chemical properties but with known health effects. Because of the difficulty of defining whether exposure to a given level of an airborne contaminant will cause harm or not, the standards described as airborne concentrations of substances represent conditions under which it is believed that nearly all workers may be repeatedly exposed for 8 hours a day 5 days a week without adverse effect. The standards apply strictly to 8 hour exposures or other stated time scale and they are not a dividing line between safe and dangerous concentrations. Because of the different basis upon which the standards are set, e.g. irritancy, narcosis, impairment of health or carcinogenicity, the standards do not indicate the relative toxicity of substances. The most commonly used standards are the Threshold Limit Values produced by the American Conference of Governmental Hygienists and they are intended to be used as guidelines for good practice. The 'Documentation of the Threshold Limit Values' which gives the background to the TLV can always be consulted for information relating to the setting of the standard.

In the UK, airborne standards consisting of mandatory exposure limits that must not be exceeded (Maximum Exposure Limits) have been established for 29 substances.

There are, in addition, approximately 700 substances with assigned Occupational Exposure Limits (OELs) intended to be used as guide lines when assessing time weighted eight hour exposures. Because exposures are not constant but vary up and down as a result of innumerable factors, it has been customary to use the eight hour period as the time base for exposure standards. For many materials, such as lead and mineral dusts, it is the total dose build up over long periods of time that appears to determine the health consequences of exposure, but quite clearly there are many materials that act quickly and that at high concentrations can have immediate effects on people. Irritants such as sulphur dioxide or ammonia can quickly induce irritation of the eyes and respiratory tract but many solvents can also quickly impair mental functions at high concentrations. Many materials, therefore, have in addition to the eight hour time weighted average a short term exposure limit (STEL) most commonly covering a period of 10 minutes (UK) or 15 minutes (USA). Workers should not be exposed to contaminant levels above the STEL in order to ensure their protection against acute effects.

Harmful skin responses can be expected if corrosive or irritant materials, liquid or dust, come into contact with the skin. However, many substances, such as organic solvents, phenol, organic forms of lead, or pesticides can be absorbed directly through the skin so contributing to the body burden. Even some gases, if present in high concentration, can be absorbed through the skin and protection of the skin as well as the respiratory tract is required. Cuts and abrasions act as a short circuit into the body for pollutants. To draw attention to this particular property of some materials, a skin notation is given with the standard for airborne concentrations (Table 13.1).

However well founded a standard, there always remains the problem of whether it is valid when the contaminant in question is mixed with other

Table 13.1. *OEL example with skin notation*

Substance	Formula	Long-term exposure limit (8 h TWA value)		Short-term exposure limit (10 min TWA value)		Notes
Allyl alcohol	$CH_2{=}CHCH_2OH$	ppm	$mg\,m^{-3}$	ppm	$mg\,m^{-3}$	
		2	5	4	10	

materials. Will the effects be additive, multiplicative or will they neutralise each other? For example, readily soluble gases such as sulphur dioxide would normally be removed from the air stream in the upper reaches of the respiratory tract and not reach the more distant and sensitive areas of the lung. However, some dust particles will adsorb gases and so enable the gas to hitch a ride and penetrate further into the lungs and to be delivered to the lining of the respiratory system as a concentrated packet. Exposure to mixtures of solvents which cause narcosis are, at the least, likely to summate their effects and in some cases one substance may potentiate the effect of another by interference with metabolic pathways involved in degrading the substances. Even the use of drugs for medical treatment or alcohol may make the worker more vulnerable to, not only chemical pollutants, but to the physical stress of heat and physical work. Exposure to creosote vapour and contamination of the skin will make the skin more sensitive to sunlight and the development of sunburn. The use of biological monitoring, that is, measurement of the amount of a contaminant accumulating in the body from samples of exhaled air, blood or urine, may help to indicate what is happening in complex airborne contaminant situations but normally an expert judgement is required. The response of industry is normally to try and keep exposures simple and as low as possible. The Health and Safety Executive provide some guidance on how to deal with mixtures in their Guidance Note No. EH40.

Although the majority of people in factories, offices and other workplaces are mainly concerned with airborne contaminants, contamination of surfaces with which people come into contact by pesticides, heavy metals or dusts can present risks to health. Deposited material can become airborne if disturbed or transferred to the hands, skin and clothing whence they enter the body by ingestion, through the skin or inhalation. Samples of dust can be collected from surfaces by scraping or brushing into containers and then

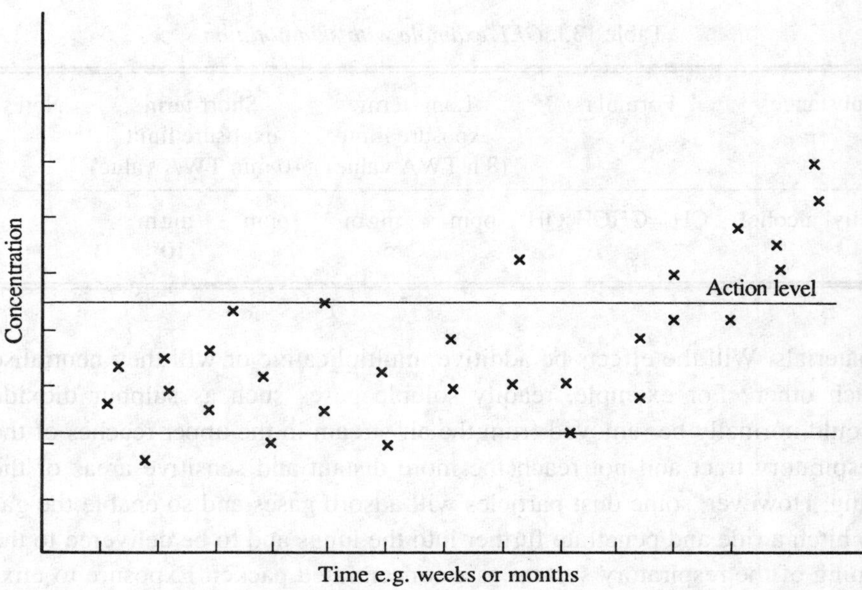

Fig.13.1. Environmental quality control monitoring. When a series of measurements of an airborne contaminant are made, the results will indicate the performance of the control systems. Any departure from a selected action level, e.g. 1 standard deviation, should lead to the control system being checked and repaired.

analysed. Filter papers can be wiped over surfaces to pick up sufficient contamination for analysis. With very toxic chemicals and radioactive materials the aim is to maintain the highest state of cleanliness but it is also important to know what services maintenance and contract personnel are exposed to as they go about their jobs and ensure that they are adequately informed and protected.

In-house or company standards

To achieve a given quality of the working environment and meet standards in terms of airborne contaminant levels, control measures have to be used. These measures include engineering control such as containment, local exhaust ventilation, work procedures designed to limit emissions into the environment, cleaning up spills of materials and many other techniques designed to control contaminant levels in the air and the exposure of people to materials. The performance of the control system can therefore be defined in terms of what it can achieve and its performance monitored over long periods of time. The control system can, in fact, be subjected to a quality control programme with the performances of the system, in terms of contaminant level, being used as an 'in company standard' (Fig.13. 1).

This standard relates to the control system not to published standards such as an OEL and for reasons to be discussed in later sections is normally well below published recommended standards.

Measuring the environment

The key to the application of standards in the occupational environment is the ability to measure the amount of a substance in air at any one time or over a period of time. Standards are therefore normally accompanied by a recommended method of measurement, some of which are enshrined as standard methods and published by the British Standards Institute (BSI) or the Health and Safety Executive (HSE) in their Guidance Note Series – Methods for the Determination of Hazardous Substances. Before the figure produced by a measurement can be interpreted, the sampling strategy and the limitations of the sampling and analytical methods need to be understood. Because of the many factors determining the concentration of a dust or gas in the air such as the quantities being emitted, whether the windows and doors are open, the direction of the wind, the work procedures, contribution of secondary sources such as clothing, analytical and sampling variables and many other factors, repeated measurements of a contaminant in air will show a range of values which follow a distribution curve with a bit of a tail at the high concentration end. The curve is a lognormal distribution and it is indicating that if workers are exposed to say, oil mist from cutting machines, solvent vapour from degreasing tanks, welding fume or sawdust, the exposure levels either as short-term exposures over a few minutes or as 8 hour daily exposures will vary and a decision has to be made as to what percentage of the readings or exposures above the standard are going to be acceptable. Clearly, if the figure is going to be 5% or 10%, workers are going to spend the majority of their time working in contaminant levels which are well below the standard. The use of an environmental standard by an organisation does not therefore mean that employees are exposed to time weighted averages equivalent to the standard: in fact quite the reverse, for the majority of exposures will be well below the standard.

Returning to the problems of measuring the environment, there are many recommended and standardised procedures for measuring gases, vapours, mists, dusts and fumes in the air but the majority of techniques involve a

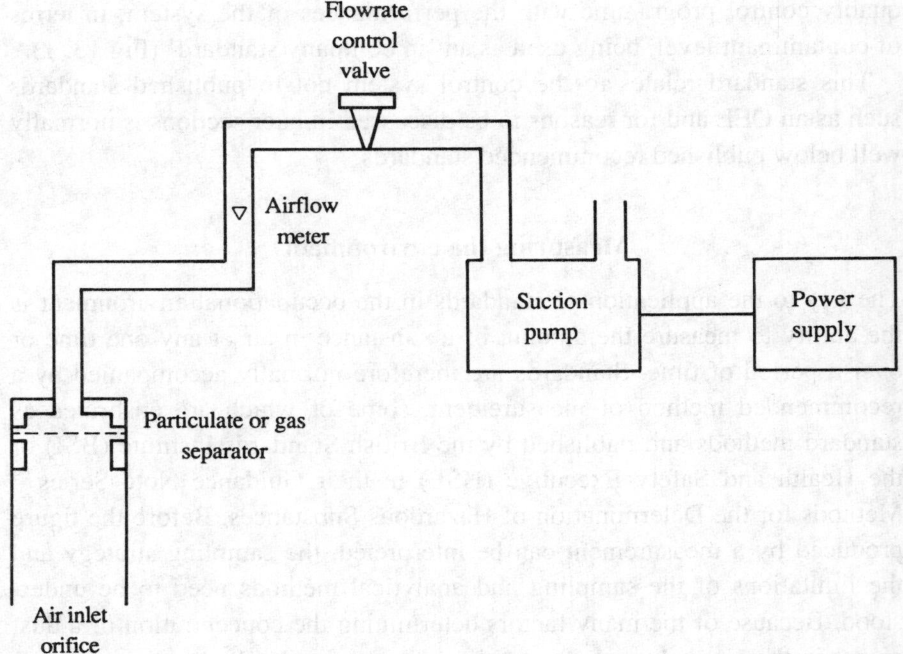

Fig.13.2. The basic elements of a sampling train.

device for capturing the contaminant, a pump of some sort to pull air through the sampling head, a method of determining the volume of air sampled and the time span over which the sample was taken (Fig.13.2).

The sample is normally analysed for the contaminant or contaminants of interest in a laboratory although many field methods do exist. The analyst is able to calculate the quantity of contaminant in a cubic metre of air from knowledge of the volume of air sampled. The value is a time weighted average over the period of sampling. However, on site analysis is often preferred and can be done using colorimetric tubes, devices which show the presence of a chemical by developing a stain that extends down the tube according to the amount present. Instrumental methods also can, in some cases, give an immediate readout.

The results of sampling are normally expressed as either parts per million of air (when a gas or vapour is involved 1 ml by volume per cubic metre is one part per million), or as milligrams per cubic metre for dusts, fumes and mists. With fibrous materials such as rock wool, glass fibre or asbestos, standards are also expressed as so many fibres per ml of air which are within the respirable size range. These are the fibres which are sufficiently small in actual diameter, 3 μm or less, to penetrate to the alveoli. It is as well to bear in mind that, whilst synthetic fibrous materials such as glass fibre and rock wool only break across the fibre so that the fibre diameter never changes, as it is fixed at the manufacturing stage, fibres of asbestos can and do break along their length to produce very long thin fibres with diameters down to about 0.2 μm. This property enables asbestos in its three main forms to produce dense dust clouds with fibres in the respirable size range.

The respirable sized dust particles can be separated from the larger particles by size selectors such as a cyclone but this is not possible with fibres and the respirable size range has to be determined by looking down a microscope and counting those fibres which fit the dimensions for respiration, i.e. an aspect ratio of 3 : 1, longer than 5 μm and 3 μm or less in diameter. Unfortunately there are many other materials, particularly organic ones, that produce fibres and these tend to get counted at the same time so that fibre counts, particularly if they are low, may be in error in terms of indicating how many asbestos fibres, if any, are present. If it is necessary to count only asbestos fibres or mineral fibres, an electron microscope is required which can identify individual asbestos fibres by their X-ray emissions, when bombarded by an electron beam. The accurate measurement of particulates in airstreams presents a number of problems, one of which is the under or over sampling, resulting from a mismatch between the speed with which the air stream impinges on the sampling head and the speed with which the air is drawn through the sampling head. If the air approaches the head faster than the face velocity of the sampler, the larger particles carry straight on and hit the sampler instead of going round. If the air approaches at a lower speed than the face velocity, the smaller particles are drawn in with the larger ones, carrying on and failing to be captured. For this reason the ideal is isokinetic sampling with the face velocity equal to the airstream velocity. For most sampling situations met in practice, it is not possible to do isokinetic sampling and the error has to be accepted. Fortunately particles below 10 μm in diameter tend to follow the airstream and it is accepted that isokinetic sampling is not necessary for this size range.

Another sampling problem is trying to sample the air that the operatives are breathing. Very large differences can exist between the concentration of a contaminant in the general environment and that in the air being breathed. This is for a number of reasons: the operative may be generating the pollutant cloud as a result of work activities, contaminated clothing may be releasing dust and vapour, the operative may get very close to the source, or the convection current produced by body heat may draw contaminant into the breathing zone.

Sampling errors produced by a mismatch between the position of the sampling head and the operative's nose and mouth can be very large, several hundred per cent. Because of these problems, care has to be exercised in sampling and adequate numbers of samples obtained to define the exposure concentration envelope. If the performance of the engineering control of the environment is to be measured, samplers are placed in selected positions in the working area. This form of monitoring is called area or static monitoring.

The monitoring techniques used are often similar to those used for personal monitoring. However, direct reading instruments which will give a second by second account of concentration and keep a record of the readings are also used quite extensively. Using such instruments continuous monitoring is possible for toxic gases and particulates in many working areas such as oil refineries, blast furnaces, and the pharmaceutical and electronics industries. Direct reading instruments enable detailed studies of emissions to be made which are very valuable in identifying the emission points, and the most effective and practical methods of control.

Accuracy of measurement is important but not easy to achieve with very large errors being possible at the sampling stage and smaller ones at the analysis stage. The latter are controlled to a certain extent by quality assurance schemes which enable laboratories to compare their performance against that of other laboratories. The quality of the sampling is not so easily checked and so sampling requires skill and experience to obtain the best results. Normally the sensitivity of the analytical methods available are adequate for most materials. It is with carcinogens and some sensitisers that the exposure limit may be set so low that analytical sensitivity may be pushed to its limit. Results may be obtained but when they are plotted against sample volume or weight of sample no relationship is found. This indicates that the figures obtained are saying more about the limitations of the analytical method than about the quantity of pollutant present. In considering the figures and values obtained from the laboratory, the possibility of errors and bias must not be ignored.

The pollutants

Whenever materials are handled, worked on or processed, there is the possibility of the air and workplace becoming contaminated by gases, vapours, fume, dusts or mists derived from the materials. The gases and vapours dilute in the air and eventually disperse but the dust, fume or mists may settle out on surfaces and present a health hazard at a later date when they are disturbed by some activity. The pollutant has to enter the air at some point, normally referred to as the emission point. This may approximate to a single point such as a grinding wheel or milling machine or be very diffuse in nature such as a conveyer belt, articles drying after printing or from some other operation such as fungal spores from a mushroom bed. All environments appear to have their problems whether a purpose-built factory, offices, quarries, sewers, shipbuilding, shops, hotels, transport or schools. Human and mechanical activity will always generate some airborne contaminant, if it is only the dust from books in a library vault.

The quantities of material required to bring airborne levels up to the occupational exposure limits can be very small indeed. A wine bottle of benzene vapour (OEL 30 mg/m^3) for example, if released into a sealed room 5 m × 5 m × 3 m is sufficient to raise the benzene vapour level of the room to the OEL. For sulphur dioxide with an OEL of 5 mg/m^3 a space 15 m × 10 m × 2.5 m would be required if a full wine bottle of the gas was to be released. It is easy to see why workers in enclosed spaces are so vulnerable to the small quantities of material trapped behind scale and in debris. With dusty material the quantities of material are equally small. A gram of finely divided sawdust could bring the dust level in a 8 m × 5 m × 2.5 m room to OEL whilst only 10 mg of respirable-sized sand (quartz) particles would be required to do the same.

With such small quantities of materials involved, environments can very easily be contaminated with dangerously high levels of particulates, gases and vapours, unless some control is exercised over emissions and ventilation maintained at high levels. Any enclosed space or a space in which air exchange is limited can rapidly become contaminated by life- or health-threatening concentrations.

Historically the most dangerous pollutants have been the silica-containing dusts generated in a large number of industrial processes and activities. Today it is recognised that any process involving the release of respirable-sized crystalline silica has to be very tightly controlled. Mineral fibres are also very dangerous if in the respirable-size range of less than 3 μm in diameter and very stringent regulations are in force for asbestos. At the moment the standard for glass fibre dust and other man-made mineral fibres is 5 mg/m^3. However, research is going on to determine if the MMMF are carcinogenic when inhaled and, although the present standard places the material in the same category as mineral oil mist and iron oxide, dust users of MMMF are advised to control exposures to as low a level as possible.

The heavy metals such as lead and mercury have traditionally produced extensive and severe health problems but other metals such as cadmium, nickel, chromium, arsenic and platinum also cause serious problems if not adequately controlled. Platinum and its salts are powerful sensitisers. Even exposures to copper and zinc, two metals essential for the health of the body, have to be controlled and exposures limited. Metals appear in a wide range of processes from refining, foundry work, alloying, machining, cutting and welding, and as catalysts and they are an important group of pollutants. The fumes of metals such as lead and cadmium are particularly dangerous due to the very high surface area to weight ratio of fumes.

Chemicals

One group of widely used chemicals are the organic solvents although they have many other uses. These are often complex mixtures produced by the distillation of petroleum fractions but can also be pure chemicals, as is the case with 1,1,1-trichloroethane, a chlorinated organic compound used widely as a solvent, for example in typing correction fluid.

Amongst the organic solvents are such materials as petroleum naphtha, petrol, paraffin and mineral spirits of the type used for paints.

Common effects produced by exposure are narcosis, although the vapour can be so concentrated that asphyxiation can also be a potential hazard. Neurotoxic effects are associated with n-hexane, a chemical that can appear in many materials such as naphtha. Benzene, a carcinogen, also appears in many products including petrol. If exposures are low there should be no health risk but good ventilation is required to minimise exposure. Organic solvents are widely used in adhesives and so are readily introduced into the built environment, along with other chemicals such as plasticisers, formaldehyde and many others, through the furnishings, tiles, paints, wallpaper and furniture. Once within the building, even very minor emissions can lead to exposure standards being reached. A basic hygiene rule that applies to all chemicals that produce a gas or vapour is always to ventilate and not to recycle unless the air can be cleaned of all the contaminants, which cannot normally be done. Because of the long time-scale (years) over which degassing can take place, a technique being used to scour buildings of volatiles is to raise the temperature and ventilate for a period, often over a weekend, a technique referred to as 'baking'.

Other materials quite widely used in manufacturing processes but also finding their way into factories to do specific tasks are such chemicals as phenol and methylene chloride for paint stripping. The latter chemical is also used very widely as a solvent for oil, fats, waxes, bitumen and as a degreasing agent. Toluene di-isocyanate, released when using two pack polyurethane paints, foam manufacture and the use of certain adhesives, may sensitise and then produce an asthmatic reaction in some people and so the vapour has to be very tightly controlled.

Another group of chemicals used within buildings are pesticides. These are manufactured and blended within factories, used in the treatment of timber and so find their way into the allied industries making such things as timber fittings for buildings. They are also incorporated into paints and surface finishes both within buildings and transport vehicles. The structure of vehicles can reach 60 °C, possibly more, when exposed to direct sunlight, with the consequent possibility of toxic emissions, as can happen for example if tributyl tin is used as a fungicide. For most work situations, the main problems will arise from handling materials impregnated with pesticides and possibly the release of contaminated dust during machining.

The chlorinated hydrocarbon insecticides are the more persistent of the organic pesticides and their use is declining because of this. Some of the fungicides contain metals such as chromium, copper, tin and zinc.

The biological materials used in industry present very widespread problems whether they are food products, woods, textile fibres or extracts of plants. In some cases it is the bacteria or fungi growing on the materials that cause the main problems; in others it is the material; for example, some hard-wood dusts produce nasal cancer and certain fruits produce contact dermatitis, e.g. citrus fruits.

When considering any material there is a general rule that says that if enough gets inside a person it will cause ill health. However benign the material may appear, it must be treated with respect and emissions and exposures of employees controlled.

Control

Control starts by selecting, when possible, the least toxic material for the task and using materials with low volatility or having dust-suppressed forms. Dust suppression is achieved by pellatisation, wetting with water or other liquids and encapsulating hazardous materials in less hazardous ones. An example of this approach is to bag the chemicals to be used in a process in the weight ranges required and, if possible, to use a plastic or material for the bag which is acceptable to the process so that it does not have to be opened. Micro-encapsulation for particularly hazardous materials is also available. Emission points should be identified and emissions limited as far as is practical. Small emissions of gases and vapours are normally of no consequence as first dilution and then removal from the area by air movements ensures that no health risk develops. With dust and fume emissions, the particles settle out and this leads to extensive surface contamination and the development of secondary sources that put at risk maintenance and service staff. Fine dust can travel for considerable distances in factories and lead to secretarial, packaging and domestic staff being exposed. Surface monitoring for lead, other metals and pesticides, for example, may be required at sites distant to the production areas to ensure adequate protection of staff.

Air sampling cannot be relied upon to identify movement of a contaminant into non-production areas due to the low levels and spasmodic nature of the contaminant transfer. Good housekeeping is, however, required to ensure that over the months and years secondary sources do not build up which could put at risk the health of staff both in the production areas and in areas distant to them such as offices and amenity areas. All engineering control whether by containment, enclosures, local exhaust ventilation (LEV), dilution or ventilation has to be subjected to an inspection and maintenance programme designed to ensure that it remains in effective working order. This is a legal requirement in the UK and most industrialised countries. As engineers cannot check equipment every day it is important that those using it can identify when it fails or its performance declines. This requires that staff understand the basic principles upon which local exhaust systems and dilution systems work and the critical importance of following laid-down work procedures to ensure that control is effective. This clearly applies to staff using LEV but it should also apply to office managers and others who work in areas subjected to engineering control of the environment.

Due to the high cost of heating the air that comes in to replace that taken away by LEV and extraction systems, the recycling of exhausted air is widely practised in industrial and commercial buildings. This practice has some drawbacks, particularly in offices where the ventilation system is not normally fitted with efficient particulate and gas filters. The absence of gas filters allows the gases and vapours from processes and materials to recycle and build up to levels that cause discomfort such as eye irritation and various subjective effects referred to as building related symptoms. During dry weather when the humidity may fall to very low levels in the building, dry throats, sore eyes, skin complaints such as dryness, itching and possibly dermatitis may develop. These effects may be caused by the low humidity but they are also partly due to increased dust levels as dust is more easily disturbed to become airborne as the humidity drops. Air conditioning can deal with this problem but is not very widely used in the UK due to costs and maintenance requirements. More widely used are the electrostatic air filters which remove particulates, particularly tobacco smoke, pollen, fungal spores and dust raised by human activity. They are also fitted with charcoal filters to remove high molecular weight gases such as acrolean. Charcoal filters are ineffective against the smaller molecules such as carbon monoxide. Frequent maintenance of such filters is critically important if they are to remain effective. This is normally done by the supplier. The mechanisms by which particulates and gases are removed from air streams, for example particulates by mechanical filtration, cyclones and electrostatic filters and gases by adsorption onto charcoal, chemical reactions, incineration and catalysed reactions, differ considerably in cost and ease of maintenance with gas removal being the more difficult. For this reason air vented from a building is seldom cleaned of pollutant gases. Such pollutants are therefore free to re-enter buildings downwind or back into the same building if the effluent is trapped in the boundary layer round the building. The re-entry of contaminants is not normally identified by the occupants of the building unless pungent or highly irritant gases are involved but, even when this is not the case, the indoor environment will deteriorate with low levels of pollutants being present.

Maintaining control of the environment

All industrial environments have some form of environmental control built into them even if it is only the work procedures and a reliance on dilution ventilation.

The first step in maintaining control of the environment is to know where it should be, i.e. its quality in terms of dust levels, gas and vapour levels and the range over which they vary. Such data are gathered by environmental monitoring over a period of time when it is known that the engineering controls and work procedures are working properly. Such monitoring defines the performance of the control systems and defines the 'in company standards' for the pollutants involved. Any departure from the operating range would then be taken to indicate a failure in the system and warrant investigation and correction.

The main approach to maintaining control is to have a planned and regular inspection and maintenance programme for all forms of engineering control. The inspection must be of such aspects that relate to environmental control, such as face velocities, volume flows, and pressure drops across filters. The work procedures and quality of supervision should also be reviewed at intervals as should the use of any protective clothing and respiratory equipment. It is now a legal requirement in the UK and many other countries to have written inspection and maintenance procedures that are able to ensure that engineering control and personal protective clothing and equipment is maintained in effective working order. Adequate records of performance tests and maintenance must be maintained.

The skill to work safely

The substances, materials and processes used in industry will always be increasing, always presenting additional challenges to employers, managers and operatives. The knowledge and ability to work safely, to control present and future hazards effectively and at low cost is an essential skill for all organisations and one that can be developed. In general, systems go wrong because of failures with the administrative aspects of health and safety programmes. The provision of information, training, supervision and maintenance, for example for staff performing operations in different environments and failing to recognise the critical importance of good ventilation whenever pollutants get into the air. Elementary, but lives are lost because of this lack of understanding. Staff at whatever level within an organisation, chief executive or line operative, must be competent regarding those aspects of the company's health and safety programme with which they are involved. Such competence can only come from involvement with a continuing education and training programme geared to update and maintain the necessary skills to assure health and safety at all levels within the organisation.

References

CHERNIACK, M. (1986). *The Hawk's Nest Incident*. Yale University Press, ISBN 0-300-03522-5.

EUROPEAN ECONOMIC COMMUNITY (1980). Directive 80/1107 on the Protection of Workers from the risks related to exposure to chemical, physical and biological agents at work. Official Journal of the EEC 1980 L327, p. 8.

HARPER, M. (1987). Possible toxic metal exposure of prehistoric bronze workers. *British Journal of Industrial Medicine*, **44**, 652–6

HEALTH AND SAFETY EXECUTIVE (1989). *Occupational Exposure Limits 1989*. Guidance Notes EH40. London, HSE.

HEALTH AND SAFETY EXECUTIVE (1989). *Monitoring Strategies for Toxic Substances*. Guidance Notes EH42. London, HSE.

HEWSTONE, R.K. User Labelling of Dangerous Substances, Section 11.4, pp 1-13. User Labelling of Dangerous Preparations, Section 11.5. In *Croner's Handbook of Occupational Hygiene*, instalment 25. Kingston-upon-Thames: Croner Publishing Ltd.

14

EXTREMELY LOW FREQUENCY ELECTROMAGNETIC RADIATION

L. H. HAWKINS

The nature of non-ionising radiation

Electromagnetic radiation (EMR) can cause damage to the body by both ionising and non-ionising mechanisms. Most of the factors leading to this damage are outside the scope of this book but there are two aspects which are closely linked with indoor air quality or the indoor environment which need to be considered. The ionising effect of radon and its daughter products is dealt with in Chapter 7. This chapter deals with extremely low frequency (ELF) fields in the frequency range of normal electrical power supplies. It does not deal with higher frequency fields due to radar, radio transmissions, and the possibility of such emissions from electronic apparatus.

Ionising radiation undoubtedly causes tissue and cellular damage and creates a risk to health that is well documented and the mechanisms for which are reasonably well understood. In this chapter we consider the possibility that electromagnetic radiation with lower levels of energy can have biological effects which occur through mechanisms that are different from those arising from ionising radiation.

In order to understand the interactions of biological systems with electromagnetic fields it is important to have some understanding of the physics of electromagnetism.

Isaac Newton first showed that a light beam could be split by a glass prism into different colour bands but that these bands could not be further split or converted into other colours. We know now, of course, that each colour is a discrete frequency band. The fact that we relate particular frequencies to a particular colour is a feature of the way the retina of the eye responds to these frequencies and the way that the brain interprets the signals

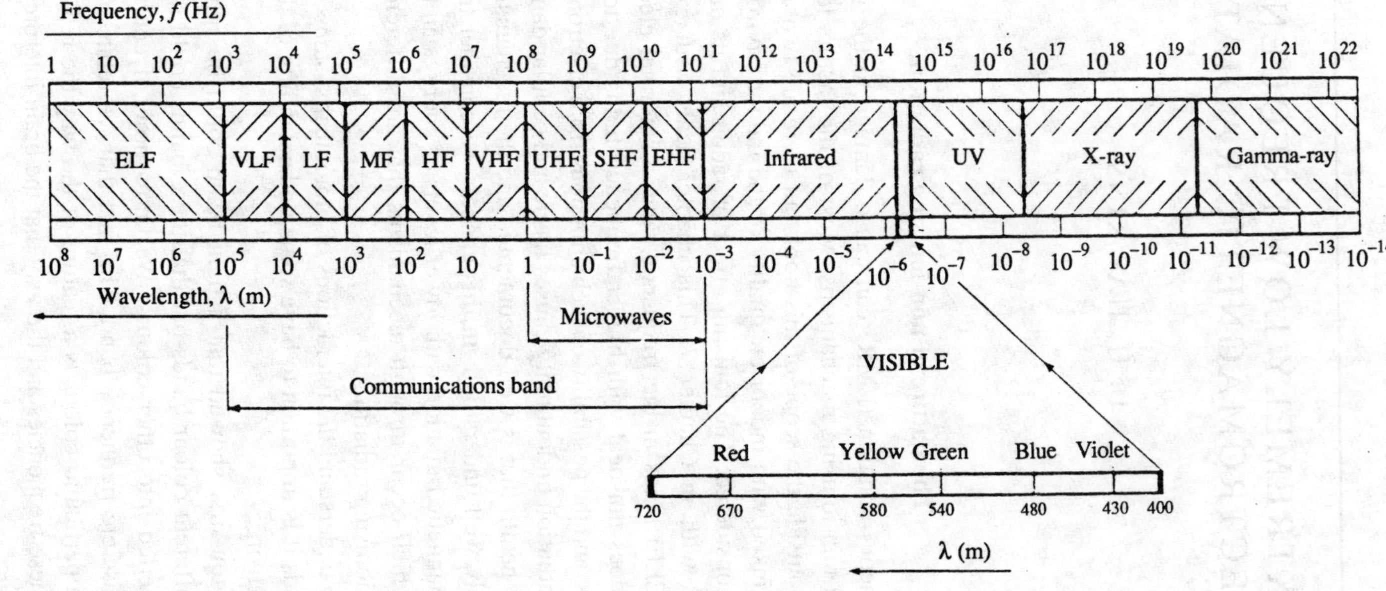

Fig.14.1. Electromagnetic spectrum.

from the retina. If we did not have visual mechanisms to interpret electromagnetic frequencies as colour, we would live, as some animals do, in a black and white world, unaware that colour existed. In the same way, we are unaware that most frequencies outside of the narrow band of visual light are present around us, since we have no physiological mechanisms to detect them. The exception to this is that some frequency bands, the ultraviolet, the infrared and microwave bands will cause some heating of the body. Visual light is one small band of frequencies in a spectrum that ranges from near zero to over 10^{24} cycles per second (Hz).

Figure 14.1 shows the full spectrum and the names we have arbitarily given to bands within it. The highest frequencies (above 10^{18} Hz) have sufficient energy to displace electrons from neutral atoms to create charged atoms or molecules. Charged atoms or molecules are referred to as ions, and for this reason, this high frequency (high energy) end of the electromagnetic spectrum is known as ionising radiation.

An electrical field is a force that exists between two charged objects. The strength of the field will depend on the potential difference between the objects and the distance between them.

$$E = V/D$$

Potential difference (volt) = V

D = Distance (metres) E = volts per metre (V m^{-1})

Where E is the electrical field strength.

If this field is unvarying with time it is said to be an electrostatic field. This would be the type of field associated with the electrical charge on a computer visual display unit (VDU) or nylon carpeting for example.

The physics of a time-varying field is more complex. It is, however, these time-varying fields that constitute the electromagnetic spectrum.

Suppose we twang a guitar string. The string will distort and a wave of motion will pass along it, much like a ripple on a pond.

The size of the ripple is the amplitude and the distance between one ripple and the next is the wavelength. The number of complete waves which pass along the string in a given time is the frequency (Fig.14.2).

The same thing can occur with an electrical field. If the source of the field is oscillating in magnitude, then the field it produces will oscillate and have a frequency and wavelength. However, there is something special about oscillating fields. They can travel long distances through space and possess energy. They radiate from a charged source and do not have to exist

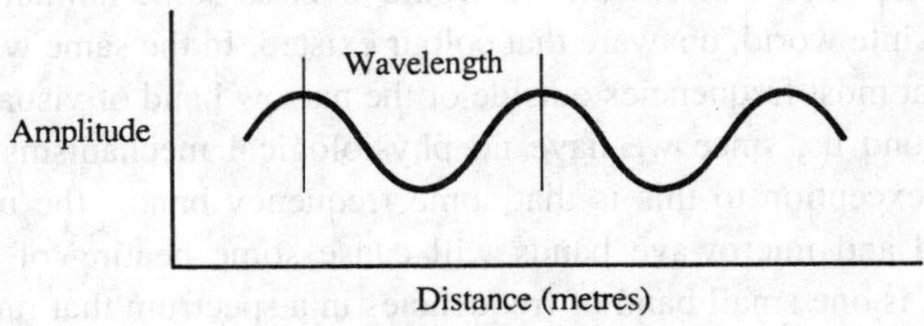

Fig. 14.2

between two charged objects. The energy they possess is inversely proportional to the wavelength (or the higher the frequency the more energy the field exerts).

It is easy to see how waves pass through an elastic medium (like a guitar string). A displacement at one point in the medium produces a force that acts upon neighbouring points so that it is repeated and propagated through the medium. The same is true of the transmission of sound waves through air. Air is able to be displaced just like any other elastic medium. But what about transmission of electromagnetic waves through a vacuum – that is much more difficult to understand since there is no medium to displace.

The problem is solved by considering that electromagnetic fields have two components, E and B. As we have seen E is the electrical field. B describes the magnetic field component. Maxwell showed that electrical fields that are time varying (i.e. which oscillate) produce a magnetic field. If this magnetic field oscillates (which it will if the electrical field oscillates) then this will in turn produce an oscillating electrical field. There is therefore a close interaction between the electrical field and magnetic field – each producing the other.

The electrical field acts as a sort of elastic medium which allows the magnetic field to propagate. Electromagnetic fields are therefore very special in being able to propagate through a vacuum by creating and carrying with them their own elastic medium. This is just as well, since without this phenomenon light and heat (infrared radiation) would not travel through space and the sun would not light or heat the earth. Electromagnetic fields propagate themselves through space at the velocity of light ($c = 2.998 \times 10^8$ m/s).

Sources of electromagnetism

Electromagnetic fields, both ionising and non-ionising are a natural feature of our environment. The sun radiates electromagnetic (EM) fields in the infrared, visible and ultraviolet regions. Cosmic rays arriving at the earth from outer space are another source of extra-terrestrial EM radiation. The earth itself has both an electrical and magnetic field. The magnetic field is about 0.35 oersteds (Oe) at the equator and 0.6–0.7 Oe at the poles. The earth also has a vertical electrical field of 120–150 Vm^{-1} in fine weather conditions, although this can increase up to 10 000 Vm^{-1} during a thunderstorm. These fields are direct current (DC), but in addition there are a number of alternating current (AC) fields originating from electrical storm activity. These so-called Schumann resonances at 9 Hz, 15 Hz, 20 Hz, 26 Hz and 32 Hz can travel long distances and are a source of extremely low frequency (ELF) radiation which are present most of the time in the atmosphere.

The relationship between natural atmospheric electricity and biological effects is interesting because it suggests that biological systems are sensitive to non-ionising radiation. It is sometimes claimed that man does not possess the sense organs to detect EM fields (with the exception of the eye and heating effects), and that for this reason such fields could not cause physiological effects. (Other animals are able to detect minute electrical and magnetic fields and use this for detecting prey and navi-gation.) The evidence that man does respond to natural atmospheric electrical events is beyond the present discussion, but a review of this can be found elsewhere (Hawkins, 1989).

In addition to natural sources of EM radiation, man is now exposed to an ever-increasing power and complexity of electromagnetic frequencies. Since the beginning of the century we have been exposed to an exponential rise in the power of radiated fields – but perhaps more importantly the range of frequencies of these fields is now being greatly extended. Since the advent of the transistor and microelectronics the range of frequencies that we can generate has dramatically increased to fill most of the region from 10^{11} Hz downwards (see Fig.14.1). These frequencies are being used for radar, television, radio, radio-telephones, military communications and other forms of communication.

We are exposed to the fields generated from overhead power line distribution, and to the fields created by the cabling and electric appliances in our homes and at work. The two major man-made sources of electrical and electromagnetic fields are therefore power distribution and usage and telecommunications.

Extremely low frequency electromagnetic fields and health

The question as to whether exposure to radiated ELF fields poses a health risk is a fundamental one in environmental and public health, and yet is one of the most controversial and difficult to answer. Evidence that humans are affected by such fields has come from two types of study – epidemiology of health risks, and laboratory-based studies of the effects of EM fields on tissues and cells.

Epidemiological evidence has been sought from people living near to sources of EM fields (usually power lines) and secondly from people working in the electrical transmission industry or with electrical equipment. In these studies it is difficult to separate the electrical and magnetic components of exposure and, indeed, in many studies there is no measure of exposure at all. Where there is, it is often an indirect measure such as the distance of dwellings from the source of exposure or the configuration of domestic wiring. For this reason and others, epidemiological studies have never been able conclusively to show an association between EM exposure and a health effect although many have pointed strongly in that direction.

The first well controlled Western study linking cancer to EM exposure was reported by Wertheimer and Leeper in Colorado in 1979. They compared 328 cases of cancer deaths in young people aged 18 or under with age-matched healthy controls. Their results showed a relative risk of childhood cancers (especially leukaemia) of about 2.0 when the subject lived near to a 'high current configuration' (HCC). HCC is an indirect measure of exposure which identifies transmission cables, transformers and substations and their proximity to the dwelling. The causes of childhood leukaemia are complex, however, and there may be a genetic element involved not allowed for in the selection of the controls. See, for example, Gardner *et al.* (1990).

Tomenius *et al.* (1982) studied 716 young people aged 18 or under who had developed tumours between 1958 and 1973. They used a control group matched for age, sex and district of birth and found that overhead power lines were in close proximity to the homes of over twice the number of cancer patients as compared to controls. In a later analysis of the data, Tomenius (1986) calculated that the relative risk for malignant disease was 2.7 in males and 1.5 in females whose homes were near to 200 kV transmission lines.

In a follow-up to their earlier study, Wertheimer and Leeper (1979) measured residential exposure to 60 Hz ELF fields in the houses of 1179 adults with cancer and in the houses of an equal number of age, sex and socioeconomic matched controls. Exposure was again estimated by HCC index. They found a statistically significant excess mortality of cancers of the central nervous system, uterus, breast and lymph glands associated with high current configuration. Several other authors have demonstrated that exposure to EM fields is associated with an increased risk of central nervous system tumours. Dutta and Millis have made an excellent review of the literature on electromagnetic fields and brain tumours (Dutta & Millis 1986). Preston-Martin *et al.* (1982) reported an excess of brain tumours among electricians and engineers in the aircraft industry. Using data extracted from the Swedish Cancer Registry, Englund *et al.* (1982) found an excess of brain tumours among welders and other occupational groups judged to have been exposed to ELF fields in their job. Many other authors have demonstrated a similar effect and an increased relative risk of various types of central nervous system tumours with ELF exposure seems, at least by epidemiological evidence, to be fairly well established. The other main type of malignancy associated with ELF field exposure is leukaemia. All the above investigations need a feasible causal link to be convincing.

As well as the studies by Wertheimer and Leeper (1979), and Tomenius *et al.* (1982), several other investigators have shown a positive association between either occupational or domestic exposure to EM fields and various types of childhood and adult leukaemia (Wright *et al.*, 1982; Coleman *et al.*, 1983; Coleman, 1985; Savitz *et al.,* 1988). However, the evidence for leukaemia reported by these investigators is not as strong as for central nervous system tumours, i.e. relative risks of between 1.05 and 1.9. Several authors have reported no association between leukaemia and EM fields (McDowall, 1986; Severson *et al.*, 1988; Fulton *et al.*, 1980).

Although a possible relationship between ELF fields and cancer has received the most study a number of authors have looked for association with other conditions. There is evidence from at least two studies that depression and headache are more common among those living near to overhead power lines (Dowson & Lewith, 1988; Perry *et al.*, 1989). Reichmanis and her colleagues showed a link between suicides and proximity to overhead power lines (Reichmanis *et al.,* 1979) and other studies have shown an association between EM exposure and heart disease (Perry & Pearl, 1988).

The mechanism of biological interaction with EM fields

EM fields, like other stressors, may show an inverse 'U' shaped relationship between the magnitude of the stressor and the stress response. Increasing dose will not necessarily increase effect (i.e. pathological risk) and there may be an optimum dose for a response, above and below which the effect diminishes. EM fields have three components (the electrical field, the magnetic field, and the frequency) each of which may have its own individual effect. There is increasing evidence that a progressively increasing relationship between dose and response may not be an appropriate model.

Delgado *et al.* (1982) used developing chick embryos to study the effect of magnetic field strength and frequency on growth abnormalities. Fields of 100 Hz and 1.2 tesla (T) had a powerful effect on the developing embryo, either delaying or arresting growth and limiting severely the differentiation of tissues.

This particular 'window' of 100 Hz and 1.2 T produced the optimum effect whilst lower or higher frequencies (10 Hz and 1000 Hz) or lower or higher magnetic fields (0.12 T or 12 T) did not produce significant effects. Other investigators have failed to reproduce these precise results.

This 'window' phenomenon has been demonstrated in many other experiments. Adey (1980) for example has shown that the calcium efflux from cells in the chicken brain occur optimally at 6–16 Hz with fields of between 10 and 56 Vm^{-1}.

Although different experiments have identified different 'windows' (either because different experimental conditions have been used or because there may be a multiplication of critical windows), there seems little doubt that biological systems can interact with external EM fields and that the effect is optimal at certain critical combinations of power and frequency. In addition, it is evident from all of the available literature that frequency has a more important effect than power and that the critical frequencies so far identified are in the ELF region (below 300 Hz). It is also apparent that the effect of EM fields is at the cellular level. In addition to his earlier work on calcium efflux from cells, Adey and his colleagues have demonstrated that EM fields can stimulate the activity of ornithine decarboxylase, an enzyme that is essential for the growth of both normal and cancer cells (Byus *et al.*, 1987).

These types of experiments have led to the development of a model which might explain how certain EM fields relate to an increase in cancer risk. The model requires that the electromagnetic field is not the primary cause of cancer, but that other environmental initiators trigger the cellular responses leading to malignancy. EM exposure then increases the risk in two possible ways. It acts as a cell growth promotor (by an effect on cellular growth promoting enzymes) and acts simultaneously to depress the immune responses that might otherwise have detected and removed the cancer cells.

The proposal that the body's immune system is suppressed by EM exposure is given credence by studies on the effects of EM field exposure on lymphocyte behaviour. Conti *et al.* (1983), using *in vitro* methods, found that in the critical frequency window of 40–70 Hz (with an electrical field of 0.1–1 Vm^{-1}), the production of lymphocytes is significantly reduced. In perhaps a more significant paper, Lyle *et al.* (1988) has shown that the cytotoxicity of T-lymphocytes is considerably suppressed in the presence of a 60 Hz field (at between 0.1–1.0 $mVcm^{-1}$). This finding has particular relevance to the relationship between EM exposure and cancer since T-lymphocytes are known to play an important role in the elimination of certain infectious organisms, tissue graft rejection and tumour immunity.

Conclusions

Man has evolved in an environment in which natural non-ionising electromagnetic fields have become incorporated into basic physiological control systems. Fundamental control of cellular organisation, membrane function, cell reproduction and growth may all be dictated by electromagnetic forces.

Artificial electromagnetic fields, to which we are now exposed at an increasing power and frequency spectrum, may in certain circumstances interfere with these basic control mechanisms. Our evidence so far is that cellular responses occur at critical frequencies or 'windows' and that many of these windows occur in the ELF region. It is of interest that the earth's natural EM field (outside the visual spectrum) consists largely of ELF frequencies.

Cellular effects may be related to an increased risk of behavioural dysfunction (depression and suicide for example), heart disease and other illnesses. The majority of epidemiological evidence however points to a significant association between exposure to EM fields and an increased risk of cancer. The effects of EM fields on cellular growth and suppression of cellular immunity can be postulated to explain an association between exposure to ELF electromagnetic fields and cancer. The precise importance of these effects as related to the indoor environment is still very unclear.

References

ADEY, W.R. (1980). Frequency and power windowing in tissue interactions with weak electromagnetic fields. *Proceedings IEEE*, **68**, 119–25.

BYUS, C.V., PIEPER, S.E. & ADEY, W.R. (1987). The effects of low energy, 60 Hz environmental electromagnetic fields upon the growth related enzyme Ornithine Decarboxylase. *Carcinogenesis*, **8**, 1385–9.

COLEMAN, M. (1985). Leukaemia mortality in amateur radio operators. *Lancet*, ii, 106–7.

COLEMAN, M., BELL, J. & SKEET, R. (1983). Leukaemia incidence in electrical workers. *Lancet*, **i**, 982–3.

CONTI, P., GIGANTE, G.E., CIFONE, M.G., ALESSE, E., IANNI, G., REALE, M. & ANGLETTI, P.V. (1983). Reduced mitogenic stimulation of human lymphocytes by extremely low frequency electromagnetic fields. *FEBS Letters*, **162**, 156–60.

DELGADO, J.M.R., LEAL, J., MONTEAGUDO, J.L. & GRACIA, M.G. (1982). Embryological changes induced by weak, extremely low frequency, magnetic fields. *Journal of Anatomy*, **134**, 533–51.

DOWSON, D. & LEWITH, G.T. (1988). Overhead high voltage cables and recurrent headache and depression. *The Practitioner*, **232**, 435–6.

DUTTA, S.K. & MILLIS, R.M. (eds.) (1986). Biological effects of electropollution. *Brain Tumours and Experimental Models*. Philadelphia: Information Ventures Inc.

ENGLUND, A., EKMAN, G. & ZABRIELSKI, L. (1982). Occupational categories among brain tumour cases recorded in the cancer registry in Sweden. *Annals of the New York Academy of Sciences*, **381**, 188–96.

FULTON, J.P., COBB, S., PREBLE, L., LEONE, L. & FORMAN, E. (1980). Electrical wiring configuration and childhood leukaemia in Rhode Island. *American Journal Epidemiology*, **111**, 292–6.

GARDNER, M.J., SNEE, M.P., HALL, A.J., POWELL, O.A., DOWNS, S. & TERRELL, J.D. (1990). Results of case-embryo studies of leukaemia and lymphomia among young people near Sellafield nuclear plant in West Cumbria. *British Medical Journal*, **300**, 423–9.

HAWKINS, L.H. (1989). Biological effects of electromagnetic fields. IOP Short Meetings. Electric field phenomenon in biological systems. *IOP Short Meetings* Series No. 21. Bristol: Institute of Physics Publishing.

LYLE, D.B., AYOTTE, R.D., SHEPPARD, A.R. & ADEY, W.R. (1988). Suppression of T-lymphocyte cytotoxicity following exposure to 60-Hz sinusoidal electric fields. *Bioelectromagnetics*, **9**, 303–13.

McDOWALL, M.E. (1986). Mortality of persons resident in the vicinity of electricity transmission facilities. *British Journal of Cancer*, **53**, 271–9.

PERRY, S. & PEARL, L. (1988). Power frequency magnetic field and illness in multistorey blocks. *Public Health*, **102**, 11–18.

PERRY, S., PEARL, L. & BINNS, R. (1989). Power frequency magnetic fields; depressive illness and myocardial infarction. *Public Health*, **103**, 177–80.

PRESTON-MARTIN, S., HENDERSON, B.E. & PETERS, J.M. (1982). Descriptive epidemiology of central nervous system neoplasms in Los Angeles county. *Annals of the New York Academy of Sciences.*, **381**, 202–8.

REICHMANIS, M., PERRY, F.S., MARINO, A.A. & BECKER, R.O. (1979). Relation between suicide and electromagnetic field of overhead power lines. *Physiological Chemical and Physics*, **11**, 395–403.

SAVITZ, D.A., WACHTEL, H., BARNES, F.A., JOHN, E.M. & TVRDIK, J.G. (1988). Case-control study of childhood cancer and exposure to 60 Hz magnetic fields. *American Journal of Epidemiology*, **128**, 21–38.

SEVERSON, R.K., STEVENS, R.G., KAUNE, W.T., THOMAS, D.B., HEUSER, L., DAVIS, S. & SEVER, L.E. (1988). Acute non-lymphocytic leukaemia and residential exposure to power frequency magnetic fields. *American Journal of Epidemiology*, **128**, 10–20.

TOMENIUS, L. (1986). 50 Hz electromagnetic environment and the incidence of childhood tumours in Stockholm County. *Bioelectromagnetics*, **7**, 191–207.

TOMENIUS, L., HELLSTROM, L. & ENANDER, E. (1982). Electrical constructions and 50 Hz magnetic field at the dwellings of tumour cases (0–18 years of age) in the County of Stockholm. *Proceedings of the International Symposium on Occupational Health and Safety in Mining and Tunnelling*. Prague.

WERTHEIMER, N. & LEEPER, E. (1979). Electrical wiring configuration and childhood cancer. *American Journal of Epidemiology*, **109**, 273–83.

WRIGHT, W.E., PETERS, J.M. & MACK, T.M. (1982). Leukaemia in workers exposed to electrical and magnetic fields. *Lancet*, **ii**, 1160–1.

15

CONTRIBUTIONS FROM OUTDOOR POLLUTANTS

G. W. CROCKFORD

Introduction

Some authorities such as the American Environmental Protection Agency consider that the risks to health and comfort from airborne pollutants may on occasions be worse inside buildings than outside them and that indoor air pollution could represent, for many people, a major portion of their exposure to air pollution. Although a large part of the indoor airborne pollutants arise from sources within the building, the outdoor air is by no means clean and in some situations can be heavily polluted. As the air within a building is in continual exchange with outdoor air, pollutants of outdoor origin are going to play some part in determining what the occupants of buildings are exposed to, at what concentrations and for how long. In at least two important respects the outdoor air is also going to influence the concentration of pollutants of indoor origins. First, low humidities give rise to more dust generation and high humidities to mould growth (Burge & Hoyer, 1990), while, second, high temperatures increase emissions of volatile materials from such things as furnishings. Awareness that outside sources and conditions could influence the quality of indoor air goes back to Roman times and probably earlier, when, in Londinium, tanneries, which used human urine, were moved downwind from the town. Probably the most dramatic examples of outdoor air influencing indoor conditions were during the smogs that occurred in many big cities until the 1960s and which penetrated buildings both as a visible aerosol and as invisible gases (Spedding, 1974). The smogs caused many thousands of deaths, and bronchitis in a considerable proportion of the population. Today, such pollution is on a much smaller scale but can nevertheless be seen in valleys during the autumn and winter where the smoke from bonfires and the few remaining houses that burn coal and wood can be seen slowly moving along the valleys, engulfing the dwellings in its path. In certain weather conditions, pollutants do not quickly dilute and disperse but just accumulate in the area around the source or slowly move across the ground, the cloud retaining its identity possibly for miles down wind.

The location of a building in relation to local sources of pollution is a major factor in determining what is going to enter the building, and the weather conditions influence the concentrations that are going to develop and the frequency of exposure. The pollutants take the form of particulates such as dust, smoke, fume and mists. The size of the particles varies from the submicron size range associated with fumes, smokes and the hazes resulting from chemical reactions within the atmosphere, to large particles many microns in size arising from traffic, industrial activity of various types and natural sources such as pollen and spores. Because the larger particles (>75 μm – aerodynamic diameter) fall out quickly (Spedding, 1974) they are normally only found in the vicinity of the source unless vigorous air movements keep them airborne. The small particles are eventually washed out by rain and snow. On entering a building, particulates will settle out on surfaces including the inside of ventilation ducts where, because many will be organic in nature or contain a percentage of organic matter, they provide a medium for mould growth within the ducts. The particles that remain airborne and those that enter through windows, doors and other gaps settle on all available surfaces causing deterioration of decorative surfaces and light fittings, equipment malfunctions, and spoilage of stored products in terms of customer appeal. Fungal and bacterial spores, very often carried on other particles, are of considerable importance to some sectors of the food processing industry and with pollens can be responsible for allergic reactions such as asthma. In some cases pollutants can enter buildings on the clothing and shoes of people. Cases have been recorded where substances such as mercury, lead and asbestos have been carried into the home on clothing and caused illness in members of the family.

Outdoor pollutants can often be detected by smell, irritation of the eyes or seen emerging from their source. Other signs such as haze, damage to vegetation and buildings, sunlight picking up small particles and discolouration of surfaces may indicate the presence of pollutants in the atmosphere. Because of the importance of airborne pollutants the World Health Organisation and the American Environmental Protection Agencies have promulgated guidelines or standards for outdoor air quality, (WHO, 1987).

Pollutant sources

There are some very important pollutants such as pollen, fungal spores, dusts from soils, and organic chemicals released into the environment by natural biological activity such as decay processes. However, the sources of most concern tend to be from human activity of one form or another. Agricultural activities lead to emissions, often on a very large scale, of smoke from burning, pollens from agricultural crops such as rape and grasses, chemicals used for pest control, dust from harvesting and soil preparation, gases such as ammonia and methane from factory farming operations. Industrial operations of all sizes and kinds release gases, vapours, fumes, smokes and dusts into the air. Although the small industrial and commercial establishments may at first sight appear of minor significance they are very numerous, cannot normally afford the latest or best forms of control, tend not to be so well managed with respect to environmental concerns as larger companies, and often are very close to commercial and residential property.

Burning

Burning is widespread. Even today, domestic coal and wood fires and bonfires create noticeable and odorous pollution. Many small businesses such as grocers, butchers, and furniture removers burn waste. The small size of the average smoke particle, about 0.07–0.2 μm, renders them invisible unless in quantity and enables them to penetrate buildings as readily as gas molecules. The particles are often tarry hydrocarbons and resins with a high 'sticking' power and in areas of turbulence such as leaks from ventilation systems, edges of doors and windows can be seen to be discoloured. The size of the particle that is released into the atmosphere with the flue gases is determined by the flue-gas velocity. In an industrial chimney the velocity may be 12 m s^{-1} and be able to carry out of the chimney particles larger than 200 μm in diameter. Such large particles will settle close to the chimney. The flue gas velocity in a domestic chimney is much lower, about 1.5 m s^{-1}, and the maximum particle size is about 75 μm in diameter. The composition of the pollutants released is determined by what is being burnt, the temperature of combustion and the availability of oxygen. With modern synthetic materials being added to bonfires or being incinerated as part of waste disposal the range of combustion products can be very large. They include carbon dioxide and monoxide, nitrogen oxides, smoke containing droplets of hydrocarbons, phosgene, hydrochloric acid, acetic acid, acrolein, polycyclic aromatic hydrocarbons such as the benzopyrenes. When PVC is burnt, chlorinated PAHs such as dioxins may be produced and also 2,3,7,8 - tetrachloradibenzo-*p*-dioxin, said to be a most potent carcinogen. Chipboards release formaldehyde and polyurethane plastics release cyanide.

A tonne of garden waste can produce 30 kg of carbon monoxide (Gamlin & Price, 1988) and lead to high carbon monoxide levels in the neighbourhood. The pollutants from such fires are, moreover, being added to that from other sources such as vehicles and industry and other people's diluted bonfire smoke.

Analysis of bonfire smoke (Gamlin & Price, 1988), has shown that it contains the carcinogenic benzopyrenes at concentrations about 300 times that in cigarette smoke. Combustion products from other sources such as wood also contain carcinogens. Combustion products both as gases and small particles gain ready access to buildings of all types, as few have filters that can remove them. Though the pollution from any one source dilutes as it moves with the wind, other pollutant sources add to it and although the particles and gases may not be visible or reach the odour threshold, they very often still cause eye irritation and have a direct influence on health status, particularly asthma, bronchitis, and lung infections (Burge & Hoyer, 1990).

Waste disposal

Modern incinerators burn materials under controlled conditions and emit flue gases at a sufficient height to ensure dilution before they come down to ground level. The range of combustion products produced is considerably smaller than that produced by fires in domestic premises, building sites, small businesses, etc. However, hydrochloric acid, nitrogen oxides, particles and some unburned volatiles may be emitted, and the furnace ash may possibly contain heavy metals. The waste coming in to the incinerator has to be handled and emissions affecting the neighbouring area may therefore occur from ground level sources.

Much waste is buried in land fill sites where smells, micro-organisms, methane and dust may produce local problems. Gases such as methane can migrate underground and seep into buildings as can gas from gas mains. Explosive concentrations of methane can exist 1000 ft from a land fill site, (O'Leary *et al.*, 1988), and, in the USA, several deaths have been caused by explosions in buildings from methane migration. If methane migrates through soil it is likely that many other gases will do likewise.

Demolition sites

When the Fulham Power Station in London was demolished in 1983, it attracted a great deal of local attention because it was known that it contained asbestos.

The airborne contaminants arising from a site are going to reflect the site's previous use and history. The main problems arise from heavy metals, asbestos, and chemicals that have, over the years, soaked into the ground. Sometimes the demolition processes such as thermal cutting can produce fumes of such things as lead, zinc or any other material that may be on the steel work being cut up. Generally, the large particulates fall out quickly, and the small particles, fume and gases remain airborne and penetrate nearby buildings. In the case of the Fulham Power Station, asbestos was identified at the boundaries but at a very low fibre count, 0.06 f ml^{-1} (Cova, 1983).

Industrial activity

Airborne emissions from industrial activities have been reduced substantially over recent years in the advanced countries. Nevertheless, substantial emissions still occur and the question is whether they are of significance in terms of penetrating buildings and producing problems. Clearly, sulphur dioxide, carbon monoxide, nitrogen oxides and fine particulates are produced by many combustion processes, whether engine exhausts, furnace gases or central heating systems and these will penetrate buildings. There are also all the new chemicals as well as older ones that are now being used very widely and in considerable quantities. In a week, a flexographic printing machine working 24 hours a day may use 1600 litres of ethanol and 400 litres of methoxy propanol. Such a machine can be in a small factory and emitting about 12 l an hour of solvent as vapour into the atmosphere and then to other buildings. Alcohol may not be of great concern to neighbours but other solvents and isocyanates, a powerful sensitizer producing asthma, could be. A small garage may well have facilities for respraying vehicles using two-pack polyurethene paints. Larger facilities with a number of spray booths and ovens for baking the car bodies are also very common and both can be within yards of houses and commercial premises. In cleaning the air from such facilities, the paint spray will be filtered out but only with a certain efficiency. The smaller particles, which carry furthest, will penetrate the filters more readily than the larger ones and a good filtration efficiency will be 98% or less. The isocyanate is normally washed out using a scrubber, but many plants do not have scrubbers, which are, in any case, only 80–90% efficient. The paint solvents will be partly removed by the scrubber if water soluble; if not, most will be released into the air. The exhaust stack is very often at roof height and in the building boundary layer. The emissions are therefore being released effectively at ground level, the ideal position for gaining entry to neighbouring buildings. Even with good effluent treatment, it is estimated that 1% of total paint sprayed with its isocyanates will be emitted, plus the solvents. In terms of isocyanates the stack concentration is about 0.5–1.3 of the maximum exposure limit.

If the factory is running a two-shift or three-shift system emissions could occur in the evenings and during the night which is important because the sensitivity of people and animals to toxic chemicals depends on a number of factors but one is the time of day with the possibility of sensitivity at night being higher than during the day. This is potentially another factor that may have to be considered when evaluating the health risks of modern chemicals when they escape into the atmosphere.

One source of pollutants that from time to time give rise to problems are sources of emission just outside the building such as drums of waste materials, work generating dusts and fumes, and materials degassing from the ground after spillages. In a recent incident, petrol fumes gave rise to problems in an office building with the hydrocarbon concentration measurements showing levels of 12 to 96 mg/m^3. The sources were three abandoned petrol tanks buried beneath an adjacent building. These were removed but some vapours were still given off by contaminated ground to produce the occasional problem.

Atmosphere dynamics

The contaminants in the air at any one time are reacting with one another to form new compounds. Sometimes the reactions require the presence of sunlight.

One of the simplest reactions is the oxidation of sulphur dioxide to sulphuric acid and the production of ammonium sulphate. All the sulphuric acid present in the air is not neutralised at once and it attaches itself to particles or to surfaces where it participates in the corrosion process of both indoor and outside surfaces. It is known that sulphur dioxide and its products are also partly responsible for the deterioration of paper and books. By adsorption on to particles, sulphur dioxide, which is a highly soluble gas, can gain entry to the deeper parts of the lung where it may cause irritation and damage whereas it would normally be removed high up in the respiratory tract.

The chemicals produced by photochemical reactions involving NO, NO_2, the hydrocarbons from vehicle emissions, industrial activity and vegetation produce haze, ozone and the chemicals responsible for eye irritation, damage to vegetation, and respiratory problems in humans.

Amongst the chemicals produced by the action of sunlight are nitrogen dioxide, ozone and aldehydes. Until recently it was considered that the ozone produced by photochemical reaction did not penetrate to any significant extent into buildings and that human exposure was limited to the period spent out of doors. The idea is that indoor surfaces scavenge the ozone from the air so producing a marked lowering of concentration indoors. However, the ozone must have time to react with surfaces, now considered to be in excess of an hour, so differences in some concentrations must be expected between well ventilated and poorly ventilated buildings (Levin, 1989). Recent research indicates that indoor ozone concentrations are frequently a significant proportion of outdoor values. If this is the case and bearing in mind that most people spend about 90% of their time indoors, it may be that the indoor exposures to ozone are the more important.

The ventilation rate of a building is going to have a pronounced effect on the occupant's exposure to ozone. Where it is low, 0.4 air changes an hour or less, the ozone concentration indoors will be substantially lower than the outdoor value. Research indicates lower values of 20% of the outdoor level up to 80 or even 90% of that outdoors. Adverse health effects occur at concentrations as low as 80 ppb for periods as little as 1 hour, (Levin, 1989).

Outdoor ozone levels in the USA of 100 ppb are often exceeded and it is estimated that a large portion of the US population is exposed to unhealthy ozone levels, particularly when building ventilation rates exceed 1 air change per hour at which value indoor ozone levels of 50% of outdoor levels are reached. In London, ozone levels of 140 ppb (1 h mean) have been recorded with 210 ppb having been reached in 1976 (Parker & Ball, 1983; Glover & Lightman, 1983). The London values could be exceeded down wind of the city as photochemical pollution takes 1 to 2 hours to form by which time the air mass is in the surrounding countryside.

With ozone as an example it is possible to see why outdoor air quality standards have attracted attention, the one for London is 80 ppb and for the US 120 ppb. The American Society of Heating Refrigeration and Air Conditioning Engineers have a Standard 62–89 'Ventilation for Acceptable Indoor Air Quality' (ASHRAE, 1989) which, if it is obeyed, will mean removing ozone from outdoor air when levels are sufficiently high to enable indoor levels to breach the standard before supplying air to the building. The standards for indoor contaminant levels and information on what values can be reached in outdoor air tell the engineer when appropriate air cleaning is required.

The recognition of health risks from ozone are of relatively recent origin. Pollen, fungal spores, and combustion products have been a problem for, probably, several thousand years but now it is possible to control the concentrations of them entering a building. It may be possible in some areas to design the outdoor environment so as to minimise pollution indoors. For example, care should be taken in selecting trees and plants for urban development as pollen can so easily be carried into the building. In addition, common sense should be applied in construction and planning, so that, for example, factories are not sited next to dwellings.

Transport to the building

The dominant factor in determining what contaminant and how much gets into a building is its proximity to a source. Although large highly visible industrial plants may be the source of many emissions they are normally at some distance from residential and commercial property. The stack emissions from such plants are likely to be carried clear of property and diluted by the time the plume comes down to ground level. The fugitive emissions, i.e., the leaks and minor emissions, produced by the normal activity of the plant, dust from roads, emissions from treatment tanks, and roof level emissions are all being released into the air essentially at ground level. The plume of contaminated air from these sources therefore moves across the ground with the prevailing wind, its concentration depending on activity at the plant, wind speed and any local features that aid mixing or restrain it. Air does not normally mix very quickly, as can be seen by plumes of smoke from stacks and aircraft vapour trails, unless there is an energy input that creates turbulence. It is therefore possible for pollutants to reach distant buildings at very high concentration unless precautions have been taken at the source to dilute and disperse. This can be the case when dilution or general ventilation is being used in the factory but local exhaust ventilation systems will often lead to the emission of contaminants at a high concentration. With poor dispersion, high concentrations of pollutants build up locally and in the plume downwind of the source, and any building in its path is likely to be exposed to concentrations well in excess of ambient air quality standards.

Due to the vagaries of the weather, such situations when neighbours are exposed to high concentrations may be infrequent with the result that engineers do not normally build the necessary air cleaning equipment into the ventilation system of the building. The result is that the ventilation system will quickly disperse polluted outdoor air throughout a building.

Entry to the building

A building normally has many gaps in its structure: windows, doors, cracks in masonary and panelling, ventilation system, fans, and air bricks. Without mechanical ventilation, there is therefore an air exchange rate which enables outdoor pollutants to gain entry and to gradually build up the internal concentration with time. The indoor concentration lags behind the outdoor level both during any rise and the consequent fall back after the outdoor levels have returned to normal. Aerosols that enter will tend to settle and will require mechanical energy to get them airborne again. Gases and vapours that have been adsorbed onto surfaces will de-gas and retard the return of the indoor air to the level of the ambient air. This phenomenon is observed, for instance, on clothing which has been worn in the presence of cigarette smoke. The smell of cigarette smoke emanates from the clothing for 24 hours or more after the exposure. Some gases, of which ozone is an example, are destroyed by the adsorbing surface.

Although the fabric of the building and furnishings control, to a certain extent, the indoor environment, it is possible to clean the air in a number of ways. A filter on air intakes will keep out a percentage of the particles although the smaller ones will tend to penetrate as will gases and vapours. Unfortunately houses and many offices rely on natural ventilation or simple fans to blow air in or out. Even when mechanical ventilation is installed, gas filters, for example charcoal, are seldom fitted, so there is only limited control over the internal environment.

Many occupants are therefore faced with the problem of trying to achieve some degree of control over pollutants of outdoor origins without expensive modification to the building. One of the most popular ways of doing this is to recycle the indoor air through electrostatic filters which also tend to be fitted with charcoal filters for the removal of gases and vapours. They can be extremely effective with, for example, pollen, many other particulates and ozone. They do not distinguish between pollutants of indoor or outdoor origins and can be justified on the basis of both. It is important, however, to recognise that as they are only cleaning and recycling the air, they can only reduce the level of contaminant, not remove it. The rate at which the pollutant enters, and the rate at which it is removed, brings the pollutant concentrations down to an equilibrium point.

Conclusions

The pollutants in the ambient air, whether of natural or man made origin, gain entry to buildings because of the normal air exchange between environment and building. The concentration that is reached within the building depends on reactions of the air exchange rate and the adsorption and chemical mechanisms which occur at the surface of walls and furnishings. These mechanisms for removal can become saturated and concentrations of some contaminants reach levels close to those outside. As the harmful effects of outdoor pollutants are identified and standards derived to protect people, there will be a growing need for air cleaning equipment to remove both particles and gases from air entering buildings.

The most satisfactory course of action is to control at source, one example being the use of catalytic converters to treat vehicle emissions, but this is not always possible. Emissions from many agricultural processes, for example, are difficult to control. The multitude of small sources, both industrial and domestic, make enforcement difficult yet their proximity to houses may be putting many people at risk. It may be that the enforcement approach to environmental pollution from these smaller but very important sources is unsuitable and bound to fail if a nuisance or direct harm has to be proved. Perhaps with any activity that is going to result in pollution, however small, the perpetrators should be able to demonstrate that they are using the best practical means of control, and to keep adequate records. Failure on either count could then lead to prosecution without having to demonstrate the occurrence of pollution.

In the immediate future, however, the best approach is to install air cleaning equipment. It will protect the decor and furnishings, reduce cleaning costs and hopefully protect the occupants.

References

ASHRAE (1989). *Standard Ventilation for Acceptable Indoor Air Quality*. Report no. 62–89. American Society of Heating Refrigerating and Air Conditioning Engineers, Inc.

BURGE, H.A. & HOYER, M.E. (1990). Indoor air quality. *Applied Occupational and Environmental Hygiene*, **5**, 84–93.

COVA, B.C. (1983). *London Environmental Bulletin*, **1** (2). The demolition of Fulham Power Station.

GAMLIN, L. & PRICE, B. (1988). Bonfires and brimstone. *New Scientist*, **5**, 48–51.

GLOVER, G.M. & LIGHTMAN, P. (1983). Airborne measurements of atmospheric pollutants in London and the South-East. *London Environmental Bulletin*, **1** (2).

LEVIN (1989). Ozone and indoor air: myth and reality. *Indoor Air Quality Update*, **2** (7).

O'LEARY, P.R., WALSH, P.W. & HOM, R.K. (1988). Managing solid waste. *Scientific American*.

PARKER, J. & BALL, D.J. (1983). Summer air pollution in London. *London Environmental Bulletin*, **1** (2).

SPEDDING, D.J. (1974). *Air Pollution*. Oxford Chemistry Series, Oxford Clarendon Press.

WHO (1987). *Air Quality Guidelines for Europe*. WHO Copenhagen. WHO Regional Publications, European Series no. 23.

16

VENTILATION FOR CONTROL OF AIRBORNE POLLUTANTS

P. APPLEBY

Introduction

Ventilation can be defined as the movement of air through a space whereby vitiated, or used, air is continually replaced with air from outside that space. This air may all originate from an external source, i.e. 'fresh' air, or some proportion of the air leaving the space may be recirculated.

The forces that set up this throughput of air may be natural, mechanical, or some combination of the two. Naturally induced forces are generated from wind pressure, internal convection currents or the two forces working together, via cracks or openings in the fabric of the building. The resultant air movement may be adventitious, in which case it is called infiltration, or deliberate and controllable, in which case it is known as natural ventilation.

Mechanical, or forced, ventilation is generally employed when natural forces cannot be relied upon to satisfy building or user requirements. The prime mover is a fan, comprising of a bladed impeller rotated by the action of an electric motor and thus creating additional capital and running cost when compared with a naturally driven system. There are many different types of forced ventilation, depending on the sophistication of control required. They range from a simple propeller fan mounted in a window through to full air conditioning, with separate supply and extract air distribution systems and facility to recirculate room air, filter, cool, heat, control humidity and even cater for different load patterns in different zones of a large building.

Regardless of the level of sophistication, every space has certain basic ventilation needs. These needs are dictated by the nature of the contaminant-producing processes which occur within the space. Contaminants evolve continuously, even in empty buildings where fabric and furnishings may emit a cocktail of constituent substances, such as formaldehyde, and temporary lodgers which have been adsorbed onto surfaces during occupancy, such as odours. When the space is occupied, additional 'fresh' air is required to dilute body odours and tobacco smoke as appropriate (see below). In addition there may be other processes requiring either dilution of low levels of contaminant emission or make-up air to replace that removed by a local exhaust system or combustion process.

Heat can be looked upon as another form of contamination and temperature rise in a space can be limited by the movement of outside air through that space. Depending on the dynamics of heat flow at the internal surfaces and cyclic variations in outdoor temperature, it is usual for peak internal space temperature to decrease with an increase in throughput of outdoor air. Furthermore, ventilation can generate air movement which can create a beneficial cooling effect at the surface of the body. Another potential source of contamination is water vapour and in some applications, particularly domestic, ventilation may be used to limit condensation. Ventilation may also have an important part to play in the control of smoke and protection of escape routes during a fire.

Since outdoor air may require treatment for much of the year and fan energy is related to the total volume flow, ventilation strategy is very closely tied into economic strategy for a particular project. Over-ventilation wastes energy, may result in over-sized hardware and can lead to draught or process problems. Under-ventilation may lead to some combination of odour, contamination, heat or moisture problems, and has been a major contributing factor in cases of 'sick building' syndrome, and chronic or acute illness and even death in some industrial applications.

The ventilation options available are characterised by the degree of control they offer over air movement and their degree of control over indoor air quality. For example natural ventilation provides limited control over air movement, dependant on careful positioning of openings in the fabric to make use of wind or convection forces as available. These forces have been observed to cause the ventilation rate to vary by a factor of 5 with changing weather conditions. Control over indoor air quality will fluctuate with natural forces, although the purity of the incoming air cannot be controlled at all. The ventilation efficiency (see below) will depend on the position of openings and the prevailing forces. Internal buoyant forces can produce a 'stack' effect which induces low level input and moves contaminants upwards. If there is enough incoming air then the contaminants will stratify above the occupied zone. This is displacement or buoyancy-assisted ventilation, where make-up and exhaust can either occur by natural or mechanical means.

Mechanical ventilation provides the designer with far more choices, although at greater cost. High-emission and toxic contaminants (including heat) can be controlled at source. General contamination and heat gain can be either diluted or displaced, and air can be distributed to wherever it is most needed. Contaminants can be removed from outdoor air and extract air, as appropriate, and the building can be sealed to reduce infiltration of air and noise from outside. Supply air condition can be automatically controlled to provide constant internal conditions if desired.

Fresh air for sedentary occupants

Leaderer & Cain (1983) have shown that the minimum ventilation rates for CO_2 control adopted in Britain and North America, of 5 and 2.5 litres/s per person respectively, were inadequate when considering odour levels perceived by people when first entering an already occupied space, and that the density of occupancy has no bearing on fresh air requirements. It is widely thought that the high incidence of complaints in North American air conditioned buildings can be largely attributed to the adoption of the 2.5 litres/s recommendation, leading to their coining of the phrase 'tight building syndrome'. Minimum fresh air requirements have been increased accordingly in both countries, to 8 litres/s per person.

Carbon dioxide dilution

Exposure to CO_2 at low concentrations is not harmful to human health. An acceptable concentration is considered to be 2500 mg/m^3. An individual's rate of respiration varies with activity, but on average each person requires about 0.2 litres/s of fresh air to provide sufficient oxygen for the metabolic processes. Carbon dioxide is expired at a rate which is primarily dependent upon diet.

It can be estimated that a man standing quietly with a metabolic rate equivalent to 1.2 met (1 met = 58.5 W/m^2) expires about 0.005 litres/s of CO_2. If the outside air contains 300 mg/m^3 of CO_2, it can be shown that a dilution air flow rate of 2.3 litres/s would prevent the concentration in the room rising above 2500 mg/m^3, assuming perfect mixing. After the oil crisis of the 1970s a fresh air rate of 2.5 litres/s per person was adopted by the North Americans as a minimum for sedentary activity with no smoking.

Odour control

Odours represent one of the biggest sources of complaint amongst sedentary workers, and are thought to be a significant contributing factor in many 'sick' buildings. There are many sources of odours in a building, including the ventilation or air conditioning system (Fanger *et al.*, 1988). The recently revised British and North American recommendations (CIBSE, 1986*a*; ASHRAE, 1989) (8 litres/s per person) result from concern over the incidence of ventilation-related problems in air conditioned buildings and are based on recent work in Denmark (Fanger, 1986) and N. America (Leaderer & Cain, 1983) which have concentrated on odour perceptions and reducing the percentage of dissatisfied occupants to an acceptable level. This work showed that there is a difference in perception between people entering a space and those who are acclimatised. Even with no smokers present, up to 40% of people entering a room supplied with 2.5 litres/s of air per person will be dissatisfied.

Unfortunately no standard has so far taken account of odours released by the building and services, or ventilation efficiency (see below). Fanger *et al.* (1988) have addressed the first deficiency by devising a unit of odour intensity which they called the 'olf', defined as the 'emission rate of air pollutants (bioeffluents) from a standard person'. Using the subjective assessment of a large group of people of odour levels in a variety of naturally and mechanically ventilated and air conditioned spaces Fanger has found that for every occupant and associated odours there may be another 4 to 5 odour equivalents (olfs) released from building materials, furnishings and the air handling system. Auditoria are particularly prone to soiling of internal surfaces because of the extent of soft furnishings and acoustically absorbent finishes. Within the air handling systems examined, the odours emitted from cooling coil drip trays and spray humidifiers were found to be particularly significant.

Environmental tobacco smoke

Leaderer & Cain (1983) have shown that allowances for environmental tobacco smoke (ETS) have also been inadequate in the past, particularly when catering for non-smokers entering the space. The issue is complicated by the fact that smoke evolves at a rapid rate and there may be high local concentrations which depend on room air movement. For low level supply systems smoke may rise vertically into the high level extract and dilution will occur at the central air handling unit. The plume will be very sensitive to cross-draughts from opening doors, to the wakes generated by moving people, and to other disturbances.

Leaderer found that, in a room with perfect mixing, 75% of non-adapted occupants are satisfied if a fresh air rate of 85 m^3 per cigarette is provided. If each cigarette takes 7.5 minutes to burn and if it is assumed that 10% of people in a large space may be smoking at any one time, a fresh air rate of 17.5 litres/s should suffice. The 1989 ASHRAE standard is based on work carried out on adapted occupants (Janssen, 1986) who only require 8 litres/s for an 80% acceptance level.

The most recent British recommendations (CIBSE, 1986*a*) refer to situations with some heavy and very heavy smoking, corresponding ventilation rates of 16, 24 and 32 litres/s per person respectively.

Fanger *et al.* (1988) found that a further 2 olfs were contributed to the emission of odours by cigarette smoking: this is with 30% of their observers being smokers, although they provide no information on smoking rates.

Intermittent and transitory occupancy

It has already been suggested that in many spaces outgassing from building and furnishing materials may occur, but in other spaces, those with mainly hard surfaces, sparse furnishings or where potentially contaminating building products are avoided, there may be insignificant release of contaminants when the building is unoccupied. For applications in which odours are likely to accumulate it is useful to provide pre-dilution by bringing in the ventilation plant some time before the occupancy arrives. For this second category it may be possible to hold back fresh air provision until the odour levels are approaching unacceptable levels.

Control methods are available which can be used to adjust fresh air rates according to the prevailing contamination levels. CO_2 or odour (air quality) sensors can be installed in the space or in extract ductwork and arranged to adjust the mixing dampers or switch fans accordingly. These methods can be used to provide automatic control over lag and lead operation and adjust fresh air volumes for transient and intermittent occupancy. However, Fanger *et al.* (1988) found no correlation between CO_2 concentrations and odour perception, which may question the validity of using CO_2 sensors for control of ventilation systems.

At perhaps an eighth of the cost of a CO_2 sampling-based system, the 'air quality' sensor based controller is a relatively cheap system. The 'air quality' sensor uses the same principle as an adsorption filter to attract gases to the porous surface of a semiconductor, the conductivity of which changes with the amount of gas adsorbed. Gases are alternately adsorbed and desorbed (by heating the sensor). This system will respond to changes in concentration of a wide range of contaminants, although little is known about the way it responds to mixtures. For the system to make an appropriate response to the gases and vapours found in tobacco smoke an adjustment to its sensitivity is required which renders it inaccurate in its response to the contaminants which indicate air purity during non-smoking occupancy.

In tests carried out in an existing auditorium in Sweden, Sødergren & Dahlgren (1985) observed that air quality sensors continued opening the fresh air dampers for some hours after the hall emptied. They suggest that this was the result of the sensor responding to contaminants released from seating and other adsorbent internal surfaces. In contrast, the CO_2 controller closed the dampers and switched off the fans under the same conditions.

The complexity, cost and questionable accuracy of the control of mixing dampers can be overcome by the use of full fresh air plant with air-to-air heat exchangers transferring heat between the exhaust and outdoor airstreams (ASHRAE, 1988). There are a number of devices on the market, some with efficiencies as high as 70 or 80%. Transfer of both sensible heat and moisture with very little transfer of contaminants is possible if a hygroscopic thermal wheel is used. However, most other arrangements, such as the plate heat exchanger, provide sensible heat transfer only, but introduce a complete physical barrier between exhaust and outdoor air streams. For the penalty of a lower efficiency, run-around coils transfer heat from an exhaust air system in one part of a building to an outdoor air intake elsewhere in the building, using standard finned coils with pumped water/glycol circulation between.

Room air movement

It can be said with some confidence that the effectiveness of all commercial air conditioning and ventilation systems depends on occupant perceptions of room air movement more than any other single aspect of design. Along with the control sensors, the air terminal devices (atds) form the main interface between the air system and the building occupants. Atds create profiles of velocity, temperature, humidity and contamination in the room, whilst also influencing the aural environment and internal appearance of the space. Inappropriate selection or positioning of atds can result in draughts, stagnation, poor air quality, excessive temperature gradients, short circuiting and unacceptable noise. The way supply air interacts with the air in the occupied zone dictates the effectiveness of the dilution of odours by the fresh air component of the supply air or the removal of odours with the exhaust air. Poor distribution of air through the occupied zone can result in some occupants not receiving their fresh air allowance which, when combined with associated inefficiencies in heat transfer and unacceptable thermal conditions, can lead to severe discomfort. At the other end of the spectrum, excessive velocities or velocity fluctuations can lead to complaints of draught.

There are many possibly conflicting influences on the choice of air terminal type and position, such as appearance and integration with fabric, structure and other services, necessitating consideration of air movement strategy during the conceptual stages of building design. For example, the choice between high and low level supply may be influenced by computing and communications requirements and the potential for underfloor distribution, which may, in turn, alter the form and structure of the building.

The following factors define the interaction between the air terminal devices and the space:

- velocity, direction and shape of the airflow patterns leaving the supply air terminal devices in relation to the occupied zone
- positions of supply and extract air terminal device
- temperature differential between room and supply air
- influence of surfaces and projections
- influence of convection currents and disturbances (e.g. occupants, machinery, infiltration air, windows, heating surfaces, etc)
- velocity fluctuations
- quality of supply air
- position of contaminant sources and sinks
- rates of evolution and sorption of contaminants

NB in this context water vapour may be considered to be a contaminant.

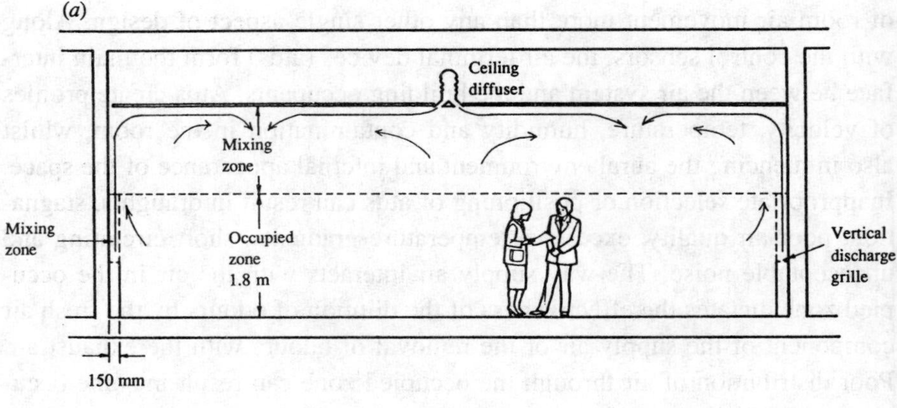

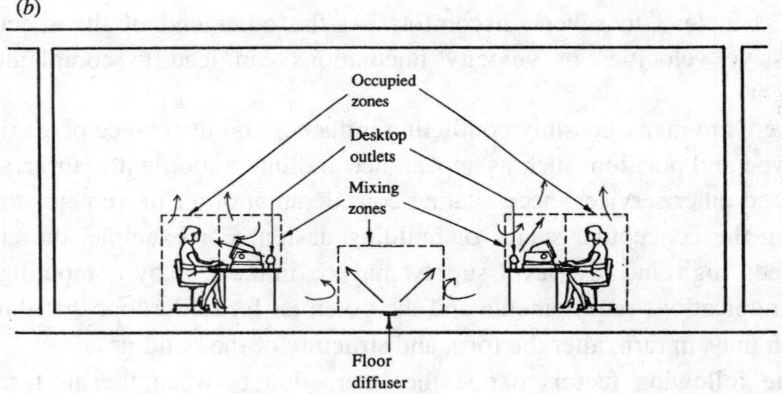

Fig. 16.1.(*a*) Supply air wall jets : mixing zones and occupied zones, (*b*) Mixing zones and occupied zones: free jet low level supply.

The occupant: supply–air interface

In room air diffusion (RAD) design it is useful to divide the space into two zones (see Fig. 16.1). CIBSE (1986*b*) define the occupied zone as 'a region, the outer limits of which are described by an envelope 1.8 m from the floor and 0.15 m from walls'. The region outside this box can be defined as the mixing zone. This definition is adequate for systems which rely on high supply air velocities and large mixing volumes or wall jets to supply air to the space. However, there are a number of methods of air supply for which this definition must be modified. For air terminal devices which supply air into the lower region of a room, such as the low velocity atds, and the rapid-diffusion floor, seat-back and desk-mounted atds associated with buoyancy-

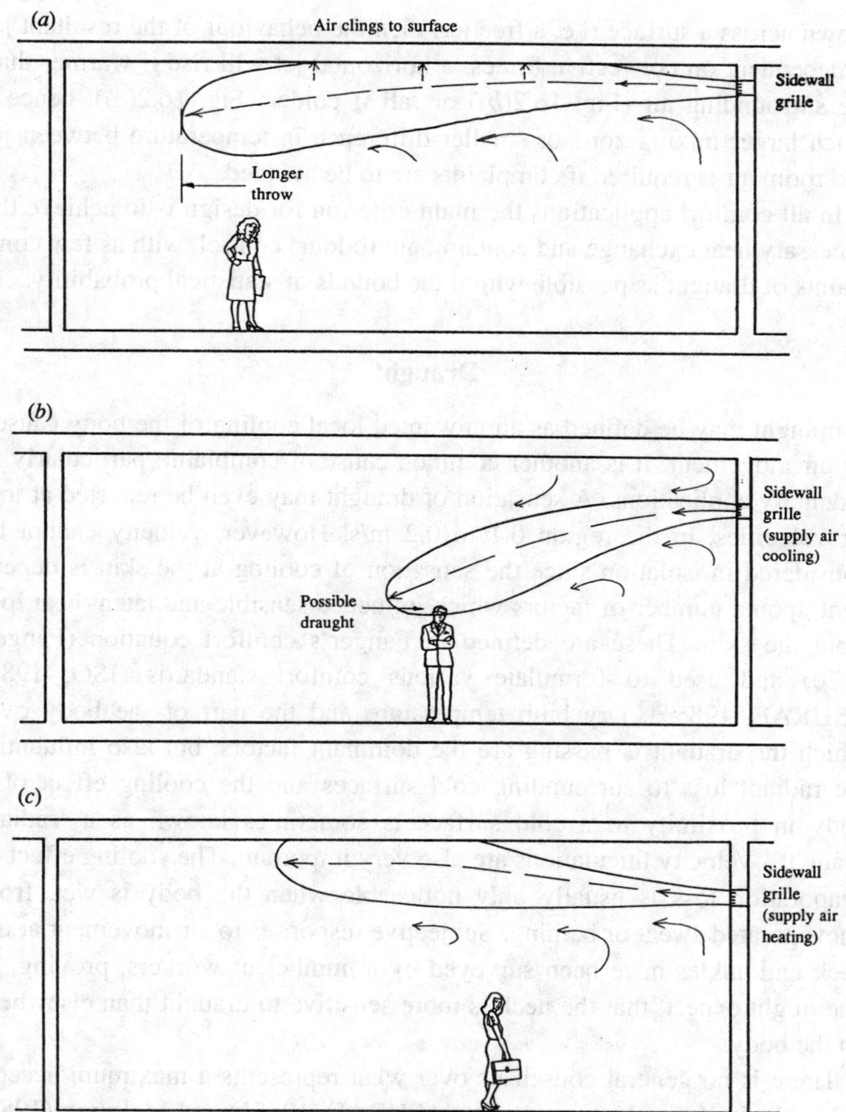

Fig. 16.2. Effect of surfaces and temperature on jets: (*a*) wall jet; (*b*) free jet (cooling); (*c*) free jet (heating).

assisted ventilation, the occupied zone is 'any region where the occupants are likely to linger for significant periods'.

Figure 16.2(*a*) shows how the supply air clings to a surface when blown across it at a velocity greater than about 1.5 m/s (i.e. a wall jet). If air is not blown across a surface (i.e. a free jet) then the behaviour of the resultant jet is dependent on convective forces: a horizontal jet will rise if warmer than the surrounding air (Fig. 16.2(*b*)) or fall if colder (Fig. 16.2(*c*)), hence a much larger mixing zone or smaller difference in temperature between jet and room air is required if complaints are to be avoided.

In all comfort applications the main criterion for design is to achieve the necessary heat exchange and contaminant (odour) control, with as few complaints of draught as possible within the bounds of statistical probability.

Draught

A draught may be defined as an unwanted local cooling of the body caused by air movement. It is another common cause of complaint, particularly in sedentary applications. A sensation of draught may even be reported at low air velocities, in the region 0.1 to 0.2 m/s. However, velocity cannot be considered in isolation since the sensation of cooling at the skin is dependent upon a number of factors which influence sensible and latent heat loss from the skin. These are defined in Fanger's comfort equation (Fanger, 1970) and used to formulate various comfort standards (ISO, 1984; ASHRAE, 1985*a*). Dry bulb temperature and the part of the body over which the draught is passing are the dominant factors, but also influential are radiant loss to surrounding cold surfaces and the cooling effect of a body in proximity to a cold surface is sometimes known as a 'radiant draught'. Velocity fluctuations are also very important. The cooling effect of evaporation loss is usually only noticeable when the body is wet, from unevaporated sweat or bathing. Subjective responses to air movement at the neck and ankles have been surveyed by a number of workers, proving, as one might expect, that the neck is more sensitive to draught than elsewhere on the body.

There is no general consensus over what represents a maximum acceptable velocity for sedentary activity. ASHRAE (1985*b*) and McIntyre (1985) quote 0.35 m/s as a desirable upper limit. However, recent Scandinavian work (Christensen *et al.*, 1984) suggests that 10% of occupants might be dissatisfied with a velocity of 0.2 m/s, even with room temperatures as high as 27 °C. At the other extreme the ISO (1984) can be used to justify a room velocity of 1.0 m/s to reduce dissatisfaction at high room temperatures. Indeed this is the principle applied in 'spot cooling' in aircraft and industrial applications or in the use of desk and punkah fans to supplement natural air movement in non-air conditioned applications. There can be a problem with papers being lifted from desks at velocities exceeding 0.8 m/s.

The convection currents, which the heat generated by our bodies create, travel upwards at a velocity of up to 0.25 m/s, which may afford some protection against downdraught, whilst walking slowly causes air to pass over the body at 1 m/s or more. However, for most sedentary applications, it is reasonable to consider a maximum of 0.25 m/s for summer cooling and 0.15 m/s for winter heating. In order to avoid complaints of 'stuffiness' it has been suggested (Jackman, 1986) that the velocity over the body should not be allowed to fall below 0.05 m/s. Since we generate higher velocities than this through convection, it is unlikely to be the low velocity which causes discomfort *per se*, but without the cooling and fresh air associated with local air movement there is a commensurate rise in temperature, humidity and odour levels.

Ventilation efficiencies

Workers using tracer gas measurement techniques (Charlesworth, 1988) to determine the actual local air change rates of outdoor air in various parts of a building have found that they tend not to be equal to the nominal air change rate for the whole building. This is due to a combination of short circuiting between supply and extract openings, uneven room air diffusion and leakage through the building fabric.

In this context, it is useful to express air change rates in their reciprocal form as the age of air, where age refers to the amount of time a parcel of air takes to be completely replaced with outdoor air. It is the time constant for air exchange.

Air exchange efficiency (E_A) relates spatially averaged local ages (T_L) to the reciprocal of the overall air change rate (N) for the building or zone under consideration ($T_O = 1/N$):

$$E_A = T_O/2T_L$$

Where N is the number of times the whole volume of air in a space is replaced with air from outside that space in 1 hour.

Ventilation efficiency or index (Skåret & Mathisen, 1983) has been defined as a ratio which indicates the difference in condition between the occupied zone and the extract air:

Efficiency for contaminant control: $E_C = C_E/C_R$ where C_E and C_R are concentration in the extract air and room (occupied zone) respectively.

It can be seen that if room and extract air are at the same condition and age then E_A is 0.5 and E_C is 1.0. For example, a space which is ventilated by a system which achieves complete mixing of supply and room air will have evenly distributed thermal and contaminant fields. Whereas if the general movement of air is upwards, and vertical gradients of temperature, concentration and moisture content are established above the occupied zone, E_C will be greater than 1.0. However, if some of the supply air short circuits the room and enters the extract system without heat exchange or dilution then E_C would be less than 1.0. This short circuiting may occur if supply and extract air terminal devices are positioned too close to each other.

Low level supply (buoyancy-assisted) systems (Appleby, 1986) promote high efficiency by employing room convection currents to create a reasonably stable upward movement of air – see Figure 16.3. Contaminants and moisture move upwards also, rather than being mixed to create a homogeneous atmosphere. For perfect piston flow E_A is 1.0 and E_C is ∞. But actual ventilation and air exchange efficiencies are not easy to predict since they depend upon the way in which contaminants, moisture and heat are picked up by the air as it moves upwards through the room. This phenomenon will depend to some extent on the characteristics of the air terminal device, particularly if it employs rapid diffusion or free jets, although the dominant influence will usually be the buoyant forces.

Supply of air from high level

The choice of supply air pattern is between free jets, wall jets and displacement downflow. A number of air terminal devices can achieve these patterns including nozzles (drum or punkah louvres), grilles, diffusers (circular, rectangular or linear), adjustable diffuser/nozzles: all of which can issue swirling or straight jets; and perforated face devices or ceilings.

Wall jets only perform satisfactorily if there are flat unobstructed surfaces available outside the occupied zone over which air can be blown without being deflected into the occupied zone or short circuiting into the extract system. Wall jets can deal with heat gains and losses with little change in room airflow patterns. They can change direction without breaking away from the surface, hence it is possible to blow air vertically up walls and across ceilings from floor or cill grilles or sidewall diffusers, or vice versa

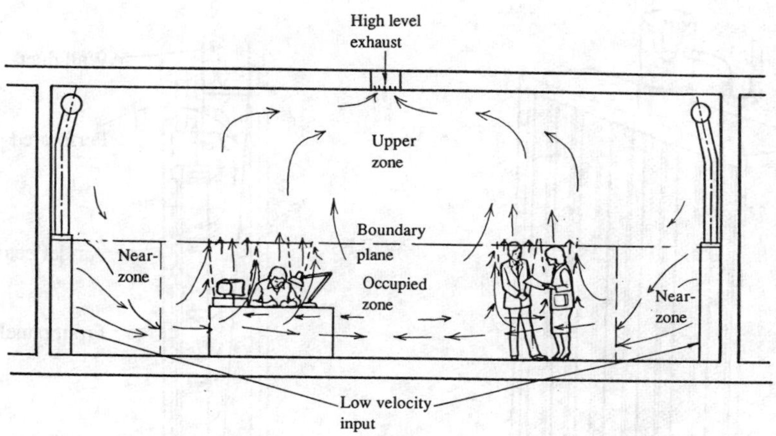

Fig. 16.3. Buoyancy-assisted room air diffusion.

from ceiling diffusers. In both cases throw will be influenced by gravitational forces when the air is travelling vertically.

Where rooms have ceilings obstructed with lighting or acoustic baffles, or geometries are not compatible with the use of wall jets, it may be feasible to use grilles, nozzles or freely suspended diffusers to produce free jets. The drop and throw of a free jet are temperature dependent and may be influenced by the momentum in rising convection currents.

Downward displacement requires very large air volume flow rates to be independent of buoyancy effects. Low heat loads, low ceiling heights and high room velocities are conducive to stability. Localised problems can occur if strong upward convective plumes force downward moving currents to be funnelled into parts of a room which have lower convective forces.

Supply of air from low level: buoyancy-assisted room air diffusion (Appleby, 1989*b*)

The supply of air from close to the floor has, in the past, been characterised either by the nature of the room air movement, as in the term 'displacement ventilation', the position of the air terminal devices, as described by the terms 'low level supply', 'personal air supply' and 'ventilation from the floor', or the type of air terminal device, as with 'low velocity supply'.

It is proposed here to differentiate between the driving forces and air terminal device. The main driving force for most of the low level supply systems

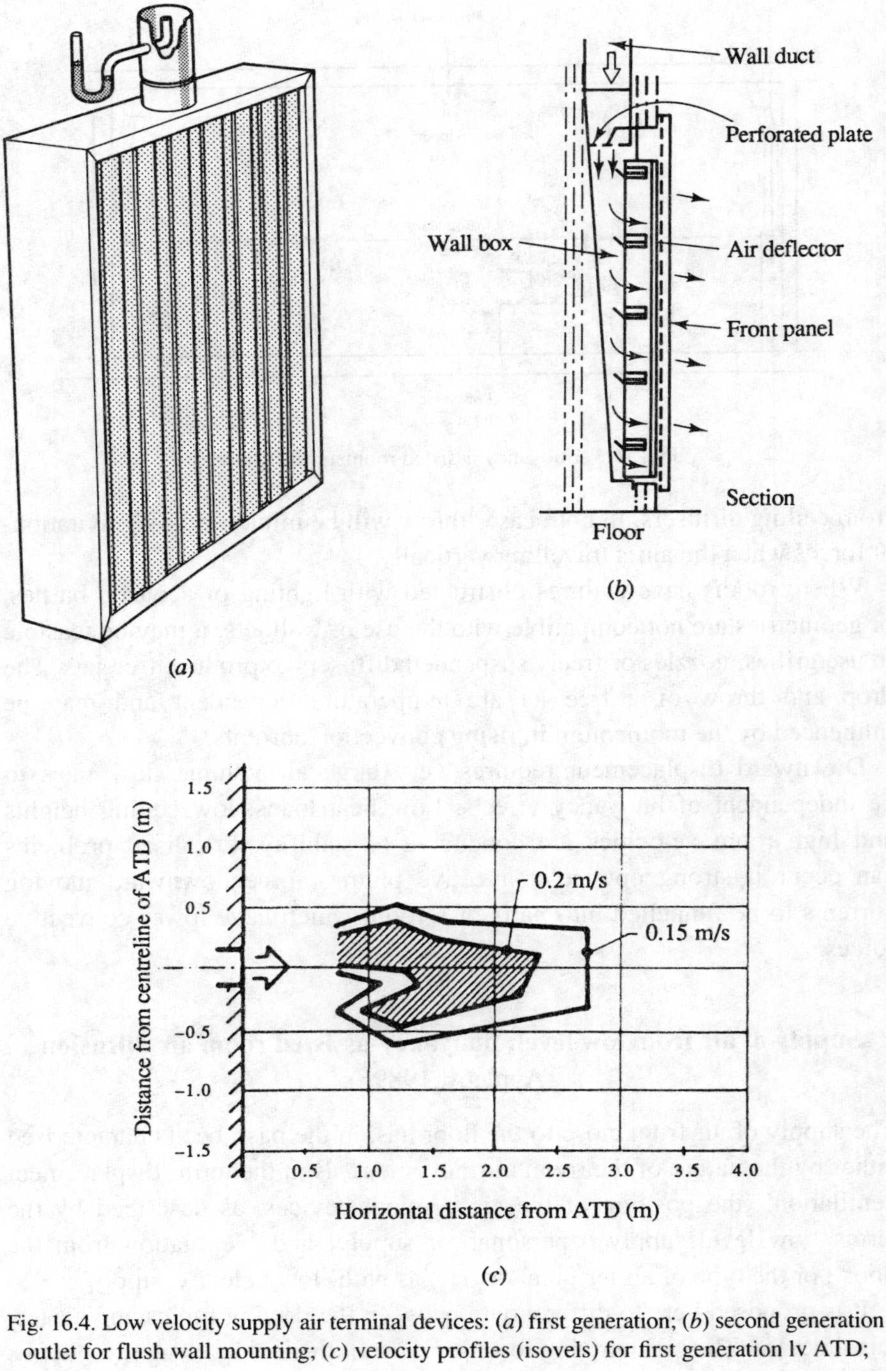

Wall duct
Perforated plate
Wall box
Air deflector
Front panel
Section
Floor

(*a*)

(*b*)

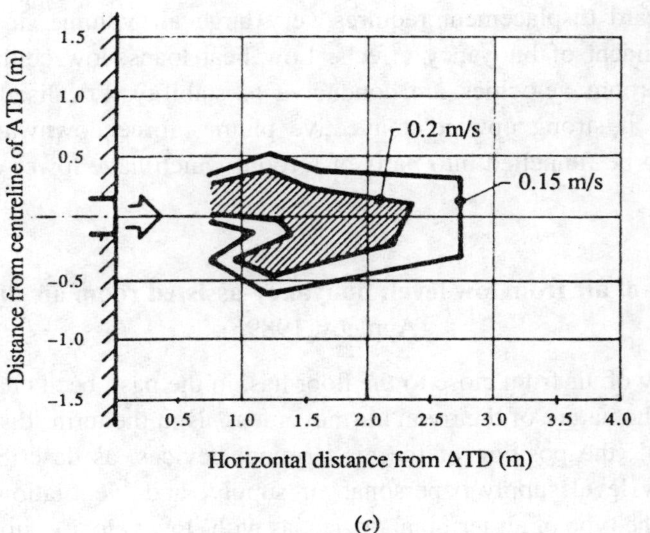

(*c*)

Fig. 16.4. Low velocity supply air terminal devices: (*a*) first generation; (*b*) second generation outlet for flush wall mounting; (*c*) velocity profiles (isovels) for first generation lv ATD;

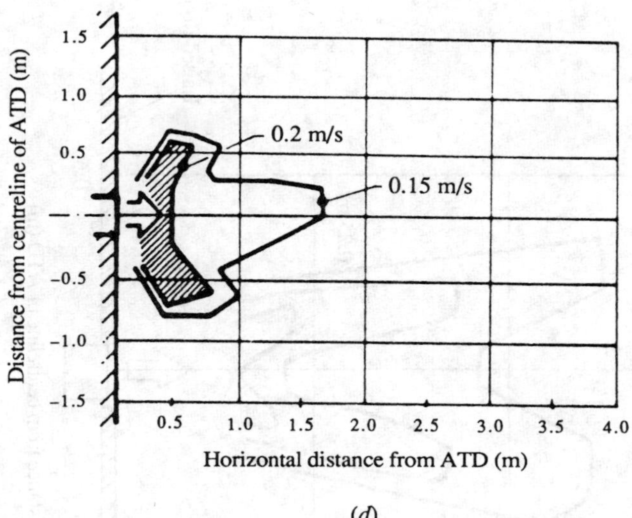

(*d*)

Fig. 16.4. Low velocity supply air terminal devices: (*d*) velocity profiles (isovels) for second generation lv ATD.

used for comfort ventilation or air conditioning comes from the convective plumes rising from the heat sources in the occupied zone (see Fig. 16.3), hence the term 'buoyancy-assisted room air diffusion'. A number of manufacturers produce supply atds which can be used with buoyancy-assisted mechanical ventilation or air conditioning. There are essentially two approaches:

- supply air at or near floor level at a low velocity, via a 'low velocity supply atd'
- supply air in the form of rapidly diffusing jets via floor, desk or seat-back atds.

The former approach is one which has been widely adopted in the Scandinavian countries, whilst floor, desk and seat-back outlets were developed primarily in Germany.

Figure 16.4 shows typical low velocity supply air terminal devices (lv supply atds) and their respective air velocity profiles measured close to the device at ankle level. Figure 16.5 shows a typical 'rapid-diffusion floor-mounted atd' and its velocity profiles. Rapid diffusion is achieved due to the highly turbulent nature of the multiple jets, which have a swirl imparted to them due to the internal geometry of the device. Figure 16.6 shows a typical seat-back outlet and corresponding velocity profiles.

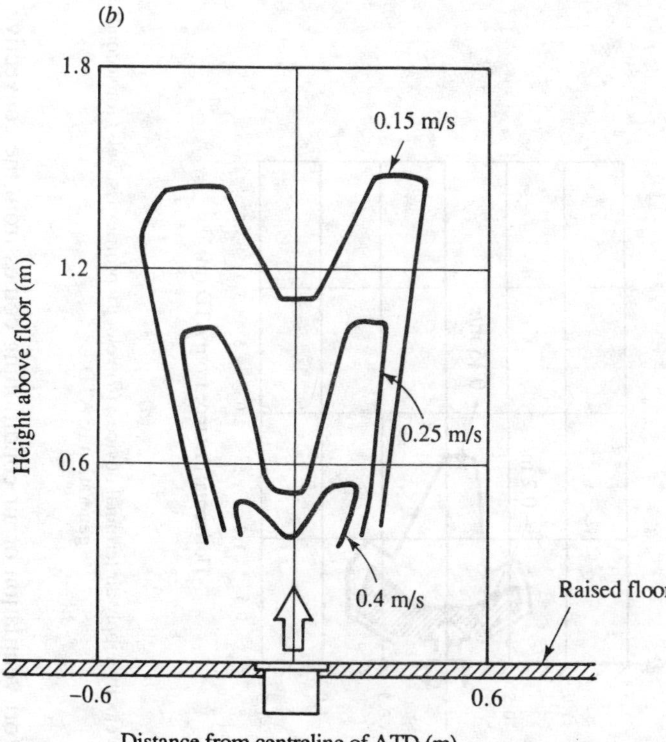

Fig. 16.5. (*a*) a floor-mounted twist outlet, note dust receptor for recessing into floor; (*b*) typical velocity profiles (isovels) for floor-mounted twist outlet.

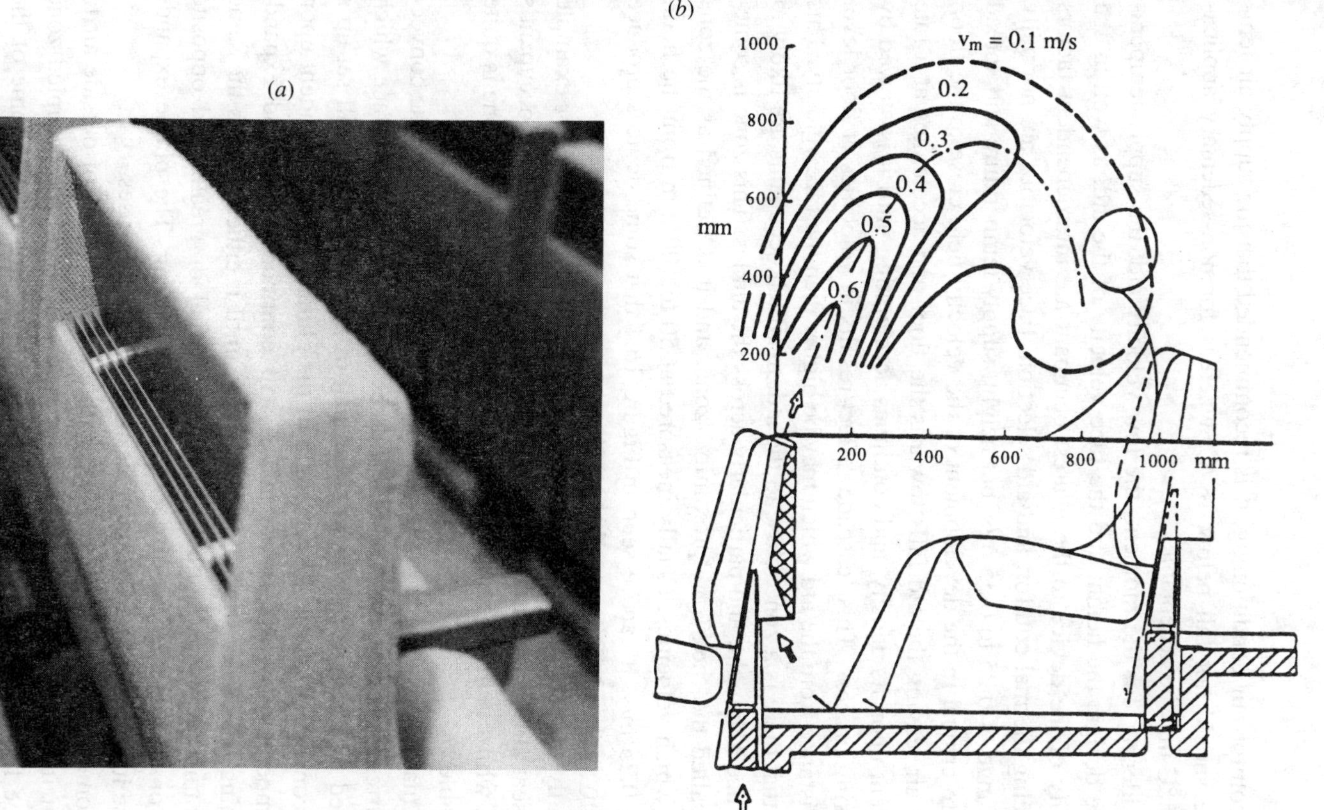

Fig. 16.6. (*a*) seat back atd for a theatre; (*b*) typical velocity profiles (isovels) for personal supply ATD.

For comfort air conditioning, it is recommended that the supply air temperature should not fall below 18 °C, whereas for non-sedentary applications 16 °C may be acceptable.

Considering, for the moment, the low velocity atd: the supply temperature, the design, the height and the face velocity of the atd, as well as the proximity of the device to the nearest occupant are interdependent factors influencing thermal comfort and atd selection. Face velocities are normally in the range of 0.25 to 0.35 m/s, but height, supply temperature and spread are more critical in the development of the velocity field close to the floor. Supply air leaving the atd falls towards the floor and accelerates at a rate dependent upon its density, and continues accelerating until it is slowed by the mixing process. This 'cascade' effect results in velocities at floor level some distance from the atd which may be higher than the face velocity. This means that there is a zone, close to the atd, in which occupants are likely to experience a draught, and particularly across the ankles. This zone is sometimes called the 'near' or 'proximity' zone, and it is defined as the zone within which velocities, usually measured at 50 to 100 mm from the floor, exceed 0.25 m/s. It can be seen in Figure 16.4 that some devices produce relatively high velocities at ankle level.

Usually the size of the near-zone is fixed by the application: for example under-seat atds in an auditorium may be only 300 mm from the occupant's ankles, whilst in many commercial applications a 1 metre near-zone is quite acceptable.

The main driving force in the majority of applications will be the convection currents generated by the heat sources in the room. Any surface which is heated to a temperature higher than that of the adjacent air will set up a rising convection plume. The strength of the plume will be dependent upon the temperature of the source and it will continue to rise and expand, entraining the cooler surrounding room air, until it either reaches the same temperature as its surroundings, meets an equal or greater and opposite force, or an impenetrable surface, such as a ceiling. The plume expands because it is fed with surrounding air continuously as it rises.

A domestic fire must be supplied with a sufficient flow of outside air to prevent it from smoking. Similarly the heat sources in the occupied zone must be fed with enough air from the low level atds or else some of the rising plume will recirculate back into the occupied zone, carrying some contaminants back with it. This may not be too much of a problem if contamination emissions are relatively low. The resultant contamination

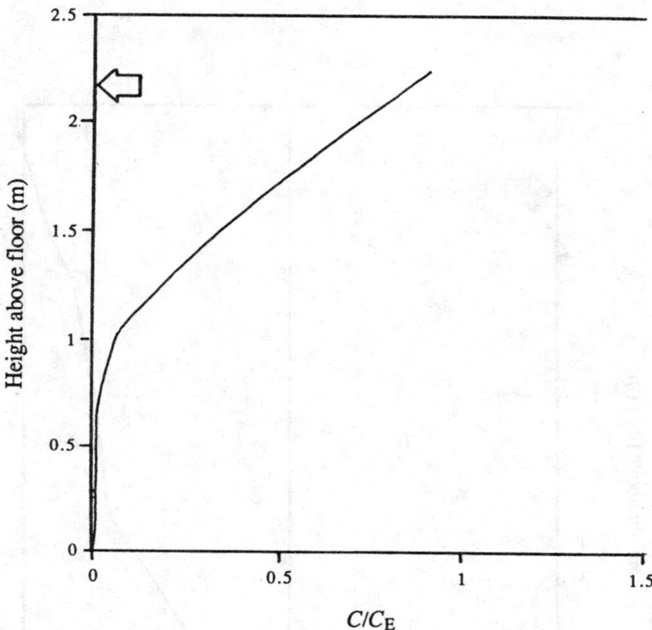

Fig. 16.7. Typical concentration gradient in a room cooled by air supplied via a low velocity atd and with buoyancy-assisted RAD.

concentrations would rarely be greater than with a high level supply system with complete mixing.

Figure 16.3 shows common room heat sources and their associated convective plumes. The 'boundary plane' is the level at which the net upward airflow is equal to that supplied into the occupied zone. If recirculation of contaminants which have entered the zone above the occupants' heads is to be avoided then the supply flow rate should be determined by calculating the net flow rate of the convective plumes as they pass through the boundary plane. This can be determined by subtracting downflow from cold surfaces from the plumes generated by warm surfaces as they pass through the boundary plane although it is hoped that downward airflows will be minimal since they will carry contaminants from the upper zone with them. For this reason buoyancy-assisted systems are only suitable for well insulated or internal spaces. Heat losses must be dealt with by locating heat emitters under windows or other cold surfaces, well away from the atds. Downflow into the occupied zone of small rooms occurs when a door is opened.

The boundary plane should be as close to the floor as possible: the higher it is taken the greater the plumes will have expanded and hence the greater

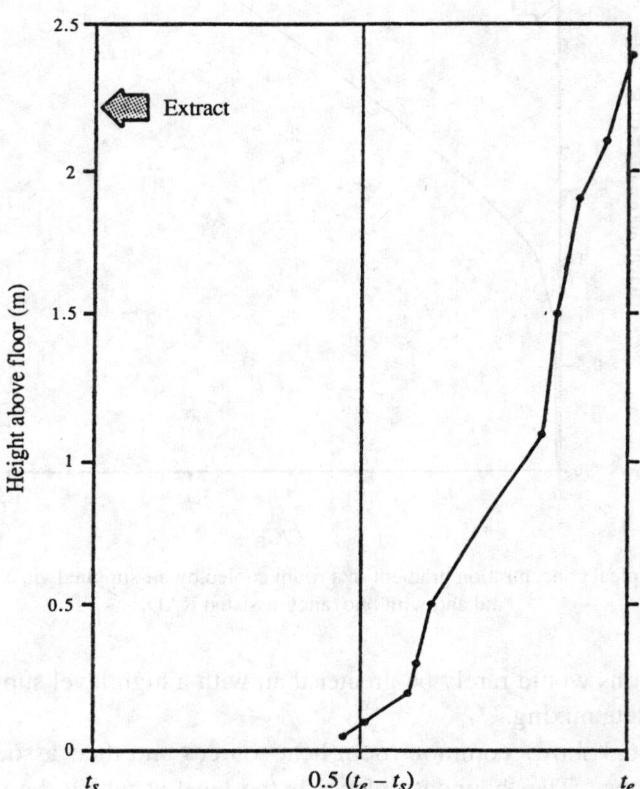

Fig. 16.8. Typical temperature gradient in a room cooled by air supplied via a low velocity atd and with buoyancy-assisted RAD.

the air volume to be supplied into the occupied zone. If most occupants are likely to be seated for most of the working day then the plane can be taken at 1.2 to 1.4 m, whereas if standing occupants are important 1.7 to 1.8 m is a reasonable height for it. Figure 16.7 shows a typical concentration profile in a room with stratification above head level, from which it can be seen that the air which is breathed is much cleaner than the extract air.

The convective heat gain from heat sources above the boundary plane, such as ceiling-mounted luminaires, will not influence the air flow between the zones. However, radiant heat from the luminaires, the sun and other sources will heat the floor and walls and establish local convective flows.

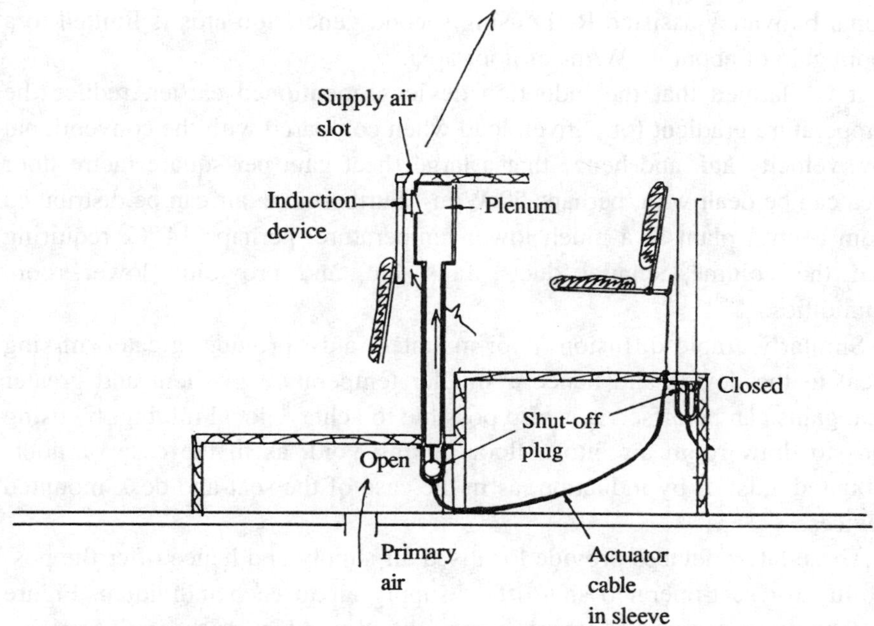

Fig. 16.9. Lecture theatre desk top supply with seat-operated shut-off.

For example, part or all of the floor may be heated and a very weak but broad convective flow established. This plume is likely to reach thermal equilibrium with the surrounding air well below the boundary plane and has the effect of raising the air temperature in the lowest part of the occupied zone and hence linearising the overall temperature gradient in the room (see Fig. 16.8). Direct solar gain through an unprotected window can cause a very strong local plume which may consume more air than is available if no allowance has been made for it.

High level mixing-type systems are designed to create uniform temperature and contamination conditions. In contrast, buoyancy-assisted systems are designed to promote stratification so that the air being breathed is nearly as pure as the air entering through the atds. Inevitably, therefore, occupants' feet will be cooler than their heads. This is further aggravated by the movement of air across the floor at ankle level and the radiant heat gain to the head from the warm ceiling and the luminaires.

A combination of the need to limit head to foot temperature difference and realistic selection of atds to give adequate usable floor area outside the near-zones means that, for the office air conditioning application, conventional buoyancy-assisted RAD using second generation atds is limited to a room gain of about 25 W/m^2 of floor area.

It is claimed that the induction devices, mentioned earlier, reduce the temperature gradient for a given load when compared with the conventional low velocity atd, and hence that a larger heat gain per square metre floor area can be dealt with, perhaps 50 W/m^2. Furthermore, air can be distributed from central plant at a much lower temperature, perhaps 14 °C, requiring half the volume, smaller ducts, fans, etc, and providing lower room humidities.

Similarly, rapid-diffusion floor-mounted atds provide greater mixing local to the device and hence a smaller temperature gradient and greater heat gains can be offset. It is also possible to achieve local mixing: by using fans to draw room air into a floor mixing void, as in the case of floor-mounted atds; or by induction, as in the case of the seat and desk-mounted devices.

These latter devices provide localised air supply and hence offer the possibility for seat-operated shut-off of supply air to each individual. Figure 16.9 shows one possible arrangement applied to a lecture theatre (Appleby, 1987), although the same principle could be applied to the underseat low velocity atds mentioned earlier. During times of low occupancy only sufficient air would be supplied into the occupied zone to feed the convective plumes rising from the occupants present and only the fresh air which is required to dilute their body odours would have to be cooled or heated in the central plant. Considerable energy savings are possible if this control arrangement is combined with a variable air volume–air handling system with accurate static pressure control.

Low velocity atds cannot be used to supply air which is warmer than the room air since its buoyancy will carry it directly to the ceiling. Floor-mounted devices can be used for heating, however: for this purpose they are normally located under windows, in which case the rapid-diffusion type would not normally be required.

Vortex-assisted room air diffusion (Appleby, 1989*a*)

Vortices occur wherever air moves and their formation is an important part of the mixing process induced by airstreams leaving air terminal devices. For example rapid-diffusion devices are designed to create a number of expanding vortices.

On a larger scale, it is possible to set up a spiralling vortex from the bulk of the air in a room by introducing air in a direction and with a momentum which induces rotation of the room air in one direction. Spiralling occurs if extract air is removed through an opening located at the apex of the spiral.

The technique has been applied full scale in Tokyo to a smoking area which was part of a larger space. Air is supplied through the four columns which supported the ceiling containing the extract opening. Despite the area being open to the surrounding space, it has been found that tobacco smoke does not escape outwards from the vortex. The Japanese are also investigating its application to specialist 'odorama' cinemas requiring rapid diffusion and removal of scents during the progress of a film.

Full fresh air systems

Whether summer cooling is to be achieved by passive or mechanical means there must be sufficient outdoor air available to meet the fresh air requirements of the occupants. In some (rare) cases it may be found that fresh air requirements exceed the air volume required to control the thermal environment, in which case a full fresh air plant will be required having a duty based on the total fresh air rate.

Natural ventilation and infiltration

Fresh air which enters the building through the action of natural forces can make up the fresh air requirements of the occupants, even during the winter months, when the movement is through cracks and temporarily opening doors. The quantity of fresh air which reaches the occupants' breathing zones will depend on the area of cracks or openings, the prevailing forces and the efficiency of diffusion through the building. For example it is generally accepted that a typical office cannot rely on natural ventilation to penetrate deeper than 6 m from the perimeter. It is also important to establish winter infiltration rates in order to calculate heat loss and hence size heat emitters and determine boiler duty. However, this usually requires some empirical allowance to account for typical door opening. Some modern forms of construction and window type do not allow sufficient infiltration to serve occupant's needs or to reduce condensation risk, and it may be necessary to incorporate deliberate crackage in the form of a trickle ventilator or an airbrick let into the window frame or wall.

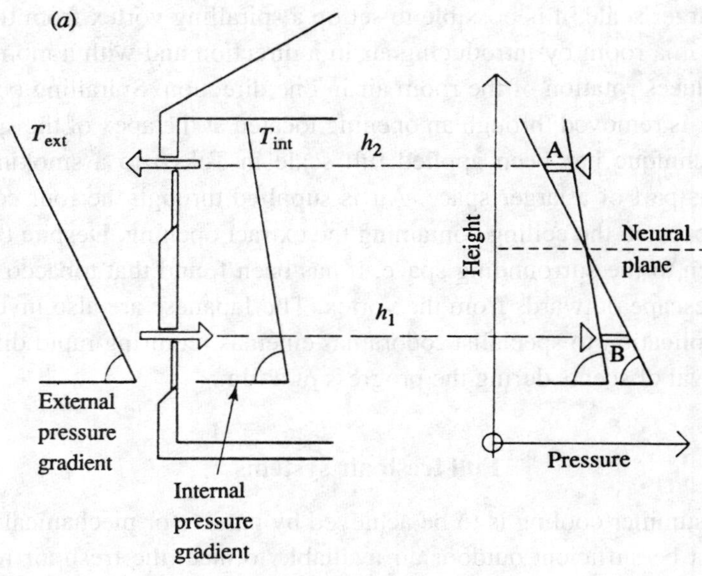

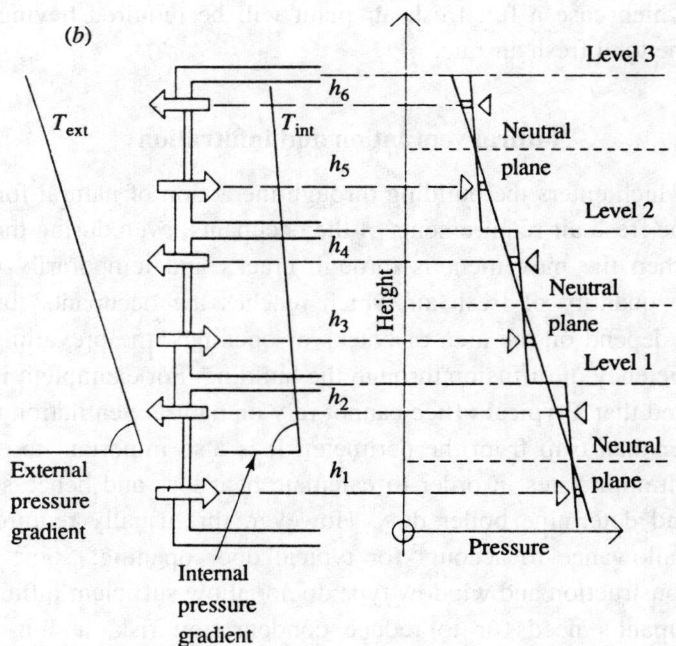

Fig. 16.10. Stack induced pressures: (*a*) between two vertically placed openings; (*b*) in multi-storey buildings with isolated floors.

Design of operable fenestration or ventilators to limit the rise in summer temperature demands a careful analysis of the climatic influences on the building and the aerodynamics of complex bluff structures and sharp-edged orifices. The dominant force in creating air movement through openings is frequently the wind, hence the conditions under which the least airflow will occur is when the wind is absent and the buoyancy forces are at their minimum. These two events may not coincide all that often and furthermore if buoyancy forces are low the heat gain will also be low. Internal peak temperatures are a function of heat gain, air flow rate and the temperature of the entering air. Buoyancy forces may dominate in spaces which are characterised by some combination of high heat gains, height and shelter from the wind. These characteristics are combined in a tall factory space which has openings incorporated into the lower walls and roof ridge, the processes emit large quantities of heat and the building is surrounded by other taller buildings. In this case quite high air change rates will be induced by the stack effect alone.

The stack effect

The stack effect arises as a result of a density difference between the air entering and air leaving a space. The most common form of the stack effect occurs when heat emission to the space warms the air to above outside temperature. This causes the air to rise and pass out through high level openings, whilst cool air enters at low level to replace it (see Fig. 16.10). Sometimes the reverse will occur, for example some spaces may be cooled below outside temperature by heat loss to the floor or other thermally massive parts of the fabric. Dense air passes out through low level openings, inducing replacement air to enter at high level. In the first case the space acts like a chimney and the heat sources are analagous to the fire. The force created by the induced 'draught' is therefore dependent upon the height of the 'chimney' and the nett heat gain and resultant temperature difference. This sets up a pressure gradient within the space such that the static pressure at low level is less than the outside pressure and the pressure at the high level opening is greater than that outside. The level at which inside and outside pressure gradients cross is known as the neutral plane (see Fig. 16.10).

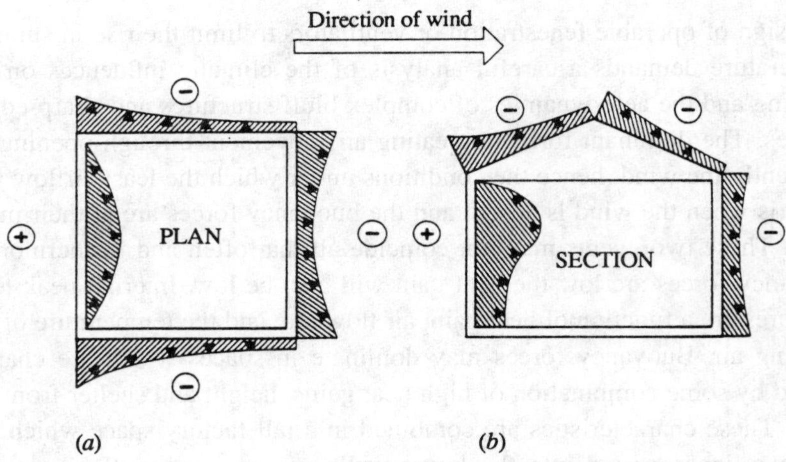

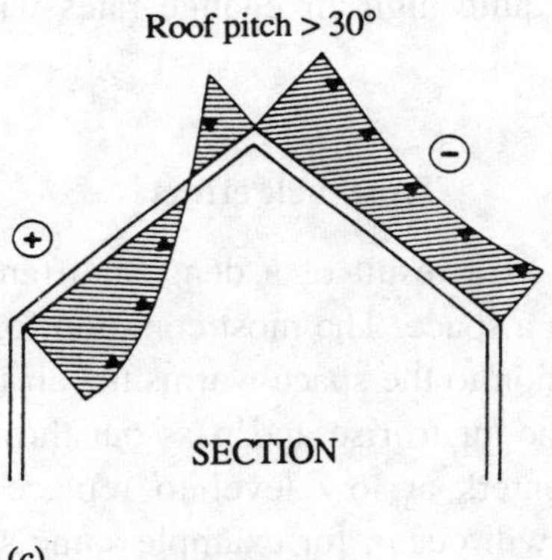

Fig. 16.11. Wind pressure distribution on a building: (*a*) plan view; (*b*) section : shallow pitch or flat roof; (*c*) section : steep pitch roof.

Wind effect

The wind which passes near to the ground is characterised by random fluctuations in velocity which, when analysed stochastically, possess a mean value of speed and direction. Wind force is also a function of height above ground, dependent upon surface obstructions and roughness and the buoyancy forces prevailing. When meeting a bluff obstruction, such as a building, the wind is deflected and induces a positive pressure on the upwind face. Flow separation produces negative pressures along the sides of the building and on the leeward face (see Fig. 16.11(*a*)). A smooth flat roof, or

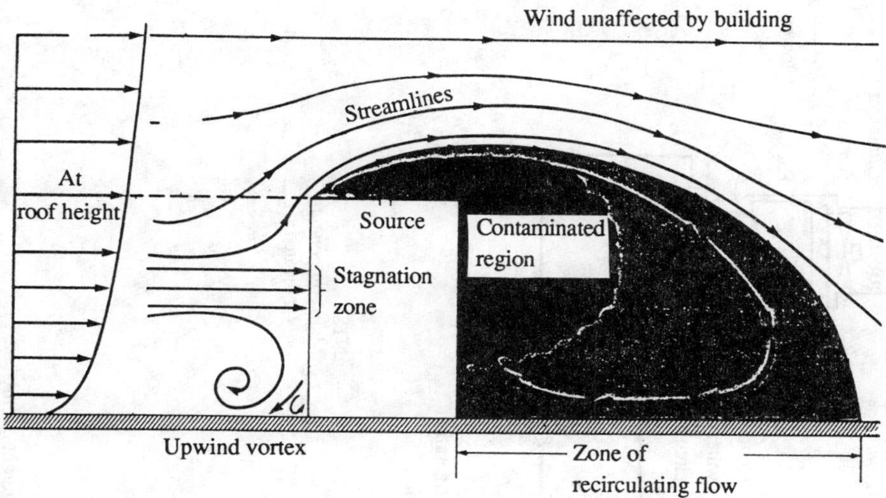

Fig. 16.12. Stylised centreline flow patterns around a rectangular building.

one with a shallow pitch (<30°) will also experience negative pressures (Fig. 16.11(*b*)), whereas a steeply pitched roof will experience positive pressures on the upwind side (Fig. 16.11(*c*)) (Liddament, 1986). Figure 16.12 shows a stylised flow pattern for a medium height building, indicating likely streamlines and recirculation zones. It can be observed from this elevation that turbulent flow regimes can occur near to the ground.

Air conditioning

Plant

Air conditioning systems incorporate, by definition, a component or components which use a mechanical refrigeration process to cool the air supplied to the space when required. This process uses a fluid which evaporates at low temperatures and pressures. There are a number of different types of refrigerant, mostly based on chlorinated fluoro-carbons (CFCs), which circulate within a closed system (see Fig. 16.13) comprising an evaporator, a compressor, a condenser and an expansion device. The medium to be cooled, either air or water, passes across the evaporator and causes the refrigerant, which may be at 3 or 4 °C, to boil, without change of temperature or pressure. This vapour, or refrigerant gas, is compressed into a high temperature high pressure gas by the compressor, hence making it hotter

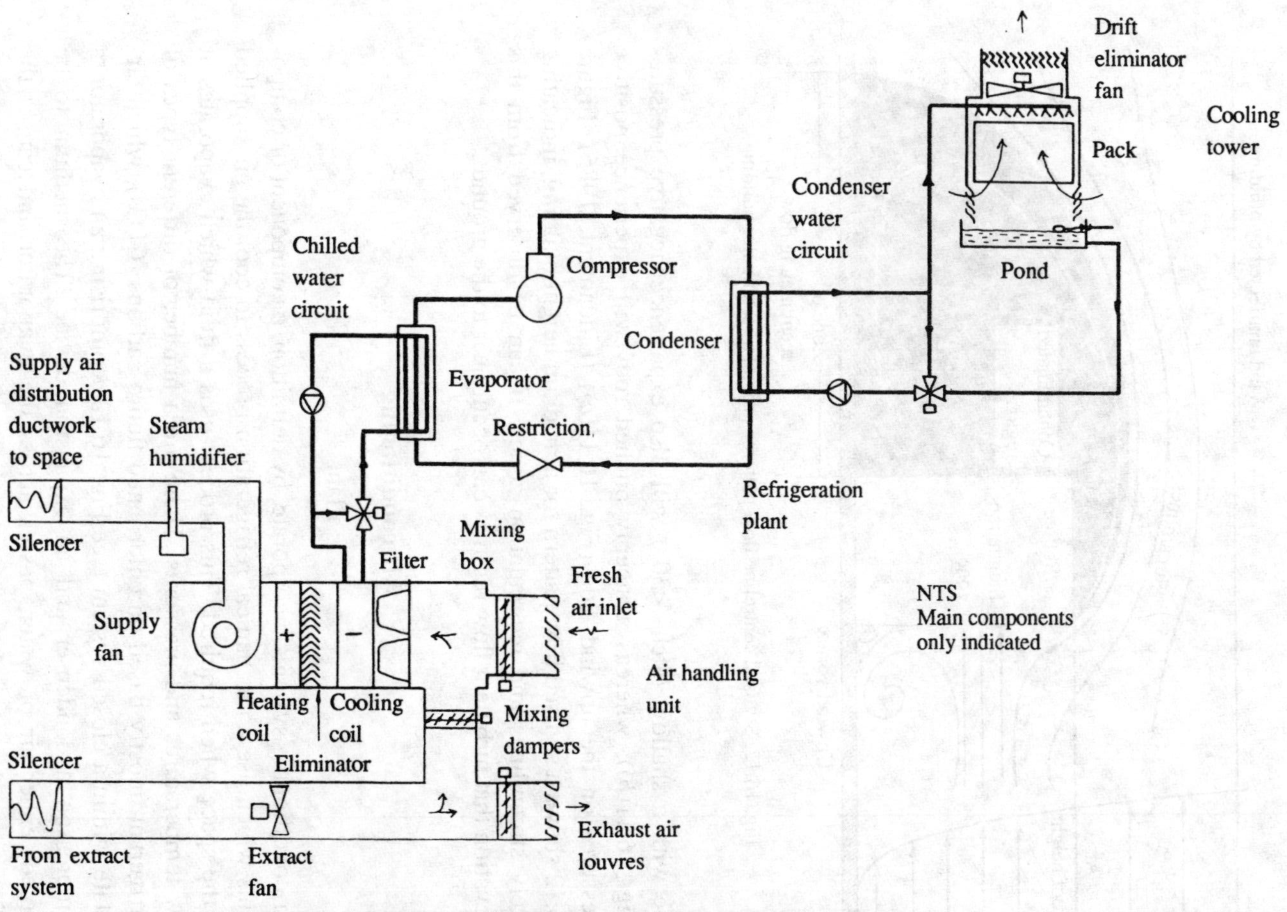

Fig. 16.13. Air conditioning system water and refrigerant circuits.

(perhaps 30 to 40 °C) than outside air. When it passes through the condenser it turns back into a liquid through direct or indirect heat exchange with outside air. The high pressure liquid refrigerant thus formed is then passed through a restriction, reducing in both pressure and temperature before entering the evaporator once more. This is the same process as that used to cool food in a domestic refrigerator or a cold store.

Small packaged air conditioners are normally installed in the wall or window of a single room with the evaporator on the room-side, cooling the air which is blown into the room. When used as a cooling coil an evaporator is usually called a direct expansion or 'd/x' coil. The condenser rejects heat to the outside air, and the compressor is usually mounted next to it, and the whole package installed within a common casing. Alternatively the condenser and compressor can be remote from the room unit, with refrigerant pipework run between the two units. This is sometimes called a 'split system'. A larger version of this uses an air handling plant which incorporates a d/x cooling coil which is connected to an air or water cooled condenser/compressor set. The largest systems generally use chilled water pumped through a cooling coil within an air handling unit. This gives finer control over output than is possible with a d/x coil.

In the past most large installations have used cooling towers to remove heat from the condensers. The cooling water is pumped over a labarynthine pack through which outside air is blown, so that air and water come into intimate contact. This enables lower condensing temperatures than is possible with an air-cooled condenser because much of the cooling is due to evaporation, and a wetted surface loses heat at a faster rate than a dry one. However, the open nature of this arrangement causes a number of problems:

(a) air entering the tower can be laden with dirt and biological matter which is scrubbed out by the water cascade, thus forming a sludge in any static parts of the system which may encourage growth of bacteria, such as *Legionella pneumophila* (CIBSE, 1987);

(b) at 20 to 35 °C the cooling water is within the optimum growth range for *Legionella pneumophila*;

(c) an aerosol forms because of the turbulent nature of the heat exchange, and if this is allowed to leave the tower it can carry bacteria for a considerable distance, perhaps into a building or the street: newer towers are equipped with very efficient 'drift' eliminators which remove most of the aerosol and bacteria, if present;

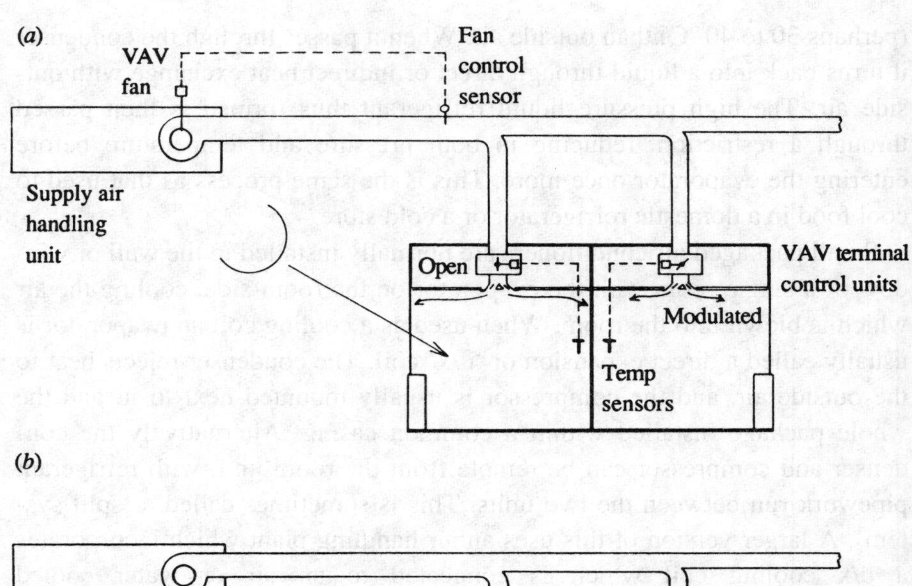

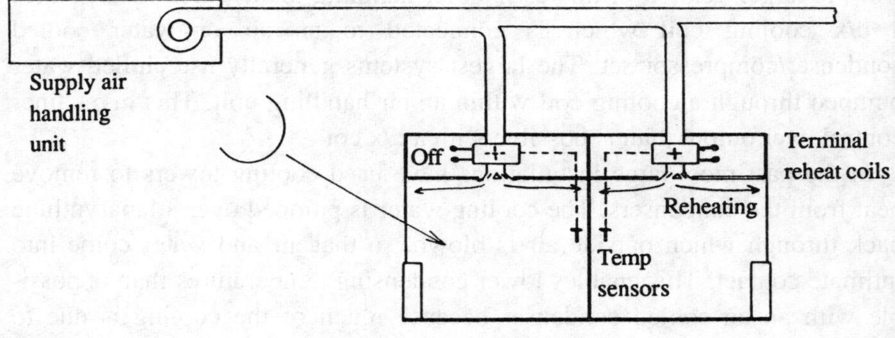

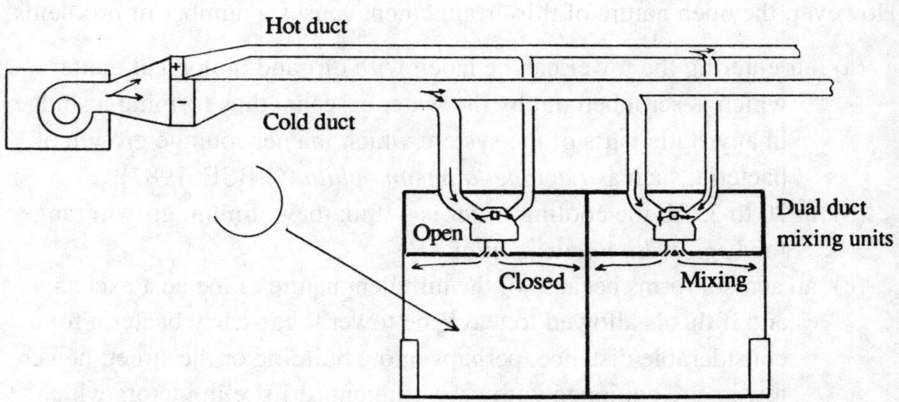

Fig 16.14. Multiple zone air conditioning systems (extract air system omitted for clarity): (*a*) variable air volume; (*b*) terminal reheat; (*c*) dual duct.

(*d*)

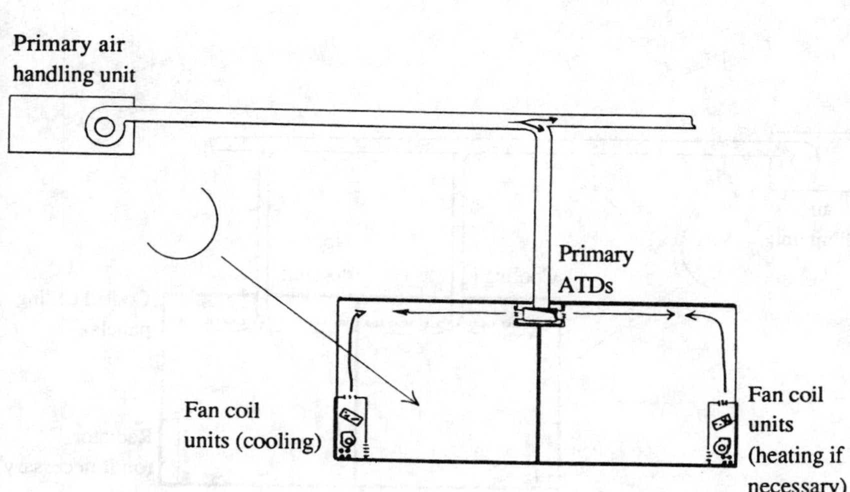

(*e*)

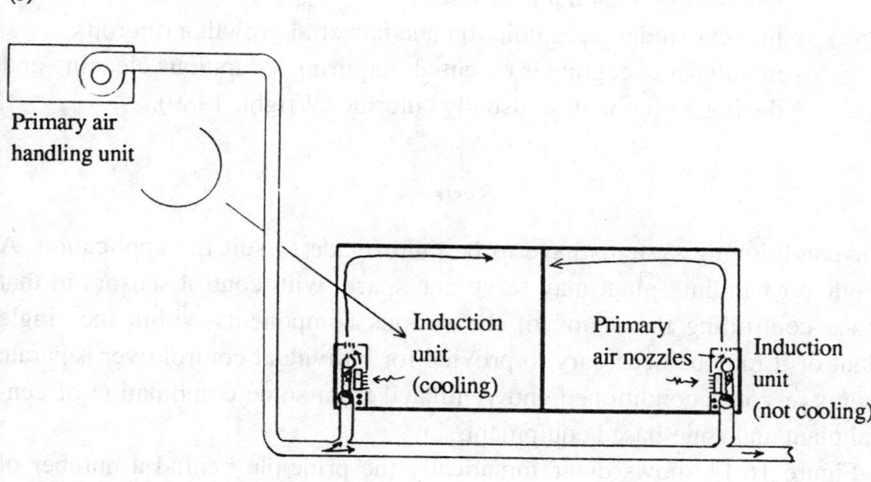

Fig. 16.14. (*d*) fan coils and primary air - room units with heating and cooling; (*e*) induction system - room units cool only.

(f)

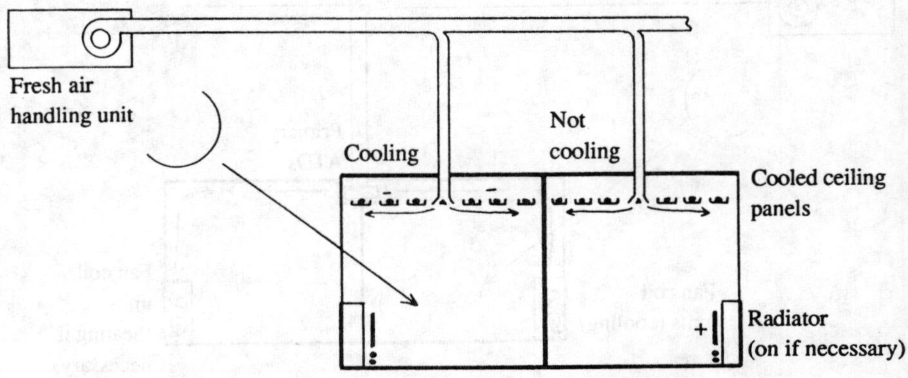

Fig. 16.14. (f) cooled ceiling and primary air.

(d) water lost due to evaporation and aerosolisation must be replaced with costly treated mains water;

(e) to prevent sludge accumulation and bacterial growth a rigorous maintenance regime is essential, requiring scrupulous cleaning and dosing with biocides, usually chlorine (Wright, 1989).

Systems

Air conditioning systems have to be tailor-made to suit the application. A single air handling plant may serve one space, with control sensors in that space controlling the action of the various components within the single plant or it may be necessary to provide for individual control over separate rooms or zones conditioned and ventilated from some combination of central plant and zone-based equipment.

Figure 16.14 shows diagrammatically the principle behind a number of multiple zone air conditioning systems. These fall into a number of categories, the most important being based on the method of zonal control, i.e. constant volume/variable temperature or variable volume/constant temperature. The most common method of dealing with zonal load variation is to modulate supply air temperature so that local room temperature changes, which occur because of alterations in solar orientation, migration of occupants or switching of apparatus, are sensed locally and used to control a zonal terminal unit. Variable air volume (VAV) systems use these sensors to modulate dampers in the supply air to each zone. This allows changes in room load to be reflected in the total air volume handled by the central plant and, provided fans are controlled efficiently, considerable savings in air handling costs can be achieved. Problems arise, however, because of:

(a) the reduction in momentum generated by atds as the volume reduces;
(b) the transition from dealing with net heat gain to net heat loss;
(c) ensuring the provision of sufficient fresh air to all parts of a building under part load conditions.

These problems can be largely overcome if:

(a) air diffusion is buoyancy-assisted (i.e. not reliant upon the momentum generated by the ATD);
(b) heat losses are dealt with separately;
(c) full-fresh air plant is employed.

There has been much effort expended on the development of high level supply atds which provide less reduction in throw on turndown than conventional devices; these generally incorporate means for discharging a constant velocity jet in parallel with the varying jet.

Air conditioning systems can also be categorised depending on the media used for the main heat exchange, i.e. all-air, air/water and unitary (fluid-based) systems.

All-air systems include all systems which use the central air handling plant to provide all of the cooling and, unless terminal heaters are provided, all of the heating, along with intake of fresh air, mixing or heat exchange with room air as necessary, filtration, humidification if necessary, and noise attenuation. This category includes VAV, terminal reheat and dual-duct.

Air/water systems use central (primary) air handling plant for the provision of fresh air, most of the cooling and, in some cases, heating being provided locally. Most air/water systems use local fans to provide room air diffusion, the exception being the induction system in which energy from the central plant is employed to induce air movement across local coils and distribute the mixture of primary and secondary air through the space. Other systems within this category employ fan coil units, reversible heat pumps and radiant ceilings to provide local heating and/or cooling.

Unitary systems use some combination of refrigerant and electricity, and/or water to provide local cooling and heating. Probably the most popular arrangement for retro-fit to small spaces is the split reversible heat pump with a 'cassette' chassis room unit which fits into a standard ceiling pattern. Recent developments in compressor technology have led to variable speed inverter control of quiet rotary devices which give much more stable control over room conditions than the traditional on–off control. A reversible heat pump incorporates a changeover valve into the refrigerant circuit which enables the condenser and evaporator to swap roles, hence, in its heat pump mode, the unit cools the outside air and heats the room air instead of vice versa.

Application

Noise and pollution sources close to a building may exclude the possibility of opening windows, and if mechanical ventilation alone cannot limit temperature rise in the occupied spaces to an acceptable level then it may be necessary to provide some form of air conditioning.

Large and complex buildings which are likely to have significant heat load phase differences between different zones will need local control over zone air temperatures. The greater the variations in load between zones the more expensive it becomes to provide satisfactory control. The ideal solution is for buildings to be constructed with external fabrics which dampen the changes in outside climate as much as possible, whilst utilising daylight, free heating from the sun and free cooling from the night sky. Furthermore, if convection currents from internal heat sources are allowed to rise, rather than mix with supply air, as with buoyancy-assisted ventilation, these sources will have a smaller influence on the internal climate.

Because of a combination of economic advantage, fashion and clever marketing, VAV systems have been the most popular type of air conditioning since the mid-1970s, particularly for offices. In some larger applications induction units or fan coil units have been used to deal with perimeter loads. Because of their greater flexibility, 4-pipe fan coil units have been most widely applied to hotels and narrow plan office buildings. Dual duct systems, although expensive and space consumptive, have been widely used in hospitals, particularly in North America. Unitary systems tend to have a shorter lifespan, can be fairly noisy and offer crude control, but are easy to install and widely used in refurbishment and tropical applications.

Conclusions

The oil crisis of the mid 1970s generated an enormous amount of research into energy conservation and led to the evolution of tight energy efficient buildings supplied with minimal quantities of outdoor air for occupants and the widespread use of variable air volume air conditioning systems. In recent years, however, many factors have combined to produce a backlash against uniform, open-plan air conditioned environments. The 'sick building syndrome' has no single cause that can be proven, but a recent British survey (Wilson & Hedge, 1987) indicates that people who work in air-conditioned buildings suffer more ailments when exposed to their workplace environment than those who work in mechanically or naturally ventilated buildings. However, in Great Britain air conditioned offices are, almost by definition, located on deep-plan sites in city centres. Over the past 20 years, this has resulted in the erection of many offices which have workspaces, remote from the window walls, which rely on artificial illumination and conditioned air to create a uniform enviroment. Because status is linked with possession of a 'defensible space' and access to a view, it is usually the management who lay claim to perimeter office space and clerical and secretarial staff are relegated to the internal spaces, in which to carry out, perhaps, repetitive and boring tasks. It is not surprising that most surveys have found a much smaller incidence of complaints from management than from other categories of office worker.

The advent of these problems, along with a significant change in the nature of the office has lead to a reappraisal of the way offices are to be designed and used. Computers and communications have rapidly become a dominant factor in the design of offices and their services. This has led to a boom in the demand for raised floors, task lighting and low level supply. Thanks to privatisation the private sector is expanding at the expense of the public sector. Furthermore the service industries are expanding as the engineering-based companies are springing up and demanding small units of office space, sufficiently flexible to meet all their computing and communications needs. This has led to an increase in the demand for multi-tenancy speculative office buildings with floor-by-floor or even suite-by-suite letting, servicing and metering, and with floor areas typically ranging from 100 to 1000 m^2 per tenant. A further trend, which is gathering pace in the City of London is the adoption of the American approach to letting 'raw', unfinished space. It is virtually standard practice in the USA to negotiate a

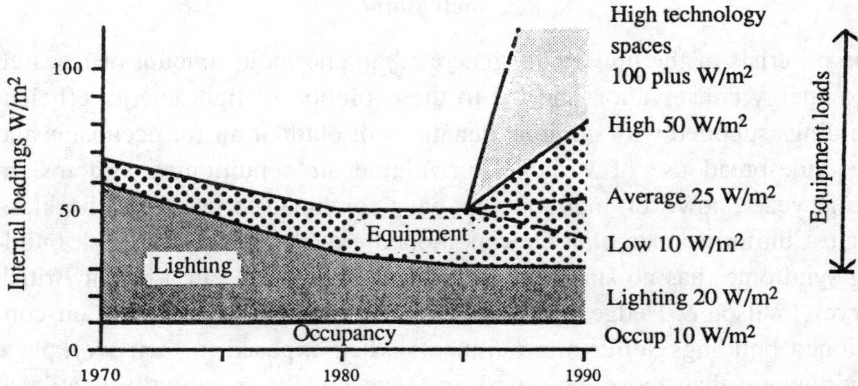

Fig. 16.15. Trends in office internal heat gains, 1970 to 1990.

customised fit-out for each tenant, including some or all of the air conditioning system. The rise in information technology (IT) has produced companies with different hierarchical structures. Because much of the repetitive work is carried out by computer, there are more executives and fewer clerical staff, whilst there is also more group working and a greater need for meetings. As a result there has been a trend towards increasing provision of offices for individuals and small teams of people, mainly occupying a band 5 m from the perimeter walls, internal spaces being used for IT workspaces, conference, office machinery, and computer and paper storage. The need for flexibility is demonstrated by comparison of the lifespan of the building and its contents. The shell might typically last 40 to 60 years, the furnishing and services 10 to 20 years, or for the duration of a tenancy, whereas the computer hardware might have an obsolescence cycle of 3 years. Therefore, during the lifetime of a structure, there may be many different internal configurations for each separately tenanted unit.

The trend towards the high technology workspace and parallel developments in lighting have resulted in very different internal heat gain patterns compared with the early 1970s. Figure 16.15 shows how the general trend in lighting gain has decreased with the evolution of high efficacy lamps and the change from the 1000 lux illuminance associated with the 'integrated environmental design' (IED) heat-from-lighting concept to the low glare, low background illuminance (300–500 lux) and task lighting necessary for vdu operation. Over the same period, typical heat gain from office machinery has risen from 5 W/m^2 to a base load of around 25 W/m^2, with many applications experiencing 50 W/m^2 during peak usage, whilst some City dealer rooms experience a 100 W/m2 peak machinery gain. This means that internal zones are now experiencing wide and unpredictable variations in heat gain which may also change as new machinery becomes available. This not only means that air conditioning is essential to ensure occupant comfort for these applications, but the designer must incorporate flexibility in plant duty and control to deal with the wide load fluctuations. Flexibility may include the facility for individuals to override the air conditioning terminals with openable windows in some areas, for individual control of air direction and temperature, and for individual adjustment of lighting and solar protection.

When considering the economic arguments relating to thermal system choice, it is worth placing the extra-over cost of air conditioning in the context of other costs that the user has to bear. Using 1985 figures, the overall cost of owning and operating an air conditioning system is about £13/m2 per annum more than for a non-air-conditioned property. This compares with an annual bill for rents and rates of about £550/m2 for the City of London, or £130–190/m2 for Greater London down to a minimum of £50/m2 in the provinces. Annual salaries and other associated overheads could come to £5000/m2, however. This makes air conditioning a far less significant overhead in the City of London than in other parts of the country.

The decision on whether to incorporate air conditioning into a particular building will depend largely on the building services designer balancing all the above factors against architect's and client's preconceptions. Many potentially catastrophic mistakes in concept or detail can be avoided by the involvement of all the building professions and the client in a design team from the conceptual stages of the project. The influence of each member of the team on the resultant form, structure, energy consumption, cost and acceptability to users can be significant, thus reducing the possibility of the building becoming 'sick' during its lifetime.

References

APPLEBY, P.H. (1986). Low level supply application and design. *Building Services Journal*, **8(11)**, 59–62.

APPLEBY, P.H. (1987). Variable volume in low level supply for a lecture theatre to give increased ventilation efficiency at low overall cost. In *Air Distribution in Ventilated Spaces Proceedings Roomvent-87*. Stockholm: Royal Institute of Technology.

APPLEBY, P.H. (1989*a*). Taming the tornado. *Building Services Journal*, **11**(**3**), 31–3.

APPLEBY, P.H. (1989*b*). Displacement ventilation – a design guide. *Building Services Journal*, **11**(**4**). 63–6.

ASHRAE (1981). Ventilation for acceptable indoor air quality. *ASHRAE Standard 62–1981*. Atlanta: ASHRAE.

ASHRAE (1985*a*) Physiological principles for comfort and health. In *Handbook: Fundamentals*, Chapter 8. Atlanta: ASHRAE.

ASHRAE (1985*b*). Space air diffusion. In *Handbook: Fundamentals*, Chapter 32; Atlanta: ASHRAE.

ASHRAE (1988). Air-to-air energy-recovery equipment, In *Handbook: Equipment*, Chapter 34. Atlanta: ASHRAE.

ASHRAE (1989). Standard 1989–62. *Ventilation for Acceptable Indoor Air Quality*. Atlanta: ASHRAE.

CHARLESWORTH, P.S. (1988). *Air Exchange Rate and Airtightness Measurement Techniques – An Application Guide*, AIC-AG-2-88. Warwick Air Infiltration & Ventilation Centre.

CHRISTENSEN, N.K., ALBRECHTSEN, O., FANGER, P.O. & TRZECIAKIEWICZ, Z. (1984). Air movement and draught, In *Indoor Air*, vol. 5: Building ventilation and thermal climate, Sweden.

CIBSE (1986*a*). *Guide Section B2:* Ventilation and air conditioning (requirements). London, Chartered Institute of Building Services Engineers.

CIBSE (1986*b*). *Guide Section B3:* Ventilation and air conditioning (systems, equipment and control). London: Chartered Institute of Building Services Engineers.

CIBSE (1987). *Technical Memorandum TM13:* Minimising the risk of Legionnaires' disease. London, Chartered Institute of Building Services Engineers.

FANGER, P.O. (1970). *Thermal Comfort; Analysis and Applications in Environmental Engineering*. Copenhagen: Danish Technical Press.

FANGER, P.O. (1986). Body odour and carbon dioxide, minimum ventilation rates, IEA energy conservation in buildings and community systems programme, *Annex IX final report*.

FANGER, P.O., LAURIDSEN, J., BLUYSSEN, P. & CLAUSEN, G. (1988). Air pollution sources in offices and assembly halls, quantified by the olf unit. *Energy and Buildings*, **12**, 7–19.

ISO (1984). 7730: Moderate thermal environments – determination of the PMV and PPD indices and specification of the conditions for thermal comfort.

JACKMAN, P.J. (1986). Parameters affecting the selection of air diffusion systems and devices, *Proceedings of the 1st European Symposium on Air Conditioning and Refrigeration*, pp. 77–84, Brussels: Expoclima.

JANSSEN, J.E. (1986). Ventilation for acceptable indoor air quality, CIBSE/ASHRAE Conference, Dublin. pp.40–9. London: Chartered Institution of Building Services Engineers.

LEADERER, B.P. & CAIN, W.S. (1983). Air quality in buildings during smoking and nonsmoking occupancy. *ASHRAE Transactions*, **89**, 2B, 601–13.

LIDDAMENT, M.V. (1986). *Air Infiltration Calculation Techniques – An Applications Guide*. Bracknell: Air Infiltration and Ventilation Centre.

McINTYRE, D.A. (1985). *Indoor Climate*. London: Applied Science.

SKÅRET, E. & MATHISEN, H.M. (1983). Ventilation efficiency – a guide to efficient ventilation, *ASHRAE Transactions*, **90**, 1B, 480–95.

SØDERGREN, D. & DAHLGREN, B. (1985). Requirement-controlled ventilation of an auditorium, *Proceedings of the International Symposium on Control and Operation of Building HVAC Systems*. Trondheim: CIB/SINTEF.

WILSON, S. & HEDGE, A. (1987). The office environment survey – a study of building sickness. London: Building Use Studies Ltd.

WRIGHT, A.E. (Chairman) (1989). *Report of the Expert Advisory Committee on Biocides*. London: HMSO.

INDEX